Ethics, Technology, and Engineering

Ethics, Technology, and Engineering

An Introduction

SECOND EDITION

Ibo van de Poel

Lambèr Royakkers

WILEY Blackwell

Registered Offices
John Wiley & Sons, Inc., 111 River Street, Hoboken, NJ 07030, USA
John Wiley & Sons Ltd, The Atrium, Southern Gate, Chichester, West Sussex, PO19 8SQ, UK

For details of our global editorial offices, customer services, and more information about Wiley products visit us at www.wiley.com.

Wiley also publishes its books in a variety of electronic formats and by print-on-demand. Some content that appears in standard print versions of this book may not be available in other formats.

Library of Congress Cataloging-in-Publication Data
Names: Poel, Ibo van de, 1966- author. | Royakkers, Lambèr, 1967- author.
Title: Ethics, technology, and engineering : an introduction / Ibo van de Poel, Faculty of Technology, Policy and Management, Delft University of Technology, Delft, NL, Lambèr Royakkers, Eindhoven University of Technology, School of Innovation Sciences, Eindhoven, NL.
Description: Second edition. | Hoboken, NJ, USA : John Wiley & Sons, Ltd., 2023. | Includes bibliographical references and index.
Identifiers: LCCN 2022057628 (print) | LCCN 2022057629 (ebook) | ISBN 9781119879435 (paperback) | ISBN 9781119879442 (epdf) | ISBN 9781119879459 (epub)
Subjects: LCSH: Technology–Moral and ethical aspects.
Classification: LCC BJ59 .P63 2023 (print) | LCC BJ59 (ebook) | DDC 174/.96–dc23/eng/20230227
LC record available at https://lccn.loc.gov/2022057628
LC ebook record available at https://lccn.loc.gov/2022057629

Cover Image: © Ander Aguirre/Shutterstock
Cover Design: Wiley

Set in 10/12.5pt GalliardStd by Integra Software Services Pvt. Ltd, Pondicherry, India

SKY10052625_080823

Contents

Acknowledgments   x
Introduction   1

1 The Responsibilities of Engineers   6

    1.1   Introduction   7
    1.2   Responsibility   9
    1.3   Passive Responsibility   11
    1.4   Active Responsibility and the Ideals of Engineers   13
        1.4.1   *Technological enthusiasm*   15
        1.4.2   *Effectiveness and efficiency* 17
        1.4.3   *Human welfare*   19
    1.5   Engineers versus Managers   22
        1.5.1   *Separatism*   24
        1.5.2   *Technocracy*   25
        1.5.3   *Whistle-blowing*   26
    1.6   The Social Context of Technological Development   30
    1.7   Chapter Summary   33
    Study Questions   34
    Discussion Questions   35

2 Codes of Conduct   37

    2.1   Introduction   38
    2.2   Codes of Conduct   39
        2.2.1   *Professional codes*   40
        2.2.2   *Corporate codes*   47
    2.3   Possibilities and Limitations of Codes of Conduct   55
        2.3.1   *Codes of conduct and self-interest* 55
        2.3.2   *The need for interpretation* 58
        2.3.3   *Can moral judgments be codified?* 60
        2.3.4   *Can codes of conduct be lived by?* 62
        2.3.5   *Enforcement*   64

2.4   Chapter Summary   66
Study Questions   67
Discussion Questions   68

3 Normative Ethics   70

3.1   Introduction   72
3.2   Ethics and Morality   75
3.3   Descriptive Statements and Normative Judgments   76
3.4   Points of Departure: Values, Norms, and Virtues   77
       3.4.1   *Values*   77
       3.4.2   *Norms*   78
       3.4.3   *Virtues*   79
3.5   Relativism   80
3.6   Ethical Theories   81
3.7   Utilitarianism   82
       3.7.1   *Jeremy Bentham*   82
       3.7.2   *Mill and the freedom principle*   87
       3.7.3   *Criticism of utilitarianism*   89
       3.7.4   *Applying utilitarianism to the Ford Pinto case*   90
3.8   Kantian Theory   91
       3.8.1   *Categorical imperative*   92
       3.8.2   *Criticism of Kantian theory*   95
       3.8.3   *Applying Kant's theory to the Ford Pinto case*   97
3.9   Virtue Ethics   97
       3.9.1   *Aristotle*   98
       3.9.2   *Criticism of virtue ethics*   100
       3.9.3   *Virtues for morally responsible engineers*   100
3.10   Care Ethics   103
       3.10.1   *The importance of relationships*   103
       3.10.2   *Criticism of care ethics*   104
       3.10.3   *Care ethics in engineering*   104
3.11   Capability Approach   106
       3.11.1   *Applications of the capability approach*   109
       3.11.2   *Criticism of the capability approach*   110
3.12   Non-Western Ethical Theories   111
       3.12.1   *Ubuntu*   112
       3.12.2   *Confucianism*   113
3.13   Applied Ethics   115
3.14   Chapter Summary   117
Study Questions   119
Discussion Questions   120

4 The Ethical Cycle   121

4.1   Introduction   122
4.2   Ill-Structured Problems   124
4.3   The Ethical Cycle   125
       4.3.1   *Moral problem statement*   127
       4.3.2   *Problem analysis*   128
       4.3.3   *Options for actions*   130

	4.3.4	*Ethical evaluation*	131
	4.3.5	*Reflection*	134
4.4	An Example		135
	4.4.1	*Moral problem statement*	135
	4.4.2	*Problem analysis*	135
	4.4.3	*Options for actions*	136
	4.4.4	*Ethical evaluation*	137
	4.4.5	*Reflection*	138
4.5	Collective Moral Deliberation and Social Arrangements		140
4.6	Chapter Summary		142
Study Questions			143
Discussion Questions			143

5 Design for Values   **145**

5.1	Introduction		146
5.2	Embedding Values in Technology		147
5.3	Designing for Values		149
5.4	Stakeholder Analysis and Value Identification		150
	5.4.1	*Stakeholder analysis*	150
	5.4.2	*Sources of value in design*	153
5.5	Conceptualization and Specification of Values		153
	5.5.1	*Conceptualization*	154
	5.5.2	*Specification*	156
5.6	Value Conflicts		159
	5.6.1	*Cost-benefit analysis*	161
	5.6.2	*Multiple criteria analysis*	163
	5.6.3	*Thresholds*	166
	5.6.4	*Respecification: reasoning about values*	167
	5.6.5	*Innovation*	170
	5.6.6	*A comparison of the different methods*	171
5.7	Prototyping and Monitoring		172
5.8	Chapter Summary		178
Study Questions			179
Discussion Questions			180

6 Ethical Aspects of Technical Risks   **182**

6.1	Introduction		183
6.2	Definitions of Central Terms		186
6.3	The Engineer's Responsibility for Safety		187
6.4	Risk Assessment		191
	6.4.1	*The reliability of risk assessments*	193
6.5	When Are Risks Acceptable?		194
	6.5.1	*Informed consent*	197
	6.5.2	*Do the advantages outweigh the risks?*	198
	6.5.3	*The availability of alternatives*	199
	6.5.4	*Are risks and benefits justly distributed?*	200
6.6	Risk Communication		201
6.7	Dealing with Uncertainty and Ignorance		203
	6.7.1	*The precautionary principle*	203
	6.7.2	*Engineering as a societal experiment*	206

6.8	Chapter Summary	209
	Study Questions	210
	Discussion Questions	212

7 The Distribution of Responsibility in Engineering — 213

7.1	Introduction	214
7.2	The Problem of Many Hands	217
7.2.1	*The CitiCorp building*	218
7.2.2	*Causes of the problem of many hands*	221
7.2.3	*Distributing responsibility*	221
7.3	Responsibility and the Law	222
7.3.1	*Liability versus regulation*	224
7.3.2	*Negligence versus strict liability*	224
7.3.3	*Corporate liability*	227
7.4	Responsibility in Organizations	227
7.5	Responsibility Distributions and Technological Designs	231
7.6	Chapter Summary	236
	Study Questions	237
	Discussion Questions	238

8 Sustainability, Ethics, and Technology — 241

8.1	Introduction	242
8.2	Environmental Ethics and the Responsibility of Engineers	244
8.3	Sustainable Development	246
8.3.1	*The Brundtland definition*	246
8.3.2	*Moral justification*	248
8.4	Engineers and Sustainability	252
8.4.1	*Design for sustainability*	252
8.4.2	*Specifying sustainability*	253
8.4.3	*Life cycle analysis (LCA)*	256
8.4.4	*Value conflicts in design for sustainability*	258
8.5	Chapter Summary	262
	Study Questions	263
	Discussion Questions	264

9 Responsible Innovation — 265

9.1	Introduction	267
9.2	Opening up the Innovation Process	269
9.2.1	*Science, technology, and society*	269
9.2.2	*The strategic importance of innovation*	270
9.2.3	*The need for responsible innovation*	271
9.3	What is Responsible Innovation?	272
9.3.1	*Responsible innovation as process*	273
9.3.2	*Responsible innovation as product*	273
9.3.3	*Responsible innovation and societal challenges*	274
9.4	Process Criteria for Responsible Innovation	274
9.4.1	*Anticipation*	274
9.4.2	*Inclusiveness*	276
9.4.3	*Reflexivity*	278
9.4.4	*Responsiveness*	279

9.5   Responsible Innovations and Societal Challenges   280
9.6   Responsible Innovation in Industry   283
9.7   Disruptive Innovation   285
        9.7.1   *Market disruption*   285
        9.7.2   *Social disruption*   286
        9.7.3   *Regulatory disruption*   287
        9.7.4   *Conceptual and normative disruption*   288
9.8   Responsible Innovation and the Responsibility of Engineers   289
9.9   Chapter Summary   290
        Study Questions   291
        Discussion Questions   292

Appendix I: Engineering Qualifications and Organizations in a Number
            of Countries   294
Appendix II: NSPE Code of Ethics for Engineers   301
Appendix III: ENGINEERS EUROPE Position Paper on Code of
             Conduct: Ethics and Conduct of Professional Engineers   306
Appendix IV: Examples of Corporate Codes of Conduct   308
Appendix V: DSM Alert Royal DSM Whistleblower Policy   314
Appendix VI: Cases   321
References   336
Index of Cases   352
Index   353

Acknowledgments

This book is based on our Dutch textbook Royakkers, L., van de Poel, I. and Pieters, A. (eds) (2004). Ethiek & techniek. Morele overwegingen in de ingenieurspraktijk, HBuitgevers, Baarn. Most of the chapters have been thoroughly revised. Some chapters from the Dutch textbook are not included and this book contains some new chapters.

The second edition contains new chapters compared to the previous edition, while some other chapters have been left out. All other chapters have been revised and updated.

Section 1.4 contains excerpts from Van de Poel, Ibo. 2007. De vermeende neutraliteit van techniek. De professionele idealen van ingenieurs, in *Werkzame idealen. Ethische reflecties op professionaliteit* (eds J. Kole and D. de Ruyter), Van Gorcum, Assen, pp. 11–23.

Section 3.13 and large parts of Chapter 4 are drawn from Van de Poel, I., and Royakkers, L. (2007). The ethical cycle. *Journal of Business Ethics*, 71 (1), 1–13.

Subsection 5.2.4 contains excerpts from Van de Poel, I. (2015a). Design for values, in *Social Responsibility and Science in innovation Economy* (eds P. Kawalec and R.P. Wierzchoslawski), KUL, Lublin, pp. 115–164.

Section 5.5 contains excerpts from Van de Poel, I. (2021). Values and design, in *Routledge Handbook to Philosophy of Engineering* (eds D.P. Michelfelder and N. Doorn), Routledge, New York, pp. 300–314.

Section 5.6 contains excerpts from Van de Poel, I. (2009a). Values in engineering design, in *Handbook of the Philosophy of Science. Vol. 9: Philosophy of Technology and Engineering Sciences* (ed. A. Meijers), Elsevier, Amsterdam, pp. 973–1006.

Subsection 5.6.4 contains excerpts from Van de Poel, I. (2015b). Conflicting values in design for values, in *Handbook of Ethics, Values, and Technological Design* (eds J. van den Hoven, P.E. Vermaas and I. van de Poel), Springer, Dordrecht, pp. 89–116.

The case of ride-sharing platforms in Section 5.7 is drawn from De Reuver, M., Van Wynsberghe, A., Janssen, M. and Van de Poel, I. (2020). Digital platforms and responsible innovation: Expanding value sensitive design to overcome ontological uncertainty, *Ethics and Information Technology*, 22, 257–267.

Section 6.7 contains excerpts from Van de Poel, I. (2009b). The introduction of nanotechnology as a societal experiment, in *Technoscience in Progress. Managing the Uncertainty of Nanotechnology* (eds S. Arnaldi, A. Lorenzet and F. Russo), IOS Press, Amsterdam, pp. 129–142.

Sections 7.1 and 7.2 contain excerpts from Van de Poel, I., Royakkers, L.M.M. and Swart, S.D. (2015). *Moral Responsibility and the Problem of Many Hands*), Routledge, New York.

The V-Chip case in Section 7.5 contains excerpts from Van de Poel, I. (2007). Ethics in engineering practice, in *Philosophy in Engineering* (eds S. Hylgaard Christensen, M. Meganck and B. Delahousse), Academica, Aarhus, Denmark, pp. 245–262, and Fahlquist, J. N., and Van de Poel, I. (2012). Technology and parental responsibility: the case of the V-chip. *Science and Engineering Ethics*, 18 (2), 285–300.

Chapter 8 is co-authored with Michiel Brumsen. It was originally written by Michiel Brumsen and has been revised for the second edition by Lambèr Royakkers and Ibo van de Poel.

Section 8.2, and subsections 8.4.2 and 8.4.4 contain excerpts from Van de Poel, I. (2017a). Design for sustainability, in *Philosophy, Technology, and the Environment* (ed D.M. Kaplan), MIT Press, Cambridge, pp. 121–142.

The box on values in the energy transition in Section 8.4.4. contains excerpts from Van de Poel, I. and Taebi, B. (2022). Value change in energy systems. *Science, Technology, & Human Values*, 47 (3), 371–379.

Sections 9.3 and 9.8 contain excerpts from Van de Poel, I., and Sand, M. (2021). Varieties of responsibility: two problems of responsible innovation, *Synthese*, 198, 4769–4787.

Section 9.6 contains excerpts from Van de Poel, I. Asveld, L., Flipse, S., Klaassen, P., Scholten, V. and Yaghmaei, E. (2017). Company strategies for responsible research and innovation (RRI): A conceptual model, *Sustainability*, 9 (11), 2045, DOI:10.3390/su9112045.

Introduction

One of the main differences between science and engineering is that engineering is not just about better understanding the world but also about changing it. Many engineers believe that such change improves, or at least should improve, the world. In this sense engineering is an inherently morally motivated activity. Changing the world for the better is, however, no easy task and also not one that can be achieved on the basis of engineering knowledge alone. It also requires, among other things, ethical reflection and knowledge. This book aims at contributing to such reflection and knowledge, not just in a theoretical sense but also more practically.

This book takes an innovative approach to engineering ethics in several respects. It provides a rather unique approach to ethical decision-making: the ethical cycle. This approach is illustrated by an abundance of cases studies and examples, not only from the US but also from Europe and the rest of the world. The book is also innovative in paying more attention than most traditional introductions in engineering ethics to such topics as ethics in engineering design, the organizational context of engineering, the distribution of responsibility, sustainability, and new technologies such as nanotechnology.

There is an increasing attention to ethics in the engineering curricula. Engineers are supposed not only to carry out their work competently and skillfully but also to be aware of the broader ethical and social implications of engineering and to be able to reflect on these. According to the Engineering Criteria 2000 of the Accreditation Board for Engineering and Technology (ABET) in the US, engineering graduates must have "an understanding of professional and ethical responsibility" and "the broad education necessary to understand the impact of engineering solutions in a global and societal context" (Herkert, 1999).

This book provides an undergraduate introduction to ethics in engineering and technology. It helps students to acquire the competences mentioned in the ABET criteria or comparable criteria formulated in other countries. More specifically, this book helps students to acquire the following moral competencies:[1]

- *Moral sensibility*: the ability to recognize social and ethical issues in engineering;
- *Moral analysis skills*: the ability to analyze moral problems in terms of facts, values, stakeholders, and their interests;

Ethics, Technology, and Engineering: An Introduction, Second Edition. Ibo van de Poel and Lambèr Royakkers.
© 2023 John Wiley & Sons Ltd. Published 2023 by John Wiley & Sons Ltd.

- *Moral creativity*: the ability to think out different options for action in the light of (conflicting) moral values and the relevant facts;
- *Moral judgment skills*: the ability to give a moral judgment on the basis of different ethical theories or frameworks including professional ethics and common sense morality;
- *Moral decision-making skills*: the ability to reflect on different ethical theories and frameworks and to make a decision based on that reflection;
- *Moral argumentation skills*: the ability to morally justify one's actions and to discuss and evaluate them together with other engineers and non-engineers;
- *Moral design skills*: the ability to consider how values, as well as modes of use and inter-action, can be inscribed into engineering artefacts at the design stages;
- *Moral agency and action skills*: the ability to respond wisely and responsibly to situations in a way that satisfies as many potential competing constraints as possible;
- *Moral situatedness skills*: the ability to acknowledge the social dimension of engineering practice and to understand the social relations of expertise in connection with technology management and decision-making.

With respect to these competencies, our focus is on the concrete moral problems that students will encounter in their future professional practice. With the help of concrete cases we show how the decision to develop a technology, as well as the process of design and production, is inherently moral. The attention of students is drawn toward the specific moral choices that engineers face. In relation to these concrete choices students will encounter different reasons for and against certain actions, and they will discover that these reasons can be discussed. In this way, students become aware of the moral dimensions of technology and acquire the argumentative capacities that are needed in moral debates.

In addition to an emphasis on cases – which is common to most other introductory textbooks in engineering ethics as well – we would like to mention three further characteristics of the approach to engineering ethics we have chosen in this textbook. The first two characteristics focus on the last three competencies.

First, we take a broad approach to ethical issues in engineering and technology and the engineer's responsibility for these. Some of the issues we discuss in this book extend beyond the issues traditionally dealt with in engineering ethics like safety, honesty, and conflicts of interest. We also include, for example, ethical issues in engineering design (Chapter 5) and sustainability (Chapter 8). We also pay attention to such technologies as the atomic bomb and nanotechnology. While we address such "macro-ethical" issues (Herkert, 2001) in engineering and technology, our approach to these issues may be characterized as inside-out, that is to say: we start with ethical issues that emerge in the practice of engineers and we show how they arise or are entangled with broader issues.

The second characteristic of our approach is that we pay attention to the broader contexts in which individual engineers do their work, such as the project team, the company, the engineering profession and, ultimately, society. We have devoted a chapter to the issues this raises with respect to organizing responsibility in engineering (Chapter 7). Where appropriate we also pay attention to other actors and stakeholders in these broader contexts. Again our approach is mainly inside-out, starting from concrete examples and the day-to-day work of engineers. It is sometimes thought that paying attention to such broader contexts diminishes the responsibility of engineers, because it shows that engineers lack the control needed to be responsible.[2] Although there is some truth in this, we argue that the broader contexts also change the content of the responsibility of engineers and in some respects increase their responsibility. Engineers, for example, need to take into account the viewpoints, values, and interests of relevant stakeholders (Chapter 1). This also implies including such stakeholders, and their viewpoints, in relevant discussion and decision-making, for example in design (Chapter 5), and responsible innovation (Chapter 9). Engineers also need to inform managers, politicians, and the public not only of technological risks but also of uncertainties and potential ignorance (Chapter 6).

The third characteristic of our approach is our attention to ethical theories and focuses on the first six competences. We consider the ethical theories important because they introduce a richness of moral perspectives, which forces students to look beyond what seems obvious or beyond debate. Although we consider it important that students get some feeling for the diversity and backgrounds of ethical views and theories, our approach is very much practice-oriented. The main didactical tool here is what we call the "ethical cycle" (Van de Poel and Royakkers, 2007). This is an approach for dealing with ethical problems that systematically encourages students to consider a diversity of ethical points of view and helps them to come to a reasoned and justified judgment on ethical issues that they can discuss with others. The ethical cycle is explained in Chapter 4, but Chapters 2 and 3 introduce important elements of it.

The development of the ethical cycle was largely inspired by the ten years of experiences we both have in teaching engineering ethics to large groups of students in the Netherlands, and the didactical problems we and our colleagues encountered in doing so (Van de Poel, Zandvoort, and Brumsen, 2001; Van der Burg and Van de Poel, 2005). We noticed that students often work in an unstructured way when they analyze moral cases, and they tend to jump to conclusions. Relevant facts or moral considerations were overlooked, or the argumentation was lacking. Ethical theories were often used in an instrumental way by applying them to cases in an unreflective way. Some students considered a judgment about a moral case as an opinion about which no (rational) discussion is possible.

The ethical cycle is intended as a didactical tool to deal with these problems. It provides students with a guide for dealing with ethical issues that is systematic without assuming an instrumental notion of ethics. After all, what is sometimes called applied ethics is not a straightforward application of general ethical theories or principles to a practical problem in an area. Rather, it is a working back and forth between a concrete moral problem, intuitions about this problem, more general moral principles, and a diversity of ethical theories and viewpoints. This is perhaps best captured in John Rawls' notion of wide reflective equilibrium (Rawls, 1971). (For a more detailed discussion, the reader is referred to Chapter 4.)

The ethical cycle provides a tool that does justice to this complexity of ethical judgment but at the same time is practical so that students do not get overwhelmed by the complexity and diversity of ethical theories. By applying the ethical cycle students will acquire the first six moral competencies that are needed for dealing with ethical issues in engineering and technology (see Figure 1.1).

This book consists of two parts. Part I introduces the ethical cycle. After an introductory chapter on the responsibility of engineers, it introduces the main elements of the ethical cycle: professional and corporate codes of conduct (Chapter 2), and ethical theories (Chapter 3). Chapter 4 then introduces the ethical cycle and offers an extensive illustration of the application of the cycle to an ethical issue in engineering.

Part II focuses on more specific ethical issues in engineering and technology. Chapter 5 deals with ethical issues in engineering design. It focuses on ethical issues that may arise during the various phases of the design process and pays special attention to how engineers

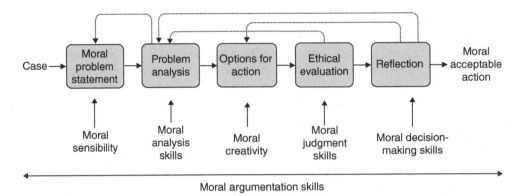

Figure 1.1 Ethical issues in engineering and technology.

are confronted with and can deal with conflicting values in design. Chapter 6 deals with technological risks, and questions about how to assess such risks, the moral acceptability of risks, risk communication, and dealing with uncertainty and ignorance. Chapter 7 discusses issues of responsibility that arise due to the social organization of engineering. It discusses in particular the problem of many hands, the difficulty of pinpointing who is responsible if a large number of people are involved in an activity, and it discusses ways of dealing with this problem in engineering. Chapter 8 discusses sustainability, both in more general terms and how it affects the work of engineers and can be taken into account in, for example, the design process. The last chapter, Chapter 9, deals with responsible innovation. The core idea of responsible innovation is to better align technological innovation with the values, needs, and expectations of society. We will discuss in particular how industry can practice responsible innovation, the disruptive character of innovations, and the responsibility of engineers for responsible innovation.

To a large extent, Parts I and II can be used independently from each other. Teachers who have only limited course hours available can, for example, choose to teach a basic introduction and only use the first four chapters. Conversely, students who have earlier followed some basic introduction to engineering ethics can be offered a course that uses some or all of the chapters from Part II. Although the chapters in Part II are consistent with the ethical cycle introduced in Part I, they contain hardly any explicit references to it and most of the necessary background would also be covered by any other basic course in engineering ethics. In fact, the chapters in Part II can also largely be used independently of each other, so that they could be used for smaller teaching modules.

Teachers, who want to offer their students an introduction to engineering ethics without discussing the various ethical theories and the ethical cycle, could choose to use the first two chapters and a selection of the chapters from Part II that deal with more specific issues. Any set-up that aims at introducing the ethical cycle should, we feel, at least include Chapters 2, 3, and 4.

Each of the chapters starts with an illustrative case study that introduces some of the main issues that are covered in the chapter. Each chapter introduction also indicates the learning objectives so that students know what they should know and be able to do after reading the chapter. Each chapter also contains key terms and a summary that provides a further guide for getting to the core of the subject matter. Study questions provide further help in rehearsing the main points and in applying the main notions to concrete examples.

A book like this is impossible without the help of a lot of people. First of all, we like to thank everybody who contributed to the composition of the Dutch textbook *Ethiek en Techniek. Morele overwegingen in de Ingenieurspraktijk* that formed the basis for this book. In particular, we would like to thank Angèle Pieters, our co-editor of the Dutch textbook and Stella de Jager of HB Uitgevers. We also like to thank Michiel Brumsen for contributing a chapter to this book. We thank Steven Ralston and Diane Butterman for translating parts of our Dutch texts. Jessica Nihlén Fahlquist, Tiago de Lima, Sjoerd Zwart, and Neelke Doorn were so kind to allow us to use a part of a common manuscript in Chapter 7 of this book. We would like to thank Michael Davis, Elena Ziliotti, Cindy Friedman, and Ilse Oosterlaken for their comments and the people of Wiley-Blackwell for their comments and support, in particular Nick Bellorini, Ian Lague, Louise Butler, Tiffany Mok, Dave Nash, Mervyn Thomas, Laura Matthews, Swetha Kodimari, Vijayalakshmi Saminathan, and Will Croft. Finally, we would like to thank the anonymous reviewers and the people who anonymously filled in a questionnaire about the scope of the book for their comments and suggestions.

Ibo van de Poel and Lambèr Royakkers

Notes

1 The last three competencies for engineering ethics education are identified by Martin et al. (2021).
2 Michael Davis, for example has expressed the concern that what he calls a sociological approach to the wider contexts that engineers face may in effect free engineers from any responsibility (see Davis, 2006).

1

The Responsibilities of Engineers

Having read this chapter and completed its associated questions, readers should be able to:

- Describe passive responsibility, and distinguish it from active responsibility;
- Describe the four conditions of blameworthiness and apply these to concrete cases;
- Describe the professional ideals of engineers: technological enthusiasm, effectiveness and efficiency, and human welfare;
- Debate the role of the professional ideals of engineering for professional responsibility;
- Show an awareness that professional responsibility can sometimes seem to conflict with the responsibility as an employee and how to deal with this;
- Discuss the impact of social context of technological development for the responsibility of engineers.

Contents

1.1 **Introduction** 7

1.2 **Responsibility** 9

1.3 **Passive Responsibility** 11

1.4 **Active Responsibility and the Ideals of Engineers** 13

    1.4.1 Technological enthusiasm 15
    1.4.2 Effectiveness and efficiency 17
    1.4.3 Human welfare 19

1.5 **Engineers versus Managers** 22

    1.5.1 Separatism 24

    1.5.2 Technocracy 25
    1.5.3 Whistle-blowing 26

1.6 **The Social Context of Technological Development** 30

1.7 **Chapter Summary** 33

Study Questions 34

Discussion Questions 35

Ethics, Technology, and Engineering: An Introduction, Second Edition. Ibo van de Poel and Lambèr Royakkers.
© 2023 John Wiley & Sons Ltd. Published 2023 by John Wiley & Sons Ltd.

1.1 Introduction

Case Dieselgate

In September 2015, Volkswagen's (VW) Dieselgate was revealed. Dieselgate is one of the biggest scandals of corporate environmental violations worldwide that directly affected consumer health and safety. Volkswagen had not only run their new engine on diesel, but also on cheating software. It had implemented a software manipulation device in the 2009–2015 vehicle models with 2.0-L diesel engines that recognized when a vehicle was undergoing an emissions test and automatically adjusted emissions to the legal thresholds.

The scandal actually began in 2006, when Volkswagen decided to widely promote its diesel car in the US by capitalizing on its success in Europe. At that time, European regulations favored diesel cars over petrol cars, mainly because of their 20% lower CO_2 emissions, and provided legal and financial incentives for carmakers and car dealerships to focus on diesel cars. Therefore, dieselization emerged as a cost-effective strategy to tackle the task of CO_2 emissions reduction. Volkswagen's innovative "turbocharged direct injection" engine (TDI), developed in 1989 was a "cleaner" diesel model with improved fuel efficiency and better performance in CO_2 emission tests. This engine system placed Volkswagen in an ideal position to emerge as a market leader in Europe, and, because of a new version of the TDI, technically known as the VW EA189, introduced in 2008, Volkswagen was ready to grow the market for diesel cars in the US, aiming "to set new environmental standards in vehicles, powertrains and lightweight construction"[1], and "becoming the most innovative high-volume brand in the US"[2].

In its promotional campaign, Volkswagen had to fight against the rather negative image of diesel engines in the US: diesel engines were believed to lack technological sophistication, and consumers had a general impression that diesel fuel was dirty. The fact that diesel cars were considered "green" by Volkswagen, and that Volkswagen claimed that they meet the US Environmental Protection Agency's (EPA) stricter standards for nitrogen oxides (NO_x) reassured the American public. A key point in what would become the scandal of cheating on emissions is that the regulations in the US were much stricter and required NO_x emissions to be six times lower than allowed in the EU at the time. The difference is that CO_2 emissions have an effect on climate change, whereas the EPA was primarily concerned with emissions that are or would be harmful to people. NO_x emissions, in particular, can contribute to cardiovascular and respiratory diseases.

However, engineers were struggling to meet the regulations of the EPA given the available development budgets and engineering constraints. Meeting the regulations was necessary because Volkswagen had to have its diesel models approved by the American automobile authorities before they could be put on the road. Two obvious solutions for emission reduction technologies were rejected: lean NO_x trap and Selective Catalytic Reduction (SCR). SCR is very effective, but was considered too costly ($350 per car) since Volkswagen focused on the less affluent customer. The lean NO_x trap was much cheaper but less effective and negatively affected engine efficiency. This left the engineers with an emission problem they could not solve.

Consequently, they opted for the cheating software that they had come across when they had examined the engine for their own design process. This software recognizes when a Volkswagen is on the test bench, sets the emission cleaning regime to maximum in order to comply with the standards, and switches the cleaning back to low as soon as it has left the test bench. This cheating software was already developed in 1999 by Audi. Audi developed a tweak to the engine control system in its diesel engines that made the engines run more quietly. They referred to the tweak as "acoustic function," but it had a problem: it increased NO_x emissions. As a way out, they developed software that would only omit the extra fuel injection during the test.

In November 2006, a software engineer was tasked with adopting the acoustic function for the VW EA189 under development. Under the responsibility of Rudolf Krebs, the head of the engine development department, a meeting was held in the same month to decide on the fraud with the people responsible for engine development and engine electronics, some 15 in total. This led to intense debate: some attendees morally objected to the cheating software, which was clearly in violation of the law, whereas others expressed fears of getting caught. In the end, Rudolf Krebs ordered the cheating software to be built, but he strongly advised not to get caught.

Despite the described problems with meeting the regulations of the EPA, Volkswagen marketed "clean," "green" diesel models, with the new slogan "Clean Diesel," that supposedly complied with US regulations on NO_x emissions. Sustainability was central to the communication of company policy and strategy.[3] This marketing strategy was very successful, leading to, for example, the Volkswagen Passat TDI Clean Diesel being selected as a "2012 Clean Car of the Year" Finalist. In 2014, 40% of the US consumers considered purchasing a diesel car, up from only 13% in 2005.

However, Volkswagen's success in the US market did raise some concerns. In 2012, a group of American environmentalists questioned why VW's diesel car models driving on US roads would cause less pollution than in Europe. The International Council on Clean Transportation (ICCT), an environmental NGO, began testing the performance of diesel cars on the road in 2013. The results showed that NO_x emissions from the cars tested were about 35 times higher than the legal limit set by the EPA. However, when the cars were tested under laboratory conditions, the results met the EPA standards.

In 2014, ICCT shared its results with both Volkswagen and the EPA for verification. The subsequent EPA investigation took more than a year. Due to the sensitivity of the issue, the EPA was reluctant to make a public statement about the "incident" without hard evidence. However, in August 2015, the EPA obtained incontrovertible proof of the company's deliberate cheating in emissions testing from whistleblower Stuart Johnson, Volkswagen's head of engineering and environmental office. Johnson revealed that software was installed on certain VW diesel cars that could detect laboratory tests and instruct the cars to enter a special laboratory mode in which they produced lower emissions. This type of software is known as a "defeat device." Volkswagen officially acknowledged their illegal software to the EPA on September 3, 2015, and the EPA made this finding public on September 18.

As a result of the investigation, the EPA ordered the recall of 482,000 diesel vehicles sold in the US that were equipped with the defeat device, dating back to the 2009 model. In Europe, an estimated 8.5 million Volkswagen diesel cars had the defeat device installed. To date, the legal costs, fines, repairs, and product buybacks associated with the scandal have cost Volkswagen more than 30 billion euros. In addition, it is estimated that this scandal has led to 1,200 premature deaths in Europe, with an associated cost of 1.9 billion euros.

Source: Based on Zang et al. (2021), Spapens (2018), and Ewing (2017).

In this case, we see how the Dieselgate scandal was caused by the company's decision to install secret software to cheat emission tests. For the company, the scandal had huge legal implications. For example, Volkswagen pleaded guilty in 2017 to fraud, obstruction of justice, and falsifying statements as part of a $4.3 billion settlement reached with the US Justice Department. Besides the company, also some persons have been charged by US prosecutors. Among them were two former VW executives, the US-based VW executive, and engineer Oliver Schmidt, who oversaw emissions issues, and Martin Winterkorn, who was Volkswagen CEO between 2007 and 2015. Schmidt was sentenced to seven years in prison and fined $400,000 after admitting to charges of conspiring to mislead US regulators and violate clean-air laws. The same charges have been made against Winterkorn in a still-ongoing trial. Winterkorn denied any knowledge of the rigged emissions test, and that he is "not aware of any wrongdoing" on his part.[4] Engineer James Robert Liang, was the first company employee to be send to prison. He was sentenced to 40 months in prison and a $200,000 fine. According to federal judge Cox, Liang was an important member of a "massive and stunning fraud": "[b]eginning in about 2006, he and his co-conspirators began to design the new 'EA 189' diesel engine for sale in the US. When Liang and his co-conspirators realized that they could not design a diesel engine that would meet the stricter US emissions standards, they designed and implemented software to recognize whether a vehicle was undergoing standard US emissions testing on a dynamometer, versus being driven on the road under normal driving conditions (the defeat device), in order to cheat US emissions tests."[5]

The Dieselgate case raises questions about the responsibility of engineers (Davis, 2019). While some of the involved engineers acted clearly irresponsibly, others – like the whistleblower – may have acted more responsibly. But what do we exactly mean by responsibility and what is the responsibility of engineers? As we will see responsibility is not only about blame for things that have gone wrong, but also about avoiding harm in the future and doing good.

This chapter first investigates what exactly responsibility is (Section 1.2), distinguishing between passive responsibility for things that happened in the past (Section 1.3) and active responsibility for things not yet attained (Section 1.4). The final two sections discuss the position of engineers vis-à-vis managers, which was obviously important in the Dieselgate scandal, the wider context of technological development, and examine the consequences for the responsibility of engineers in this wider context.

1.2   Responsibility

Whenever something goes wrong or there is a disaster or scandal like that of Dieselgate, the question of who is responsible for it often quickly arises. Here responsibility means in the first place being held accountable for your actions and for the effects of your actions. The making of choices, the taking of decisions, and also failing to act are all things that we regard as types of actions. Failing to save a child who is drowning is therefore also a type of action. There are different kinds of responsibility that can be distinguished. A common distinction is between active responsibility and passive responsibility. Active responsibility is responsibility before something has happened. It refers to a duty or task to care for certain states of affairs or persons. Passive responsibility is applicable after something (undesirable) has happened.

Responsibility (both active and passive) is often linked to the role that you have in a particular situation. In the case described here, James Robert Liang fulfilled the role of engineer and not that of, for example, a family member. You often have to fulfill a number of roles simultaneously such as those of friend, parent, citizen, employee, engineer, expert, and colleague. In a role you have a relationship with others, for instance, as an employee you have a relationship with your employer, as an expert you have a relationship with your customers and as a colleague you have relationships with other colleagues. Each role brings with it certain responsibilities. A parent, for example, is expected to care for their child. In the role of employee, it is expected that you will execute your job properly, as laid down in collaboration with your employer; in the role of expert, it will be presumed that you furnish your customer with information that is true and relevant, and in the role of colleague you will be expected to behave in a collegial fashion with others in the same work situation. An engineer is expected to carry out their work in a competent way. Roles and their accompanying responsibilities can be formally laid down, for instance legally, in a contract or in professional or corporate codes of conduct (see Chapter 2). In addition, there are more informal roles and responsibilities, like the obligations one has within a family or toward friends. Here, too, agreements are often made and rules are assumed but they are not usually put down in writing. We will call the responsibility that is based on a role you play in a certain context **role responsibility**.

> **Role responsibility** The responsibility that is based on the role one has or plays in a certain situation.

Since a person often has different roles in life they have various role responsibilities. One role may come with responsibilities that conflict with the responsibilities that accompany another role. Liang, for example, in the Dieselgate scandal had both, the role as an employee and as an engineer. As an employee, he was expected to be loyal to his company and to listen to his superiors, who eventually ordered the cheating software to be built. As an engineer, he was expected not to develop technologies that can pose unacceptable human and environmental risks, and to avoid deceptive acts. This would imply a refusal to design and implement the defeat software.

> **Moral responsibility** Responsibility that is based on moral obligations, moral norms or moral duties.
>
> **Professional responsibility** The responsibility that is based on one's role as professional in as far it stays within the limits of what is morally allowed.

Although roles define responsibilities, **moral responsibility** is not confined to the roles one plays in a situation. Rather it is based on the obligations, norms, and duties that arise from *moral* considerations. In Chapter 3, we will discuss in more detail what we mean by terms like morality and ethics, and what different kinds of ethical theories can be distinguished. Moral responsibility can extend beyond roles. In the Dieselgate scandal, it was part of Liang's moral responsibility to care for the consequences of the used cheating software for consumers and environment. Moral responsibility can, however, also limit role responsibilities because with some roles immoral responsibilities may be associated. (Think of the role of Mafioso.) In this and the next chapter, we are mainly interested in the **professional responsibility** of engineers.

Professional responsibility is the responsibility that is based on your role as a professional engineer as far it stays within the limits of what is morally allowed. Professional responsibilities are not just passive but they also contain an active component. We will examine the content of the professional responsibility of engineers in more detail in Section 1.4, but first, we turn to a more detailed description of passive responsibility.

1.3   Passive Responsibility

Typical for **passive responsibility** is that the person who is held responsible must be able to provide an account why they followed a particular course of action and why they made certain decisions which have led to the undesired outcome. In particular, the person is held to justify their actions toward those who are in a position to demand that the individual in question accounts for their actions. In the case of the Dieselgate scandal, Volkswagen had to be able to render account for its actions to customers, to society, and to the sitting judge. We will call this type of passive responsibility **accountability**.

Passive responsibility often involves not just accountability but also **blameworthiness**. Blameworthiness means that it is proper to blame someone for an undesired outcome (partially) caused by their actions. To decide whether someone is blameworthy, four conditions need to apply: causal contribution, wrong-doing,

Passive responsibility   Backward-looking responsibility, relevant after something undesirable occurred; specific forms are accountability, blameworthiness, and liability.

Accountability   Backward-looking responsibility in the sense of being held to account for, or justify one's actions toward others.

Blameworthiness   Backward-looking responsibility in the sense of being a proper target of blame for an undesired outcome. In order for someone to be blameworthy, usually the following conditions need to apply: causal contribution, wrong-doing, foreseeability, and freedom.

foreseeability, and freedom. The extent to which you can be blamed is determined by the degree to which these conditions are fulfilled. The four conditions will be illustrated on the basis of the Dieselgate scandal.

Causal contribution

The first criterion is that the person who is held responsible must have made a causal contribution to the undesired outcome for which they are held responsible. Two things are to be kept in mind when judging whether someone made a causal contribution to a certain outcome. First, not only an action, but also a failure to act may often be considered a causal contribution, like in the case of the Dieselgate scandal: the failure to prevent Volkswagen from cheating on emissions testing. Second, one causal contribution is usually not a sufficient condition for the occurrence of the outcome under consideration. Often, a range of causal contributions will have to be present for the outcome to occur. A specific causal contribution will often be a necessary ingredient in the actual chain of events that led to the undesired outcome, that is, without the causal contribution the outcome would not have occurred.

Both the head of the engine development department and the engineers designing and implementing the cheating software made a causal contribution to the company's emissions cheating because both could have averted the scandal. The head of the engine department could have decided not to build the cheating software, and the engineers could have refused to build the software. Some engineers initially objected to the cheating software, but they did persist or lacked the power to prevent the scandal. In retrospect, they could possibly have gone public by informing the press. Perhaps, they could also have intervened earlier on in the process – before the decision had to be made about the cheating software – to ensure that the emission problem had been better tackled.

Wrong-doing

Whenever one blames a person or institution, this person or institution must have performed an action – which has contributed to the undesired outcome – that is in breach of a norm. This can be a norm that is laid down in the law. In the Dieselgate scandal, for example, Volkswagen violated the US Clean Air Act, the primary federal air quality law intended to reduce and control air pollution nationwide, and engineer James Robert Liang was guilty of fraud and conspiracy charges. In this book, we are not just interested in legal norms, but also in moral ones. We will therefore investigate different kinds of ethical frameworks that can be applied in judging the moral rightness or wrongness of actions and their consequences. This includes ethical frameworks such as your own conscience and moral beliefs but also codes of conduct (Chapter 2) and ethical theories (Chapter 3). Together these frameworks form a means of thinking about how one can arrive at what is good, how one can act in the right way. For example, the engineers who contributed to the design and implementation of the cheating software violated the rule stating that engineers shall not aid or abet the unlawful practice of engineering by a person of a firm, often mentioned in codes of conduct for engineers.

Foreseeability

A person who is held responsible for something must have been able to know the consequences of their actions. The consequences are the harm actually arising from transgressing a norm. People cannot be held responsible if it is totally unreasonable to expect that they could possibly have been aware of the undesired outcome. What we do expect is that people do everything that is reasonably possible to become acquainted with the possible consequences of their actions.

In the Dieselgate scandal, the head of the engine development department and the engineers designing and implementing the cheating software knew, or at least could have known, that the software was likely to be used in the coming VW vehicles model with diesel engines for the US market, so violating the law. They also knew, or could have known, that this might lead to additional health risks.

A question that still remains is whether former Volkswagen CEO Martin Winterkorn could have known about the cheating software. He has repeatedly denied any knowledge of the software, but according to an investigation, he failed to respond properly to signs that the company may have been using cheating software, and therefore his claim to have known absolutely nothing about irregularities is doubtful. Moreover, the relevant question is not whether he did not know but whether he reasonably could and should have known it.

Freedom of action

Finally, the one who is held responsible must have had freedom of action, that is, they must not have acted under compulsion. Individuals are either not responsible or are responsible to a lesser degree if they are, for instance, coerced to take certain decisions. The question is, however, what exactly counts as coercion. A person can, for example, be "forced" or manipulated to work on the development of a particular technology under the threat that if they do not cooperate they will sacrifice their chances of promotion. In this case, this person is strictly speaking not coerced to work on the development of the particular technology, they can still reasonably act differently. Therefore, the person remains responsible for their actions. However, since they are is also not entirely free we could say that their responsibility is somewhat smaller than in the case where they had freely chosen to be involved in the development of this technology.

The engineers of Volkswagen were under high pressure because of the corporate culture. Volkswagen's corporate culture was the old fashioned "command and control" model by overly demanding, and by rule of fear. A former VW employee pointed out that the pressure on the engineers was enormous: "Do what you are told. If you fail, expect being yelled at by your manager, or worse, getting fired."[6] So, the engineers working on the cheating software chose to look to the other side, because they feared they might lose their job otherwise. And they might if they spoke up, so the possibilities open to the engineers were limited. The only thing they could have possibly done to prevent the emission cheating was informing the press but that would have had negative consequences (e.g., dismissal) for the engineers and their families. In this case, you could argue that the potential negative personal consequences the engineers faced diminishes their responsibility.

1.4   Active Responsibility and the Ideals of Engineers

Google employees: Google should commit to not weaponizing its technology

In April 2017, Google launched a pilot program with the Department of Defense's Project Maven to develop and integrate "computer-vision algorithms needed to help military and civilian analysts encumbered by the sheer volume of full-motion video data that DoD collects every day in support of counterinsurgency and counterterrorism operations."[7] A year later, in April 2018, Over 3,000 Google employees, including many senior engineers in the area of artificial intelligence research, signed a petition in protest against this Artificial Intelligence project. In an open letter, originally published in the New York Times, addressed to CEO Sundar Pichai, Google employees expressed concern that the US military could weaponize AI and apply the technology toward refining drone strikes and other kinds of lethal attacks:

> We believe that Google should not be in the business of war. Therefore we ask that Project Maven be cancelled, and that Google draft, publicize and enforce a clear policy stating that neither Google nor its contractors will ever build warfare technology. (…)
>
> We cannot outsource the moral responsibility of our technologies to third parties. Google's stated values make this clear: Every one of our users is trusting us. Never jeopardize that. Ever. This contract puts Google's reputation at risk and stands in direct opposition to our core values. Building this technology to assist the US Government in military surveillance – and potentially lethal outcomes – is not acceptable. Recognizing Google's moral and ethical responsibility, and the threat to Google's reputation, we request that you:
>
> 1   Cancel this project immediately
> 2   Draft, publicize, and enforce a clear policy stating that neither Google nor its contractors will ever build warfare technology.[8]

Google announced on June 1, 2018 – after mounting pressure from its employees – that it would not renew its contract with the Pentagon's Project Maven when it was to expire in 2019. It also stepped out of the running for DOD's JEDI (the gratuitous Star Wars reference stands for Joint Enterprise Defense Infrastructure) cloud contract around the same time, because of ethical concerns centering on the use of artificial intelligence.

Google had also accommodated the second request of its employees. It released in June 2018 an ethical code for AI. Part of this code "Artificial Intelligence at Google: Our Principles" reads as follows:

[W]e will not design or deploy AI in the following application areas:

1 Technologies that cause or are likely to cause overall harm. Where there is a material risk of harm, we will proceed only where we believe that the benefits substantially outweigh the risks, and will incorporate appropriate safety constraints.
2 Weapons or other technologies whose principal purpose or implementation is to cause or directly facilitate injury to people.
3 Technologies that gather or use information for surveillance violating internationally accepted norms.
4 Technologies whose purpose contravenes widely accepted principles of international law and human rights.

We want to be clear that while we are not developing AI for use in weapons, we will continue our work with governments and the military in many other areas. These include cybersecurity, training, military recruitment, veterans' healthcare, and search and rescue. These collaborations are important and we'll actively look for more ways to augment the critical work of these organizations and keep service members and civilians safe.[9]

In the previous section, we have considered questions of responsibility when something has gone wrong. Responsibility is also something that comes into play beforehand, if nothing has yet gone wrong or if there is the opportunity to realize something good. We will refer to this as **active responsibility**. If someone is actively responsible for something they are expected to act in such a way that undesired consequences are avoided as much as possible so that positive consequences are realized. Active responsibility is not primarily about blame but requires a certain positive character trait in dealing with matters. Philosophers call such positive character traits virtues (see Chapter 3). Active responsibility, moreover, is not only about preventing the negative effects of technology, such as illustrated by the Google employees (see box), but also about realizing certain positive effects.

> **Active responsibility**   Forward-looking responsibility referring to a duty or task to care for certain state-of-affairs or persons.

Active Responsibility

Mark Bovens mentions the following features of active responsibility:

- Adequate perception of threatened violations of norms;
- Consideration of the consequences;

- Autonomy, that is, the ability to make one's own independent moral decisions;
- Displaying conduct that is based on a verifiable and consistent code; and
- Taking role obligations seriously. (Bovens, 1998)

One way in which the active responsibility of engineers can be understood is by looking at the **ideals** of engineers. Ideals, as we will understand the notion here, have two specific characteristics. First ideals are ideas or strivings which are particularly motivating and inspiring for the person having them. Second, it is typical for ideals that they aim at achieving an optimum or maximum. Often, therefore, ideals cannot be entirely achieved but are strived for. In the course of practicing their profession engineers can be driven by several ideals. Those can be personal ideals such as the desire to

> **Ideals** Ideas or strivings which are particularly motivating and inspiring for the person having them, and which aim at achieving an optimum or maximum.
>
> **Professional ideals** Ideals that are closely allied to a profession or can only be aspired to by carrying out the profession.

earn a lot of money or to satisfy a certain degree of curiosity but they can also be social or moral ideals, such as wanting to implement technological ends to improve the world. Those are also the types of ideals that can spur people on to opt for an engineering field of study and career. Some of these ideals are directly linked to professional practice because they are closely allied to the engineering profession or can only be aspired to by carrying out the profession of engineer. We call such ideals **professional ideals**. As *professional* ideals, these ideals are part of professional responsibility in as far they stay within the limits of what is morally allowed. Below, we shall therefore discuss three different professional ideals of engineers and we shall establish whether these ideals are also morally commendable.

1.4.1 Technological enthusiasm

Technological enthusiasm pertains to the ideal of wanting to develop new technological possibilities and take up technological challenges. This is an ideal that motivates many engineers. It is fitting that Samuel Florman refers to this as "the existential pleasures of engineering" (Florman, 1976). One good example of technological enthusiasm was the development of

> **Technological enthusiasm** The ideal of wanting to develop new technological possibilities and taking up technological challenges.

Google Earth, a program with which, via the Internet, it is possible to zoom in on the earth's surface. It is a beautiful concept but it gives rise to all kinds of moral questions inherent in the use of remotely sensed images, "as Google Earth might be seen as a panoptic viewing technology that leaves no voice to those being viewed" (Myers, 2010, p. 455). For instance, in the area of privacy, you can study the opposite neighbor's garden in great detail, and in the field of security, terrorists could use it to plan attacks. In a documentary on the subject of Google Earth one of the program developers admitted that these are important questions.[10] Nevertheless, when developing the program these were matters that the developers had failed to consider because they were so driven by the challenge of making it technologically possible for everyone to be able to study the earth from behind their computer.

Technological enthusiasm in itself is not morally improper; it is in fact positive for engineers to be intrinsically motivated as far as their work is concerned. The inherent danger of technological enthusiasm lies in the possible negative effects of technology and the relevant social constraints being easily overlooked. This has been exemplified by the Google Earth example. It is exemplified to an extreme extent by the example of Wernher von Braun (see box).

Wernher von Braun (1912–1977)

Wernher von Braun is famous for being the creator of the space program that made it possible to put the first person on the moon on July 20, 1969. A couple of days before, on July 16, the Apollo 11 spaceship used by the astronauts to travel from the earth had been launched with the help of a Saturn V rocket and Von Braun had been the main designer of that rocket. Sam Phillips, the director of the American Apollo program, was reported to have said that without Von Braun the Americans would never have been able to reach the moon as soon as they did. Later, after having spoken to colleagues, he reviewed his comment by claiming that without Von Braun the Americans would never have landed on the moon full stop.

Von Braun grew up in Germany. From an early age he was fascinated by rocket technology. Ac-

Figure 1.1 Wernher von Braun. Photo: NASA/ Wikimedia Commons/Public domain.

cording to one anecdote Von Braun was not particularly brilliant in physics and mathematics until he read a book entitled *Die Rakete zu den Planetenraümen* by Hermannn Oberth and realized that those were the subjects he would have to get to grips with if he was later going to be able to construct rockets. In the 1930s Von Braun was involved in developing rockets for the German army. In 1937 he joined Hitler's National Socialist Party and in 1940 he became a member of the SS. Later he explained that he had been forced to join that party and that he had never participated in any political activities, a matter that is historically disputed. What is in any case striking is the argument that he in retrospect gave for joining the National Socialist Party which was this: "My refusal to join the party would have meant that I would have had to abandon the work of my life. Therefore, I decided to join" (Piszkiewicz, 1995, p. 43). His life's work was, of course, rocket technology and a devotion to that cause was a constant feature of Von Braun's life.

During World War II Von Braun played a major role in the development of the V2 rocket, which was deployed from 1944 onwards to bomb, among other targets, the city of London. Incidentally more were killed during the V2-rocket's development and production – an estimated 10 000 people – than during the actual bombings (Neufeld, 1995, p. 264). The Germans had deployed prisoners from the Mittelbau-Dora concentration camp to help in the production of the V2 rockets. Von Braun was probably aware of those people's abominable working conditions.

There is, therefore, much to indicate that Von Braun's main reason for wanting to join the SS was carefully calculated: in that way he would be able to continue his important work in the field of rocket technology. In 1943 he was arrested by the Nazis and later released. It was claimed that he had allegedly sabotaged the V2 program. One of the pieces of evidence used against him was that he had apparently said that after the war the V2 technology should be further developed in the interests of space travel – and that is indeed what ultimately happened when he later started to work for the Americans. When, in 1945, Von Braun realized that the Germans were going to lose the war he arranged for his team to be handed over to the Americans.

In the United States, Von Braun originally worked on the development of rockets for military purposes but later he fulfilled a key role in the space travel program, a program that was ultimately to culminate in man's first steps on the moon. Von Braun's big dream did therefore ultimately come true.

Source: Based on Stuhlinger and Ordway (1994), Neufeld (1995), and Piszkiewicz (1995).

1.4.2   Effectiveness and efficiency

Engineers tend to strive for effectiveness and efficiency. **Effectiveness** can be defined as the extent to which an established goal is achieved; **efficiency** as the ratio between the goal achieved and the effort required. The drive to strive toward effectiveness and efficiency is an attractive ideal for engineers because it is – apparently – so neutral and objective. It does not seem to involve any political or moral choices, which is something that many engineers experience as subjective and therefore wish to avoid. Efficiency is also something that in contrast, for example, to human welfare can be defined by engineers and is also often quantifiable. Engineers are, for example, able to define the efficiency of the energy production in an electrical power station and they can also measure and compare that efficiency. An example of an engineer who saw efficiency as an ideal was Frederick W. Taylor (see box).

> **Effectiveness**   The extent to which an established goal is achieved.
>
> **Efficiency**   The ratio between the goal achieved and the effort required.

Frederick W. Taylor (1856–1915)

Frederick Taylor was an American mechanical engineer. He became known as the founder of the efficiency movement and was specifically renowned for developing scientific management also known as Taylorism.

Out of all his research, Taylor became best known for his time-and-motion studies. There he endeavored to scientifically establish which actions – movements – workers

Figure 1.2 Frederick Taylor. Photo: Bettmann/Bettmann/Getty Images.

were required to carry out during the production process and how much time that took. He divided the relevant actions into separate movements, eliminated all that was superfluous and endeavored, with the aid of a stopwatch, to establish precisely how long the necessary movements took. His aim was to make the whole production process as efficient as possible on the basis of such insight. Taylorism is often seen as an attempt to squeeze as much as possible out of workers and in practice that was often what it amounted to but that had probably not been Taylor's primary goal. He believed that it was possible to determine, in a scientific fashion, just what would be the best way of carrying out production processes by organizing such processes in such a way that optimal use could be made of the opportunities provided by workers without having to demand too much of them. He maintained that his approach would put an end to the on-going conflict between the trade unions and the managerial echelons, thus making trade unions redundant. He was also critical of management which he found unscientific and inefficient. To his mind having the insight of engineers and their approach to things would culminate in a better and more efficient form of management.

In 1911 Taylor published his *The Principles of Scientific Management* in which he explained the four principles of scientific management:

- Replace the present rules of thumb for working methods with methods based on a scientific study of the work process;
- Select, train, and develop every worker in a scientific fashion instead of allowing workers to do that themselves;
- Really work together with the workers so that the work can be completed according to the developed scientific principles;
- Work and responsibility are virtually equally divided between management and workers. The management does the work for which it is best equipped: applying scientific management principles to plan the work; and the workers actually perform the tasks.

Though Taylor was a prominent engineer – for a time he was, for instance, president of the influential American Society of Mechanical Engineers (ASME) – he only had a limited degree of success when it came to the matter of conveying his ideas to people. They were not embraced by all engineers but, thanks to a number of followers, they were ultimately very influential. They fitted in well with the mood of the age. In the United States the first two decades of the twentieth century were known as the "Progressive Era." It was a time when engineers clearly manifested themselves as a professional group capable of promoting the interests of industry and society. It was frequently implied that the engineering approach to social problems was somehow superior. Taylor's endeavors to achieve a form of management that was efficient and scientific fitted perfectly into that picture.

Source: Based on Taylor (1911), Layton (1971), and Nelson (1980).

Though many engineers would probably not have taken things as far as Taylor did, his attempt to efficiently design the whole production process – and ultimately society as a whole – constituted a typical engineering approach to matters. Efficiency is an ideal that endows engineers with authority because it is something that – at least at first sight – one can hardly oppose and that can seemingly be measured objectively. The aspiration among engineers to achieve authority played an important part in Taylor's time. In the United States the efficiency movement became an answer to the rise of large capitalistic companies where managers ruled and engineers were mere subordinate implementers. It constituted an effort to improve the position of the engineer in relation to the manager. What Taylor was really arguing was that engineers were the only really capable managers.

From a moral point of view, however, effectiveness and efficiency are not always worth pursuing. That is because effectiveness and efficiency suppose an external goal in relation to which they are measured. That external goal can be to consume a minimum amount of non-renewable natural resources to generate energy, but also war or even genocide. It was no coincidence that Nazis like Eichmann were proud of the efficient way in which they were able to contribute to the so-called resolving of the Jewish question in Europe which was to lead to the murdering of six million Jews and other groups that were considered inferior by the Nazis like Gypsies and mental patients (Arendt, 1965). The matter of whether effectiveness or efficiency is morally worth pursuing, therefore, depends very much on the ends for which they are employed. So, although some engineers have maintained the opposite, the measurement of the effectiveness and efficiency of a technology is value-laden. It presupposes a certain goal for which the technology is to be employed and that goal is likely to be value-laden. Moreover, to measure efficiency one needs to calculate the ratio between the output (the external goal) and the input, and also the choice of the input may be value-laden. A technology may, for example, be efficient in terms of costs but not in terms of energy consumption.

1.4.3   Human welfare

A third ideal of engineers is that of contributing to or augmenting human welfare. The professional code of the American Society of Mechanical Engineering (ASME) and of the American Society of Civil Engineers (ASCE) states that "engineers shall use their knowledge and skill for the enhancement of human welfare." This also includes values such as health, the environment, and sustainability. According to many professional codes that also means that: "Engineers shall hold paramount the safety, health and welfare of the public" (as, for example, stated by the code of the National Society of Professional Engineers, see Chapter 2). It is worth noting that the relevant values will differ somewhat depending on the particular engineering specialization. In the case of software engineers, for instance, values such as the environment and health will be less relevant while matters such as the privacy and reliability of systems will be more important. One of the most important values that falls

Johan van Veen (1893–1959)

Figure 1.3 Netherlands. Viewed from a US Army helicopter, a Zuid Beveland town gives a hint of the tremendous damage wrought by the 1953 flood to Dutch islands. Photo: Agency for International Development/Wikimedia Commons/Public domain.

Johan van Veen is known as the father of the Delta Works, a massive plan devised to protect the coasts of the South-western part of the Netherlands which materialized after the flood disaster of 1953 (see Figure 1.3). During the disaster 1835 people died and more than 72,000 were forced to evacuate their homes.

Before the disaster occurred there were indications that the dykes were not up to standard. In 1934 it was discovered that a number of dykes were probably too low. In 1939 Wemelsfelder, a Public Works Agency employee working for the Research Service for the Estuaries, Lower River Reaches and Coasts sector, was able to support that assumption with a series of models. Even before the big disaster of 1953 Johan van Veen had emphasized the need to close off certain estuaries.

Van Veen studied civil engineering in Delft before then going on, in 1929, to work for the Research Service which he was later to head. On the basis of his interest in the history of hydraulic engineering and his activities with the Public Works Agency, he gradually became convinced that the danger posed by storm-driven flooding had been vastly underestimated and that the dykes were indeed too low. Van Veen was quite adamant about his beliefs which soon earned him the nickname, within the service, of "the new Cassandra" after the Trojan priestess who had perpetually predicted the fall of Troy. He even adopted the pseudonym Cassandra in the epilogue to the fourth edition of his book *Dredge, Drain, Reclaim* which was published in 1955. According to Van Veen, Cassandra had been warning people about the too-low state of the dykes since 1937. In the fifth edition of his book, which appeared in 1962, Van Veen revealed that he was in fact Cassandra (Van Veen, 1962). Van Veen's reporting of the lowness of the dykes was not something that was welcomed. In fact, it was deliberately kept secret from the public. It is even said that Van Veen was sworn to silence on the matter.

In 1939 Van Veen became secretary of the newly created Storm Flood Committee. In that capacity, he was given the space to elaborate several of his plans for the further defense of the Netherlands. In public debates, he consistently based his arguments for those plans on the need to combat silting up and the formation of salt-water basins. Undoubtedly that was because even then he was unable to publicly air his views about safety.

Even though pre-1953 there was growing doubt within the Public Works Agency as to the ability of the existing dykes to be able to withstand a storm-driven flood that was not a matter that became publicly known. It was not only the Public Works Agency and the relevant minister that kept quiet about the possibility of a flood disaster. At that time the press was not keen to publish such doom and gloom stories either. As there was little or no publicity about the inadequacy of the dykes the inhabitants of Zeeland were thus totally surprised by the disaster. There are no indications that in the period leading up to 1953 steps were taken to improve the storm warning systems and the aid networks. If that had happened, then undoubtedly considerably fewer people would have lost their lives.

Source: Based on Ten Horn-van Nispen (2002), Van der Ham (2003), and De Boer(1994).

under the pursuit of human welfare among engineers is safety. One of the engineers who was a great proponent of safety was the Dutch civil engineer Johan van Veen (see box).

From a moral point of view, the professional ideal of human welfare is hardly contestable. One could maybe wonder whether serving human well-being is a moral obligation for engineers, and whether they have a social responsibility. There are some elements of these engineering social responsibilities that are widely agreed upon, as we already mentioned: human health and safety, protection of the environment, and sustainability. Other engineering social responsibilities have less consensus across countries and disciplines, such as the mandate to participate in pro bono work; to strive for social justice, relating to the distribution of wealth and privileges in society, as well as issues related to poverty and development; and to embrace diversity (Bielefeldt, 2018). But if engineers choose to have these latter engineering social responsibilities, this seems certainly laudable. Therefore, from a moral angle, this ideal related to social responsibility has another status than the other two ideals discussed above. As we have seen technological enthusiasm and effectiveness and efficiency are ideals that are not necessarily morally commendable, although they are also not always morally reprehensible; in both cases much depends on the goals for which technology is used and the side-effects so created. Both ideals, moreover, carry the danger of forgetting about the moral dimension of technology. On the other hand, the ideal of human welfare confirms that the professional practice of engineers is not something that is morally neutral and that engineers do more than merely develop neutral means for the goals of others.

1.5   Engineers versus Managers

Engineers are often salaried employees and they are usually hierarchically below managers. Just as with other professionals this can lead to situations of conflict because they have, on the one hand, a responsibility to the company in which they work and, on the other hand, a professional responsibility as engineers, including – as we have seen – a responsibility for human welfare. We will discuss here three models of dealing with this tension and the potential conflict between engineers and managers: separatism, technocracy, and whistleblowing. These three models are positions that engineers can adopt versus managers in specific situations, but they also reflect more general social frameworks for dealing with the potential tension between engineers and managers.

Case   Challenger

When, on the morning of January 28, 1986, the mission controllers' countdown began for the 25th launching of the space shuttle, it was almost four degrees Celsius below freezing point (or about 25 degrees Fahrenheit). After 73 seconds the Challenger space shuttle exploded 11 kilometers above the Atlantic Ocean. All seven astronauts were killed. At the time it was the biggest disaster ever in the history of American space travel.

After the accident an investigation committee was set up to establish the exact cause of the explosion. The committee concluded that the explosion leading to the loss of the 1.2 billion dollar spaceship was attributable to the failure of the rubber sealing ring (the O-ring). As the component was unable to function properly at low temperatures the flame burnt through the joint of the booster rocket. This led to the separation of booster rocket, which then crashed into the external tank, which caused a structural failure of the external tank and an explosion.

On the day of the fatal flight, the launching was delayed five times, partly for weather-related reasons. The night preceding the launching was very cold; it went down to minus 10 degrees Celsius (or 14 degrees Fahrenheit). NASA engineers confessed to remembering having heard that it would not be safe to launch at very low temperatures. They, therefore, decided to have a telephone conference on the eve of the launching between NASA and Morton Thiokol (a NASA supplier) representatives. Roger Boisjoly, an engineer at the Morton Thiokol also participated with some of his colleagues, and they underlined the risk of the O-rings eroding at low temperatures. The O-rings had never been tested in subzero conditions. The engineers recommended that if the temperature fell below 11 degrees Celsius (or 52 degrees Fahrenheit) then the launch should not go ahead. The weather forecast indicated that the temperature would not rise above freezing point on the morning of the launch. That was the main reason why Morton Thiokol initially recommended that the launch should not be allowed to go ahead.

The people at NASA claimed that the data did not provide sufficient grounds for them to declare the launching, which was extremely important to NASA, unsafe. A brief consultation session was convened so that the data could once again be examined. While the connec-

Figure 1.4 Challenger Space Shuttle. Photo: Bob Pearson/AFP/Getty Images.

tion was broken for five minutes the General Manager of Thiokol commented that a "management decision" had to be made. For Morton Thiokol it was too much of a political and financial risk to postpone the launch, since after the launching NASA would make a decision regarding a possible contract extension with the company. It was at least the case that Boisjoly felt that people were no longer listening to his arguments. After discussing matters among themselves the four managers present, the engineers excluded, put it to the vote. They were reconnected and Thiokol, ignoring the advice of Boisjoly, announced to NASA its positive recommendations concerning the launching of the Challenger. Several minutes after the launch someone from the mission control team concluded that there had: "obviously been ... a major malfunction."

Several months after the Challenger disaster, Roger Boisjoly said the following: "I must emphasize, I had my say, and I never [would] take [away] any management right to take the input of an engineer and then make a decision based upon that input ... I have worked at a lot of companies ... and I truly believe that ... there was no point in me doing anything further [other] than [what] I had already attempted to do" Goldberg (1987, p. 156). It is a view that fits into what might be termed separatism.

Source: Based on Vaughan (1996), and the BBC documentary *Challenger: Go for Launch* of Blast!Films.

1.5.1   Separatism

Engineer James Robert Liang who was involved in developing the cheating software in the Dieselgate scandal said in his defense that he was just doing what he was told while trying to be a loyal worker.[11] Just as in the Challenger case (see box) **separatism** is involved: "the notion that scientists and engineers should apply the technical inputs, but appropriate management and political organs should make the value decisions" (Goldberg, 1987, p. 156). Separatism is well illustrated by the **tripartite model**.

> **Separatism**   The notion that scientists and engineers should apply the technical inputs, but appropriate management and political organs should make the value decisions.
>
> **Tripartite model**   A model that maintains that engineers can only be held responsible for the design of products and not for wider social consequences or concerns. In the tripartite model three separate segments are distinguished: the segment of politicians; the segment of engineers; and the segment of users.

In the tripartite model three separate segments are distinguished (Figure 1.5). The first segment contains politicians, policy makers, and managers who establish the objectives for engineering projects and products and make available resources without intervening in engineering matters. They also stake out the ultimate boundaries of the engineering projects. The second segment relates to the engineers who take care of the designing, developing, creating, and executing of those projects or products. The final segment, the users, includes those who make use of the various technologies. According to this model engineers can only be held responsible for the technical creation of products.

The tripartite model (see, for example, Van de Poel (2001); originally based on Boers (1981)) is based on the assumption that the responsibility of engineers is confined to the engineering choices that they make. The formulation of the design assignment, the way in which the technology is used and the consequences of all of that are not thus considered to be part of the responsibility of engineers. According to this view, the responsibility of engineers limits itself to the professional responsibility that they have to their employer, customer, and colleagues, excluding the general public. The case of Werner von Braun illustrates this well. Von Braun was reconciled to the subordinate role of engineers but perpetually sought ways of pursuing his technological ideals and, in doing so, displayed a degree of indifference to the social consequences of the application of his work and to the immoral intentions of those who had commissioned the task. His creed must have been: "In times of war, a man has to stand up for his country, as a combat soldier as a scientist or as an engineer, regardless of whether or not he agrees with the policy his government is pursuing" (Stuhlinger and Ordway, 1994, p. xiii). It is a role that might alternatively be described as being that of a **"hired gun."** The dangerous side of this role can perhaps best be summed up in the words of the song text of the American satirist Tom Lehrer:[12]

> **"Hired gun"**   Someone who is willing to carry out any task or assignment from their employer or client without moral scruples.

Figure 1.5   The tripartite model.

> Once the rockets go up
> Who cares where they come down
> "that's not my department,"
> said Wernher von Braun.

According to Deborah Johnson (2020), hired guns are the antithesis of professionals: "[e]ngineering is a profession, and members of a profession have responsibilities that go beyond their own short-term self-interest and the interests of their clients" (Johnson, 2020, p. 91). In a letter of the US-based VW executive and engineer Oliver Schmidt to judge Cox, it became clear that he had acted as a hired gun. Schmidt had agreed to follow a script, or talking points, agreed by Volkswagen management and a high-ranking lawyer, to conceal from regulators the existence of the defeat device, at a meeting with Alberto Ayala, a California Air Resources Board executive. In the letter he admitted his mistake and that he should have taken his professional responsibility:

> In hindsight, I should never have agreed to meet with Dr Ayala on that day. Or better yet, I should have gone to that meeting and ignored the instructions given to me and told Dr Ayala that there is a defeat device in the VW diesel engine vehicles and that VW had been cheating for almost a decade. I did not do that and that is why I find myself here today.[13]

1.5.2 Technocracy

An alternative for the engineer as a "hired gun" is offered by Frederick Taylor. He proposed that engineers should take over the role of managers in the governance of companies and that of politicians in the governance of society. This proposal would lead to the establishment of a **technocracy**, that is, government by experts. Accordingly, the role of engineers would be that of technocrats who, on the basis of technological insight, do what they consider best for a company or for society. The role of technocrats is problematic for a number of reasons. First, it is not exactly clear what unique expertise engineers possess that permits them to legitimately lay claim to the role of technocrats. As we have seen, concealed behind the use of apparently neutral terms like efficiency there is a whole world of values and conflicting interests. Admittedly engineers do have specific technological knowledge and they do know about, for example, the risks that may be involved in a technology. When it comes to the underlying goals that should be pursued through technology or the acceptable levels of risk they are not any more knowledgeable than others. A second objection to technocracy is that it is undemocratic and paternalistic. We speak of **paternalism** when a certain group of individuals, in this case, engineers, make (moral) decisions for others on the assumption that they know better what is good for them than those others themselves. In that way, paternalism denies that people have the right to shape their own lives. That clashes with the people's moral autonomy – the ability of people to decide for themselves what is good and right. Moral autonomy is often considered an important moral value.

> **Technocracy** Government by experts.
>
> **Paternalism** The making of (moral) decisions for others on the assumption that one knows better what is good for them than those others themselves.

1.5.3   Whistle-blowing

Case   Inez Austin

Inez Austin was one of the few female engineers at the company Westinghouse Hanford, when in 1989 she became senior process engineer for that company at the Hanford Nuclear Site, a former plutonium production facility in the state of Washington in the United States. In June 1990, she refused for safety reasons to approve a plan to pump radioactive waste from an old underground single-shell tank to a double-shell tank. Her refusal led to several retaliatory actions by her employer. In 1990 she received the lowest employee ratings in all her 11 years at the company. Doubts were raised about the state of her mental health and she was advised to see a psychiatrist. In 1992, Austin received the Scientific Freedom and Responsibility Award from the American Association for the Advancement of Science (AAAS) "for her courageous and persistent efforts to prevent potential safety hazards involving nuclear waste contamination. Ms. Austin's stand in the face of harassment and intimidation reflects the paramount professional duty of engineers – to protect the public's health and safety – and has served as an inspiration to her co-workers." Nevertheless, after a second whistle-blowing incident, relating to the safety and legality of untrained workers, her job was terminated in 1996.

Source: Based on https://onlineethics.org/cases/engineers-and-scientists-behaving-well/inez-austin-protecting-public-safety-hanford-nuclear, accessed November 17, 2022.

A third role model is offered by Van Veen. He accepted, to an important extent, his subordinate role as engineer but he did endeavor to find channels, internally and externally, to air his grievances on safety. Though he never went public as such his role verges on that of whistle-blower as they report internal wrongs externally in order to warn society. An example of a whistle-blower is given in the boxed case on Inez Austin. The term **whistle-blowing** is the act of publicly disclosing information by someone about serious wrongdoing within their organization. Serious wrongdoing does not only include the endangerment of public health, safety, or the environment but also indictable offences, violation of the law and of legislation, deception of the public or the government, corruption, fraud, destroying or manipulating information, and abuse of power, including sexual harassment and discrimination. As the box shows whistle-blowing may well lead to conflicts with the employer. In fact, whistle-blowing can be severely detrimental to the whistle-blower, facing expensive court cases and even being jailed (Kenny, 2019). A study showed that two-thirds of identifiable whistle-blowers experienced physical and/or mental health issues, and three-quarters felt stigmatized (Redman and Caplan, 2015). While many countries and organizations have laws and policies to encourage and protect whistle-blowers (see Chapter 2), in practice these laws and policies are weak and are often simply ignored. So, it is not surprising that

Whistle-blowing   The act of publicly disclosing information by someone about serious wrongdoing within their organization.

the fear for the consequences of whistle-blowing stops people from blowing the whistle. This was also the case in the Dieselgate scandal, and that was the reason that it took so long before Stuart Johnson, the head of Volkswagen's Engineering and Environmental Office, revealed the existence of the cheating software in 2015. After Johnson, more engineers admitted this when they were questioned about irregularities in the emissions data. As we already mentioned the company's culture was one that was ruled by fear, especially when Winterkorn became CEO. Winterkorn expressed himself "loudly, yelling at anything that displeased him and he was also physically intimidating" (Spapens, 2018, p.100), and often "top VW managers received relentless criticism from Winterkorn in front of their peers" (ibid., p. 100). Through this culture, engineers and managers were left in constant uncertainty about their jobs. This was the main reason that, in spite of the fact that they had plenty of opportunities to protest or blow the whistle, they decided to choose obedience over professional responsibility.

Guidelines for Whistle-blowing

Business ethicist Richard De George has proposed the following guidelines, for when whistle-blowing is morally required:

1   The organization to which the would-be whistleblower belongs will, through its product or policy, do serious and considerable harm to the public (whether to users of its product, to innocent bystanders, or to the public at large).
2   The would-be whistleblower has identified that threat of harm, reported it to her immediate superior, making clear both the threat itself and the objection to it, and concluded that the superior will do nothing effective.
3   The would-be whistleblower has exhausted other internal procedures within the organization (for example, by going up the organizational ladder as far as allowed) – or at least made use of as many internal procedures as the danger to others and her own safety make reasonable.
4   The would-be whistleblower has (or has accessible) evidence that would convince a reasonable, impartial observer that her view of the threat is correct.
5   The would-be whistleblower has good reason to believe that revealing the threat will (probably) prevent the harm at reasonable cost (all things considered). (De George, 1990)

Whistle-blowers are often seen as people who are morally to be commended and are heroes (see box about Frances Haugen), however, some authors argue that the potential whistle-blower has a duty to blow the whistle. Ceva and Bocchiola, for example, argue that whistle-blowing is an issue neither of personal morality nor professional duty, and challenge the notion of whistle-blowing as heroic: "[t]o see whistleblowing as a supererogation has the effect of reducing any such report of organizational wrongdoing to a one-shot game in which an individual sacrifices her life and career for the sake of higher moral ideal and goals" (Ceva and Bocchiola, 2019, p. 49).

Frances Haugen, "21st century hero"

Figure 1.6 Frances Haugen Michael Morgernstern.

Beginning in September 2021, The Wall Street Journal published The Facebook Files, a series of news reports "based on a review of internal Facebook documents including research reports, online employee discussions and drafts of presentations to senior managers".[14] These documents indicate that Facebook was aware of various problems caused by its apps, ranging from its impact on teens' mental health and the extent of misinformation on its platforms, to fomenting ethnic violence in countries such as Ethiopia and human traffickers' open use of its services. For example, with regard to the impact on teens' mental; health, statistics from Facebook showed that one-third of the teen girls said that when they felt bad about their bodies, Instagram (one of the companies owned by Facebook) made them feel worse.

Frances Haugen's identity as the Facebook whistle-blower was revealed on *60 Minutes* on October 3, 2021.[15] Haugen is a data scientist from Iowa with a degree in computer engineering and a Harvard master's degree in business. In 2019, she was recruited by Facebook, and she wanted to work on issues around misinformation, since she was motivated by the loss of a friendship due to online conspiracies. She became a product manager within Facebook's Civic Integrity team, the unit tasked with monitoring and limiting misinformation. She and her team had lots of ideas for how to make Facebook less harmful, but her optimism that she could make a change from inside was short-lived. The team was, for example, understaffed, and the concerns she raised were repeatedly discounted by the managers. The Civic Integrity team played a key role in the company's efforts to get through the presidential election in 2020 relatively unscathed by tweaking the algorithm such that polarizing political content was given lower priority. After the election, Facebook removed these so-called safeguards, an action Haugen believes was at least partially responsible for the riots at the Capitol in Washington on January 6, 2021, and disbanded the misinformation team. It was the last straw for Haugen. She resigned, but still had access to the intranet and started secretly collecting documents about the strategy of Facebook. She passed these documents on to the Wall Street Journal, leading to The Facebook Files.

In her opening statement to the US Senate Commerce Subcommittee on Consumer Protection on October 5, 2021, Haugen, hailed as "21st century hero" by the Senate, said

> I joined Facebook because I think Facebook has the potential to bring out the best in us, but I'm here today because I believe Facebook's products harm children, stoke division, and weaken our democracy. The company's leadership knows how to make Facebook and Instagram safer, but won't make the necessary changes because they have put their astronomical profits before people.[16]

Haugen's main concern was Facebook's "engagement-ranking" system, more commonly known as "the algorithm," that chooses which posts, out of thousands of options, to rank at the top of users' feeds, which according to Haugen is doomed to amplify the worst in us. This system is essentially designed to keep people on the site/app for as long as possible, and it is this time and the data thus collected that is ultimately monetized:

> One of the things that has been well documented in psychology research is that the more times a human is exposed to something, the more they like it, and the more they believe it's true. One of the most dangerous things about engagement-based ranking is that it is much easier to inspire someone to hate than it is to compassion or empathy. Given that you have a system that hyperamplifies the most extreme content, you're going to see people who get exposed over and over again to the idea that [for example] it's O.K. to be violent to Muslims. And that destabilizes societies.[17]

Haugen came forward because she realized a "frightening truth": "almost no one outside of Facebook knows what happens inside Facebook. As long as Facebook is operating in the dark, it is accountable to no one. And it will continue to make choices that go against the common good."[18] Haugen's disclosure has prompted policymakers not only in the US, but also around the world, to intensify their calls for regulation of social media platforms to tackle harmful online content and regulate the social media giant Facebook. Besides the testimony before the US Senate Commerce Subcommittee on Consumer Protection, Haugen was invited for a hearing by the Parliament of the United Kingdom and by the European Parliament. After the meeting with the European Parliament, Chair Anna Cavazzini said:

> Whistleblowers like Frances Haugen show the urgent need to set democratic rules for the online world in the interest of users. Her revelations lay bare the inherent conflict between the platform's business model and users' interests. It shows that we need strong rules for content moderation and far-reaching transparency obligations in Europe. It also shows that corporate self-regulation has not worked. (…) All allegations in the 'Facebook Files' must be investigated. As the Internal Market Committee is currently negotiating the Digital Services Act and the Digital Markets Act, a public hearing with Frances Haugen will enrich the democratic discourse and our current legislative work in the committees concerned.[19]

It does not seem desirable to let the professional ethics of engineers – or people of any other profession – be exclusively dependent on such practices. Although whistle-blowing may sometimes be unavoidable, as a general social framework for dealing with the potential tension between engineers and managers, it is unsatisfactory. In the first place, whistle-blowing usually forces people to make big sacrifices and one may question whether it is legitimate to expect the average professional to make such sacrifices. In the second place, the effectiveness of whistle-blowing is often limited because as soon as the whistle is blown the communication between managers and professionals has inevitably been disrupted. It would be much more effective if at an earlier stage the concerns of the professionals were to be addressed but in a more constructive way. This demands a concept of the engineer's role in which the engineer as professional is not necessarily opposed to the manager. It means that engineers have to be able to recognize moral questions in their professional practice and discuss them in a constructive way with other parties.

1.6   The Social Context of Technological Development

Engineers are not the only ones who are responsible for the development and consequences of technology. Apart from managers and engineers, there are other actors who influence the direction taken by technological development and the relevant social consequences. We use the term **actor** here for any person or group that can make a decision on how to act and that can act on that decision. A company is an actor because it usually has a board of directors that can make decisions on behalf of that company and is able to effectuate those decisions. A mob on the other hand is usually not an actor. A variety of actors can be distinguished that usually play a role in technological development:

Actor   Any person or group that can make a decision on how to act and that can act on that decision.

Users   People who use a technology and who may formulate certain wishes or requirements for the functioning of a technology.

Regulators   Organizations who formulate rules or regulations that engineering products have to meet such as rulings concerning health and safety, but also rulings linked to relations between competitors.

- Developers and producers of technology. This includes engineering companies, industrial laboratories, consulting firms, universities, and research centers, all of which usually employ scientists and engineers.
- **Users** who use the technology and formulate certain wishes or requirements for the functioning of the technology. The users of technologies are a very diverse group, including both companies and citizens (consumers).
- **Regulators** such as the government, who formulate rules or regulations that engineering products have to meet such as rulings concerning health and safety, but also rulings linked to relations between competitors. Regulators can also stimulate certain technological advances by means of subsidies.

Also other actors may be involved in technological development including, for example, professional associations, educational institutes, interest groups, and trade unions

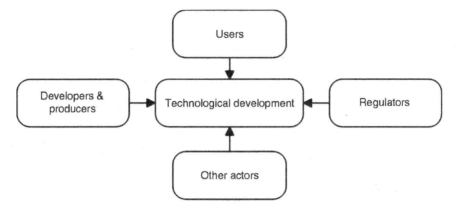

Figure 1.7 Technological development map of actors.

(see Figure 1.7). All these actors have certain **interests,** – things they strive for because they are supposed to be beneficial or advantageous for them. The interests of the various actors will often conflict, so that there is no agreement on the desirable direction of technological development.

In addition to actors that influence the direction of technological development we distinguish stakeholders. **Stakeholders** are actors that have an interest ("a stake") in the development of a technology, but who cannot necessarily influence the direction of technological development. An example is people living in the vicinity of a planned construction site for a nuclear plant. Obviously these people have an interest in what type of reactor is built and how safe it is but they may not be able to influence the technology developed. Of course, such groups may organize themselves and try to get a say in technological development and they may do so more or less successfully. Stakeholders are not only relevant because they may become actors that actually influence technological development, they are also important from a moral point of view. As we have seen, stakeholders are actors whose interests are at stake in technological development. It is often assumed that morality and ethics require that we do not just neglect the rights and interest of those stakeholders because they are powerless but that we should somehow take them into account.[20]

The possibility of steering technological development is not only restrained by the fact that a large number of actors are involved in the development of technology but also

Interests Things actors strive for because they are beneficial or advantageous for them.

Stakeholders Actors that have an interest ("a stake") in the development of a technology.

Case   The Invention of Teflon

Roy Plunkett – a 28-year-old chemist at Du Pont – was requested in 1938 to develop a new, non-poisonous coolant for fridges. He, therefore, filled a metal tube with a little-used mixture and with tetrafluorethylene that would perhaps possess cooling qualities. When he went to get the mixture out of the tube nothing came out but the tube was 60 grams heavier than normal. There, therefore, had to be something in it. After having sawn open the tube it was discovered that a pale and fatty, wax-like, white powder was stuck to the side. Nobody knew what it was so they began to experiment with the substance which turned out to be completely unique. It was given the name Teflon, after "tef" – the nickname given by chemists to tetrafluorethyleen – followed by "lon" – a suffix that Du Pont frequently used for its new products.

Du Pont devoted a great deal of time and money to discovering the exact characteristics of Teflon. It turned out to be complicated and expensive to produce Teflon. The first time that it was ever used was during World War II in order to reinforce the closing rings of the atom bomb. Teflon thus remained a state secret. It was not until 1946 that it was introduced to the general public.

Teflon has nowadays a wide range of uses. It is maybe best known as coating for non-stick frying pans. Although Teflon was long seen as a wonder material, it has recently come under some suspicion. In 2005, the Scientific Advisory Board of the Environmental Protection Agency (EPA) in the US found that perfluorooctanoic acid (PFOA), a chemical compound used to make Teflon, is "likely carcinogenic."[21] In 2005, scientists of US Food and Drug Administration (FDA) found small amounts of PFOA in Teflon cookware (Begley et al., 2005), while DuPont scientists did not detect PFOA in such pans (Powley et al., 2005). In 2004, Du Pont still maintained that "evidence from 50 years of experience and extensive scientific studies supports the conclusion that PFOA does not cause adverse human health effects."[22]

Nevertheless, in 2006 Du Pont committed itself to eliminating the release of PFOA to the environment (Eilperin, 2006).

Source: Adapted from Grauls (1993, pp. 123 ff).

Technology Assessment (TA)
Systematic method for exploring future technology developments and assessing their potential societal consequences.

Collingridge dilemma This dilemma refers to a double-bind problem to control the direction of technological development. On the one hand, it is often not possible to predict the consequences of new technologies already in the early phases of technological development. On the other hand, once the (negative) consequences materialize it often has become very difficult to change the direction of technological development.

because technological development is an unpredictable process (see Teflon box). In the course of time, a variety of methods and approaches have been developed to deal with this unpredictable character of technology development. This is done by a discipline known as **Technology Assessment (TA).** Initially TA was directed at the early detection and early warning of possible negative effects of technological development. Although such early detection and warning is important, it became increasingly clear that it is often not possible to predict the consequences of new technologies already in the early phases of technological development, as is also underscored by the Teflon example. On the other hand, it appeared that once the (negative) consequences materialize it has often become very difficult to change the direction of technological development because the technology has become deeply embedded in society and its design is more or less fixed. This problem is known as the **Collingridge dilemma**, after David Collingridge who first described it (Collingridge, 1980). Various approaches have been developed to overcome the Collingridge dilemma, one is **Constructive Technology Assessment (CTA)**. The idea behind CTA is that TA-like efforts are to be carried out parallel to the process of technological development and are fed back to the development and design process of technology (Schot, 1992; Schot and Rip, 1997). CTA aims at broadening the design process, both in terms of actors involved and in terms of interests, considerations and values taken into account in technological development. Among other things, this implies that stakeholders get a larger say in technological development.

There are also more recent approaches that aim at broadening the design process of technological development, such as **Design for Values** and **Responsible Innovation**, which will be discussed in Chapters 5 and 9. The approach of responsible innovation is to encourage innovators to work together with stakeholders during the research and innovation process, to better align the (expected) outcomes of innovation with the values, needs, and expectations of society. **Design for Values** is a systematic approach to addressing values of moral importance in (engineering) design. Both Responsible Innovation and Design for Values are in line with the demands for active responsibility, discussed in this chapter.

Constructive Technology Assessment (CTA) Approach to Technology Assessment (TA) in which TA-like efforts are carried out parallel to the process of technological development and are fed back to the development and design process.

Responsible Innovation Approach that stimulates innovators to work with stakeholders during the research and innovation process, to better align the (expected) outcomes of innovation with the values, needs, and expectations of society.

Design for Values Approach that aims at integrating values of ethical importance in a systematic way in (engineering) design.

What are the implications of the social context of technological development for the responsibility of engineers? In one sense, it diminishes the responsibility of engineers because it makes clear that engineers are just one of the many actors involved in technology development and cannot alone determine technological development and its social consequences. In another sense, however, it extends the responsibility of engineers because they have to take into account a range of stakeholders and their interests. Engineers cannot just as technocrats decide in isolation what the right thing to do is, but they need to involve other stakeholders in technological development and to engage in discussions with them.

1.7   Chapter Summary

In this chapter, we have discussed the responsibility of engineers. The notion of responsibility has different meanings. One sense of responsibility, accountability, implies the obligation to render an account of your actions and the consequences of these actions. If you are not able to give a satisfactory account, you are blameworthy. Usually four conditions need to apply in order to be blameworthy: causal contribution, wrong-doing, foreseeability, and freedom. In addition to accountability and blameworthiness, responsibility has an active component relating to preventing harm and doing good.

There are two main grounds of responsibility: the roles you play in society and moral considerations. Engineers have two main role responsibilities, one as engineers, the other as employees. As engineer you have a professional responsibility that is grounded in your role as engineer insofar as that role stays within the limits of what is morally allowed. Three professional ideals were examined as potential parts of the professional responsibility of engineers: technological enthusiasm, effectiveness and efficiency, and human welfare. The first two ideals are not always morally commendable and can in fact even become immoral when pursued in the light of immoral goals. The third ideal is morally laudable and, therefore, part of the professional responsibility of engineers.

Your professional responsibility as an engineer may sometimes seem to conflict with your responsibility as an employee. We have discussed three models for dealing with this apparent conflict: separatism, technocracy, and whistle-blowing. Separatism implies that the professional responsibility of engineers is confined to engineering matters and all decisions are made by managers and politicians. The disadvantage of this model is that engineers may end serving immoral goals and lose sight of the engineering ideal of public welfare. Technocracy means that engineers take over the decision power of managers and politicians. One disadvantage of this model is that engineers do not possess the expertise on which to decide for others what human welfare is or what is safe enough. Another disadvantage is that this model is paternalistic. Whistle-blowing means that you, as an engineer, speak out in public about certain wrong-doing or dangerous situations in a company. Although whistle-blowing may sometimes be required it is not a very attractive model for the relation between engineers and managers. Instead of any of the three models, it might be better to work on a relation between engineers and managers that is more cooperative and mutually supportive, such as a model in which engineers think about broader issues than just engineering decisions but do not decide on these issues alone.

The responsibility of engineers is further complicated by the social context of technological development. Apart from engineers, a whole range of other actors is involved in technological development. This diminishes the responsibility of engineers as their causal contribution to technology and the foreseeability of consequences is diminished. At the same time, it introduces additional responsibilities, because engineers also need to take into account other stakeholders and their interests in the development of new technologies.

Study Questions

1 What are the five features of active responsibility according to Bovens?

2 What is the difference between passive and active responsibility?

3 What criteria (conditions) are usually applied when deciding whether someone is passively responsible (blameworthy) for a certain action and its consequences?

4 According to Udo Pesch (2015), Constructive Technology Assessment (CTA) and Design for Values are methodologies that can be used to identify the contours of a framework for active responsibility of engineers. Explain why the methodologies CTA and Design for Values can help foster active responsibility.

5 Suppose one person's actions have led to the injury of another person. What additional criteria must be satisfied in order to imply that the first person is passively responsible for the injury?

6 Consider the following situation: An engineer who has been involved in the design of a small airplane for business travel, type XYZ, finds out that they used the wrong software to calculate the required strength of the wing. They have used a standard software package but now realize that their package was not fit for this specific type of airplane. The very same day they find this out, a plane of type XYZ crashes and all four passengers die. The investigation shows that the plane has crashed due to an inadequate design of the wing.

Do you consider this engineer responsible for the plane crash and the death of four people? (If you think there is not enough information to arrive at a judgment, indicate what information you would need to make a judgment and how this information would affect your judgment.)

7 In general, nobody will want to deny that engineers have an active responsibility for technologies they design and/or work with. In practice, however, many engineers find it problematic to act on this responsibility. Describe three problems for the idea that engineers should take responsibility for technologies and give a concrete example of each problem from engineering practice.

8 Explain what is meant by "separatism," and explain why the tripartite model illustrates separatism so well.

9 Why is it so difficult to steer technological development?

10 Explain why the ideal "public welfare" in professional ethics is the most important one for engineers from a moral perspective.

11 Look for an example of technological enthusiasm in your own field of study. Would you characterize this enthusiasm in this case as morally commendable, morally reprehensible, or just morally neutral? Argue your answer.

Discussion Questions

1 How serious does a wrongdoing have to be that it creates a moral obligation for an engineer to blow the whistle in spite of the very negative consequences for the engineer, such as dismissal?

2 Do you agree with Ceva and Bocchiola (2019) that whistle-blowing is an issue neither of personal morality nor professional duty, and that whistle-blowing isn't something that is heroic?

3 Can companies, as contrasted to people, be morally responsible? In what sense are companies different from people and is this difference relevant for moral responsibility?

4 What would you, as an engineer, have done if you had participated in the meeting in 2006 where the decision had to be made whether to use the cheating software in the Dieselgate scandal?

5 Give an example in engineering practice, and explain what is meant by "moral responsibility" in that example and how it extends beyond role responsibility.

6 As was already mentioned in the case Challenger (see subsection 1.5.1), Roger Boisjoly said – several months after the disaster – the following: "I must emphasize, I had my say, and I never [would] take [away] any management right to take the input of an engineer and then make a decision based upon that input … I have worked at a lot of companies … and I truly believe that … there was no point in me doing anything further [other] than [what] I had already attempted to do" Goldberg (1987, p. 156). Explain why this fits the term "separatism", and do you think Boisjoly's separatist argument is sound?

Notes

1 https://www.annualreports.com/HostedData/AnnualReportArchive/v/OTC_VWAGY_2009. pdf, accessed January 11, 2023.

2 https://www.reliableplant.com/Read/15320/vw-of-america-moves-forward-with-'strategy-2018'-plan, accessed January 11, 2023.

3 https://www.volkswagenag.com/presence/konzern/images/teaser/history/chronik/annual-report/2010-Annual-Report, accessed February 20, 2023.

4 https://www.theguardian.com/business/2015/sep/23/volkswagen-ceo-martin-winterkorn-quits-over-diesel-emissions-scandal, accessed March 22, 2022.

5   https://www.justice.gov/opa/pr/volkswagen-engineer-sentenced-his-role-conspiracy-cheat-us-emissions-tests, accessed March 2, 2022.

6   https://medium.com/swlh/the-lie-of-clean-diesel-f55b75031066, accessed March 17, 2022. See also Ewing (2017).

7   https://www.defense.gov/News/News-Stories/Article/Article/1254719/project-maven-to-deploy-computer-algorithms-to-war-zone-by-years-end, accessed March 20, 2022.

8   https://www.nytimes.com/2018/04/04/technology/google-letter-ceo-pentagon-project.html, accessed March 20, 2022.

9   https://blog.google/technology/ai/ai-principles/, accessed November 17, 2022.

10 "Google: Achter het scherm" ("Google: Behind the Screen"), Tegenlicht, https://www.vpro.nl/programmas/tegenlicht/kijk/afleveringen/2005-2006/google-achter-het-scherm.html, accessed June 1, 2022.

11 https://www.reuters.com/article/us-volkswagen-emissions-sentencing-idUSKCN1B51YP, accessed February 20, 2023.

12 Text from the number "Wernher von Braun" by Tom Lehrer that featured in his album *That was the year that was* (1965).

13 https://www.reuters.com/article/us-volkswagen-emissions/volkswagen-misused-me-accused-executive-tells-judge-idUSKBN1DX0BW, accessed March 17, 2022.

14 https://www.wsj.com/articles/the-facebook-files-11631713039, accessed March 16, 2022.

15 See https://www.youtube.com/watch?v=_Lx5VmAdZSI, accessed March 16, 2022.

16 https://edition.cnn.com/business/live-news/facebook-senate-hearing-10-05-21/h_cc8e64ccefb1b6431db1a41594cde902, accesses March 10, 2022.

17 https://time.com/6121931/frances-haugen-facebook-whistleblower-profile, accessed March 10, 2022.

18 https://edition.cnn.com/2021/10/04/tech/facebook-whistleblower-prepared-testimony/index.html, accessed March 10, 2022.

19 https://www.europarl.europa.eu/news/nl/press-room/20211011IPR14619/facebook-files-meps-to-invite-whistleblower-frances-haugen-to-a-hearing, accessed March 10, 2022.

20 https://cen.acs.org/articles/83/i27/PFOA-CALLED-LIKELY-HUMAN-CARCINOGEN.html., Accessed 2 March 2023.

21 We will discuss the reasons for this assumption in more details in later chapters.

22 https://www.latimes.com/archives/la-xpm-2004-jul-09-na-teflon9-story.html, accessed March 10, 2022.

2

Codes of Conduct

Having read this chapter and completed its associated questions, readers should be able to:

- Describe professional codes and corporate codes;
- Differentiate between three types of codes of conduct: aspirational, advisory, and disciplinary;
- Understand the role of codes of conduct with respect to the responsibility of engineers;
- Identify the strengths and weaknesses of codes of conduct;
- Evaluate the role of global codes for multinationals and for engineers.

Contents

2.1 **Introduction** 38

2.2 **Codes of Conduct** 39

    2.2.1 Professional codes 40
    2.2.2 Corporate codes 47

2.3 **Possibilities and Limitations of Codes of Conduct** 55

    2.3.1 Codes of conduct and self-interest 55
    2.3.2 The need for interpretation 58

2.3.3 Can moral judgments be codified? 60

2.3.4 Can codes of conduct be lived by? 62

2.3.5 Enforcement 64

2.4 **Chapter Summary** 66

Study Questions 67

Discussion Questions 68

Ethics, Technology, and Engineering: An Introduction, Second Edition. Ibo van de Poel and Lambèr Royakkers.
© 2023 John Wiley & Sons Ltd. Published 2023 by John Wiley & Sons Ltd.

2.1   Introduction

Case   Bay Area Rapid Transport Project

Figure 2.1   BART train. Photo: NISEE-PEER Earthquake Engineering Library, University of California, Berkeley.

In March 1972 Holger Hsortsvang, Max Blakenzee, and Robert Bruder, three engineers, working on the *Bay Area Rapid Transport Project (BART)* (see Figure 2.1) in California (United States) and responsible for the design and creation of an automatic guided train system, were dismissed. These engineers had been expressing their doubts about the safety of the system via internal memos since 1969 to their managers. The response was "don't make trouble." In 1971 they brought their concerns in confidence to members of the board of directors, thus bypassing their immediate superiors. That was unconventional for the BART organization and indeed for any hierarchical organization. The director they finally made contact with turned out to be very interested in their case and so he promised to raise it with the management. He furthermore promised to keep their names anonymous and do nothing to damage their interests. However, two days after the encounter the full story was published in the *Contra Costa Times*. At first the engineers denied having any involvement in the matter but once their involvement was confirmed they were immediately fired without cause or appeal. They subsequently took the matter to court.

In the wake of the affair one of the organizations to become involved was the Institute of Electrical and Electronic Engineers (IEEE). The IEEE decided to send what is known as an amicus curiae letter to the law courts. (An amicus curiae is an "friend of the court": someone, not a party to a case, who voluntarily offers information on a point of law or some other aspect of the case to assist the court.) The letter emphasized the fact that according to the IEEE's professional code, engineers are responsible for the "safety, health and welfare of the public." The IEEE also argued that the professional code is an implicit aspect of the employment contract. If this argument had been accepted by the judge, then it would have meant that employees who act in accordance with what is stated in the professional code may not be simply dismissed.

After the three engineers had lost their job, their concerns were decisively confirmed on October 2, 1972, three weeks after BART began carrying passengers. There was a train system accident and several passengers were injured. Despite this, the three engineers accepted an out-of-court settlement reported to be $25,000 per person. The presumed reason for this was that they had in the first instance lied about their involvement in the matter which had weakened their case. Apart from anything else, the dismissals were very detrimental for the careers of all three engineers.

Source: Based on Anderson et al. (1980), Anderson et al. (1983), and Unger (1994, pp. 12–17).

In this case, the three engineers acted out of a sense of professional responsibility. This professional responsibility was codified in the IEEE code of conduct and was related to the safety, health, and welfare of the public. Although their professional organization supported their behavior, it could not prevent them from being dismissed. In this chapter, we discuss the role of codes of conduct in engineering. In particular, we focus on professional codes as they have been proposed by professional engineering societies and on corporate codes, as they have been formulated by companies. In Section 2.2, we discuss these two types of codes, their structure and their content. In Section 2.3, we discuss a number of common objections that have been leveled against codes of conduct. This includes the problem that is highlighted by the case above, that acting according to the code, may nevertheless lead to dismissal. In Section 2.4, we will discuss codes of conduct in an international context.

2.2 Codes of Conduct

Codes of conduct are codes in which organizations lay down rules for responsible behavior of their members. Such rules may be detailed and prescriptive, but they can also be formulated more broadly and express the values and norms that should guide behavior and decision-making (Hummels and Karssing, 2007). Codes of conduct are often intended as an addition to the rules of the law. When codes of conduct are enforced this is usually done by the organization that formulated the code. For engineers, two types of codes of conduct are especially important: **professional codes** that are for-

> **Codes of conduct** A code in which organizations (like companies or professional associations) lay down rules for responsible behavior of their members.
>
> **Professional code** Code of conduct that is formulated by a professional association.
>
> **Corporate code** Code of conduct that is formulated by a company.

mulated by professional associations of engineers, and **corporate codes** of conduct that are formulated by companies in which engineers are employed.

Codes of conduct are formulated for a variety of reasons, such as: increasing moral awareness; the identification and interpretation of the moral norms and values of a profession or a company; the stimulation of moral discussion; as a way to increase accountability to the outside world; and, finally, to improve the image of a profession or company. Depending on the exact objectives of a code of conduct, a distinction can be made between three types of codes of conduct:[1]

> **Aspirational code** A code that expresses the moral values of a profession or company.
>
> **Advisory codes** A code of conduct that has the objective to help individual professionals or employees to exercise moral judgments in concrete situations.
>
> **Disciplinary code** A code that has the objective to formally enforce that the behavior of all professionals or employees meets certain moral standards.

- An **aspirational code** expresses the moral values of a profession or company. The objective of such a code is to express to the outside world the kind of values the profession or company is committed to.
- An **advisory code** has the objective to help individual professionals or employees to exercise moral judgments in concrete situations on the basis of the more general values and norms of the profession or company.
- A **disciplinary code** has the objective to formally enforce that the behavior of all professionals or employees meets certain values and norms. While also advisory – and even aspirational – codes may be aimed at ensuring that the behavior of professionals is in line with certain moral standards, only disciplinary codes are formally enforced, and involve formal sanctions if the code is not followed.

Most professional codes for engineers are advisory. Usually, they have the following more specific objectives: increasing awareness of and sensitivity for moral issues in the daily exercising of the profession; helping in analyzing such moral issues and in formulating key questions or issues with respect to these moral issues; and, finally, helping in coming to a judgment on these moral issues. Corporate codes of conduct are more often disciplinary. In such cases, their objective is to enforce that all employees act according to certain rules. The formulation of codes of conduct is only one of the activities that professional associations and companies can undertake to stimulate responsible behavior by their members. Other activities include the appointment of a confidant or committee which whom moral problems can be discussed or the organization of training sessions for dealing with moral dilemmas.

2.2.1 Professional codes

Professional codes are rules for the exercising of a profession and are formulated by a professional society. These codes are usually not intended to formally enforce behavior, but to inform, encourage, and inspire their members in particular ways. Professional codes have been formulated for a variety of professions like doctors, nurses, lawyers, the police, and corporate managers. Also engineers have professional codes of conduct.

Historically, the development of professional codes for engineers began in England in 1771 with the code of the *Smeatonian Society*. More influential for the current professional codes for engineers was the formulation of a range of professional codes for different engineering professions like civil, mechanical, and electrical engineering in the first decade of the twentieth century in the United States. The early codes regulated people's entry into the profession and the behavior of members toward each other and in relation to employers and clients. World War II gave the formulation of codes of conduct a new boost. The gas chambers and scientific experiments that had been carried out by the Germans on people during the World War II had given science and technology a bad image. The atomic bomb also showed clearly that technology gave rise to certain moral issues.

Case The Atomic Bomb

In 1932 James Chadwick discovered the neutron, which later proved the key to nuclear fission and the discovery of the atomic bomb. The Hungarian scientist Leó Szilárd as early as October 1933 realized that "a chain reaction might be set up if an element could be found that would emit two neutrons when it swallowed one neutron" (Jungk 1958, p. 54). This chain reaction would result in the production of large amounts of energy that might be used to produce energy but might also be put to bad purposes. In the same year, Hitler had come to power in Germany and Szilárd had fled to London to escape Nazi prosecution. Szilárd therefore started lobbying for not publishing the results of studies on this topic,

Figure 2.2 Atomic bomb mushroom cloud. Photo: Library of Congress/United States Department of Defence.

as he feared they could be misused by the German government; he was however not very successful.

In 1934 the research groups of both Enrico Fermi and Irene Joliot-Curie disintegrated heavy atoms by spraying them with neutrons. At this point these scientists did not realize that they had achieved fission. It took until 1938 before the experiments were properly interpreted, after another experiment with bombarding uranium with neutrons by the German physicist Otto Hahn, who is usually credited with discovering nuclear fission. On February 2, 1939, Szilárd wrote a letter to Joliot-Curie: "Obviously, if more than one neutron were liberated, a sort of chain reaction would be possible. In certain circumstances this may then lead to the construction of bombs which would be extremely dangerous in general and particularly in the hands of certain governments" (Jungk, 1958, p. 77), and "We all hope that there will be no or at least not sufficient neutron emissions and therefore nothing to worry about" (Jungk, 1958, p. 77). At that time, Joliot-Curie was just at the point of experimental realization of the mentioned chain reaction and her group published the results to the dismay of Szilárd.

As Szilárd feared that the Germans might be able to develop an atomic bomb, he began to look for ways to persuade the US government also to do so. In August 1939, he succeeded in convincing Einstein to sign a letter to President Roosevelt in which they warned for the developments in Germany and urged more American studies on the subject. The letter eventually reached Roosevelt in October 1939, and contributed to the establishment of the so-called Manhattan Project, a large research project in the US that would eventually result in the production of atomic bombs. After the war, Einstein came to regret his cooperation deeply: "If I had known that the Germans would not succeed in constructing the atom bomb, I would never have lifted a finger" (Jungk, 1958, p. 87).

Toward the end of the war, a number of scientists working on the Manhattan Project became concerned about the use of the atomic bomb they had developed by the US government. In July 1945, 69 scientists signed a petition drafted by Szilárd. This petition, among other contained the following passages (https://www.atomi carchive.com/resources/documents/manhattan-project/petition.html):

> We, the undersigned scientists, have been working in the field of atomic power. Until recently, we have had to fear that the United States might be attacked by atomic bombs during this war and that her only defense might lie in a counterattack by the same means. Today, with the defeat of Germany, this danger is averted and we feel impelled to say what follows:
>
> The war has to be brought speedily to a successful conclusion and attacks by atomic bombs may very well be an effective method of warfare. We feel, however, that such attacks on Japan could not be justified, at least not unless the terms which will be imposed after the war on Japan were made public in detail and Japan were given an opportunity to surrender.
>
> The added material strength which this lead [in the development of the atomic bomb] gives to the United States brings with it the obligation of restraint and if we were to violate this obligation our moral position would be weakened in the eyes of the world and in our own eyes. It would then be more difficult for us to live up to our responsibility of bringing the unloosened forces of destruction under control.

The signed petition never reached President Truman. On August 6, 1945, the US dropped the atomic bomb "Little Boy" on the city of Hiroshima, followed on August 9 by the dropping of the "Fat Man" nuclear bomb over Nagasaki. The bombs killed as many as 140,000 people in Hiroshima and 80,000 in Nagasaki by the end of 1945. On August 15, 1945, Japan announced its surrender to the Allied Powers.

Source: Mainly based on Jungk (1958).

One of the ways of restoring the social image of science and technology after World War II was by establishing professional codes. In 1950 the German engineers' association, the Verein Deutscher Ingenieure (VDI), drew up an oath for engineers, which was clearly inspired by the dubious role of some engineers and scientists during World War II. One of the things stated in the professional code was that engineers should not work for those who fail to respect human rights.[2] Also in the United States, most of the professional codes were reformulated after World War II: the duty of the engineer to serve the public interest was especially stressed in the new codes of conduct. Organizations like the National Society of Professional Engineers (NSPE, see Appendix II), the American Society of Civil Engineers (ASCE), and The American Society of Mechanical Engineering (ASME) formulated codes of conduct stating that engineers "should hold paramount the safety, health and welfare of the public."

The engineering profession in the United States has been a world leader in promoting engineering ethics code development and associated educational activities. Due to their leadership other nations have followed the American lead and have adopted US codes, not only in Western countries. For example, the Japan Society of Professional Engineers (JSPE) has adopted the NSPE code of conduct. The JSPE was formed in 2000, and since 2001 it is an affiliate group of NSPE. Its mission is "to encourage Japanese engineers to hold paramount the safety, health, property and welfare of the public through internationally recognized practice of professional engineering ethics."[3] The codes of conduct of the Japan Society of Civil Engineers (JSCE) and Japanese Society for Engineering Education (JSEE) also emphasize that engineers should hold paramount the safety, health and well-being of human beings. These codes show that in spite of the cultural differences there is a lot of consensus about the responsibilities of engineers and the basic moral standards for engineering. Since the globalization of the world's economics has increased the working space of engineers, engineers need, however, to be aware of the cultural differences, which can influence engineering decisions. Luegenbiehl and Clancy (2017) mentioned two basic moral principles for global engineering related to cross-cultural values:

1 Engineers should endeavor to understand and respect the nonmoral cultural values of those they encounter in fulfilling their engineering duties. If engineers fail to develop an adequate understanding of practices belonging to other cultures – and are required to communicate with engineers from or the public of these cultures – they will not be able to practice engineering competently.
2 Engineers should endeavor to refuse to participate in engineering activities that are claimed to reflect cultural practices but that violate basic moral principles for global engineering. So, engineers should make sure that they are not misled into believing that potentially immoral behavior is simply a matter of cultural practice.

In addition to national engineering societies, Europe has an overarching professional organization, the ENGINEERS EUROPE, formerly known as the European Federation of National Engineering Associations (FEANI). ENGINEERS EUROPE was established in 1951 by a group of German and French engineers. At the moment, professional associations from 33 European countries are member of ENGINEERS EUROPE. ENGINEERS EUROPE strives for a single voice for the engineering profession in Europe and wants to affirm and develop the professional identity of engineers.[4] It has formulated a universal statement regarding the conduct of professional engineers, which can be implemented by national member's societies in their code of conduct. The ENGINEERS EUROPE code (see Appendix III) thus has a quite different status than most US codes like the NSPE code which is reflected in the content of the code, in particular, the ENGINEERS EUROPE code is much more general (and vague) and contains much less detail than, for example, the NSPE code. In 1946, NSPE released its Canons of Ethics for Engineers and Rules of Professional Conduct, which evolved to the current Code of Ethics, adopted in 1964. Although these statements of general principles provided guidance, many engineers requested interpretations of how the Canons and Rules would apply to specific circumstances. These requests eventually led to the establishment of the NSPE Board of Ethical Review (BER) in 1954. The BER considers cases involving either real or hypothetical matters submitted from NSPE members, other engineers, public officials, and members of the public, and reviews each case in the context of the NSPE Code of Ethics and earlier BER opinions. The compendium of cases, in the meantime more than 650 cases, offers a unique catalog of ethics interpretation and application of the code of ethics for the profession. Box NSPE Board of Ethical Review provides an example of such a review.

NSPE Board of Ethical Review

Case: Public Health, Safety, and Welfare—Drinking Water Quality (Case 20-4, Year 2020)

Facts:

Engineer A is a professional engineer who serves as the superintendent and chief engineer for the Metropolitan Water Commission (MWC). In order to reduce municipal expenditures and lower water rates, the MWC has been considering changing its water supply source from purchasing water from remote reservoirs from another regional authority to using the local river as the MWC's source. Engineer B, a consulting engineer retained by the MWC charged with evaluating water treatment needs for the change in water source, provided a report to Engineer A recommending extensive capital investments and a three-year timeline for further evaluation of water quality, design, and construction of improvements. The improvements are needed prior to the change in water source to ensure that sufficient corrosion control is provided so that old service pipes in the MWC service area don't leach lead at levels in excess of drinking water standards. Both Engineer A and Engineer B met with the MWC at a meeting sparsely attended by the public and recommended that the change in water source be substantially delayed until improvements could be completed. Despite those recommendations, the MWC voted to proceed simultaneously with the accelerated evaluation and design of needed water treatment improvements and the change in water source.

Question(s):

1   What are the ethical obligations of Engineer A and Engineer B in this circumstance?
2   What should Engineer A and Engineer B do?

NSPE Code of Ethics References:

Section II.1.: Engineers shall hold paramount the safety, health, and welfare of the public.

Section II.1.a.:If engineers' judgment is overruled under circumstances that endanger life or property, they shall notify their employer or client and such other authority as may be appropriate.

Section II.1.c.:Engineers shall not reveal facts, data, or information without the prior consent of the client or employer except as authorized or required by law or this Code.

Section III.1.b:Engineers shall advise their clients or employers when they believe a project will not be successful.

Discussion:

(…) The engineering judgments of Engineer A and Engineer B were overruled by the MWC. If Engineers A and B believe life or property is endangered, Section II.1.a. provides that not only shall the employer or client be notified, but also all other appropriate authorities. It appears that the state regulatory agency has been contacted; however, there should be a formal presentation of the facts, findings, and recommendations. This action may also address Section II.1.c. As Engineers A and B are required to hold paramount the safety, health, and welfare of the public, and as this duty is a fundamental canon of the NSPE Code of Ethics, the consent of the MWC is not required.

Additionally, if project success is defined as "the public will not be endangered at all," then Engineers A and B should advise their client that they believe the project will not be successful. Again, as with the state regulatory agency, this advisement should proceed in a formal manner.

The formal presentations satisfy Engineer A's and Engineer B's duty to report. However, in the event that these formal presentations fail to sway the MWC to change its plans, given the gravity of the danger to public health and safety, Engineers A and B have an obligation to further pursue the matter.

Conclusions:

1 In fulfillment of their ethical obligations under the Code, Engineers A and B should formally communicate their concerns to the MWC, including that they believe the project will not be successful.
2 Both Engineers A and B have ethical obligations to notify the MWC and other appropriate authorities that prematurely changing the water source puts the public health and safety at risk. Furthermore, Engineers A and B have independent obligations to formally and in writing report their concerns to the state regulatory agency. While they may provide a joint and cooperative report, each has an independent obligation. Neither the consent nor opposition of the client is a factor in their fulfillment of this obligation.

Source: Reproduced with permission National Society of Professional Engineers https://www.nspe.org/resources/ethics/ethics-resources/board-ethical-review-cases/public-health-safety-and-welfare-3/, accessed November 17, 2022.

Professional codes for engineers provide content to the responsibility of engineers. They express the moral norms and values of the profession. Most modern professional codes relate to three domains: (1) conducting a profession with integrity and honesty, and in a competent way; (2) obligations toward employers and clients; and (3) responsibility toward the public and society.

Integrity and competent professional practice

All professional codes include the obligation to practice one's profession with **integrity** and **honesty**, and in a competent way. This is the traditional core of all professional codes. To practice one's profession in a competent way means that the practitioner must be competent and the professional practice must be conducted skillfully. This implies that practitioners must be well enough educated, must keep up to date in their field and must take only work in their field of competence. With integrity and honesty, we mean that the profession must be conducted in an honest, faithful, and truthful manner. This entails, for instance, that facts may not be manipulated and agreements must be honored.

> **Integrity**   Living by one's own (moral) values, norms, and commitments.
>
> **Honesty**   Telling what one has good reasons to believe to be true and disclosing all relevant information.

Engineers shall perform services only in the areas of their competence. (NSPE)

Engineers shall issue public statements only in an objective and truthful manner. (NSPE)

Engineers shall not be influenced in their professional duties by conflicting interests. (NSPE)

Engineers shall maintain their relevant competences at the necessary level and only undertake tasks for which they are competent. (ENGINEERS EUROPE)

Engineers shall carry out their work in full understanding of all laws, rules, and regulations as well as of well-founded principles, actively and willingly taking the lead in the observance of societal standards and seeking to improve them in response to both social and technological change. (JSCE)

> **Conflict of interest**   The situation in which one has an interest (personal or professional) that, when pursued, can conflict with meeting one's professional obligations to an employer or to (other) clients.

Sometimes it is also stipulated that the profession must be practiced in an independent and impartial way. Usually this is meant to imply that engineers should avoid conflicts of interests. You have a **conflict of interest** if you have an interest that, when pursued, conflicts with meeting your obligations to your employer or clients. This may be a personal interest, like when you have stocks in a company that produces a certain kind of measuring apparatus and you have to advise a large client about what measuring apparatus to use. It can also be an interest that derives from another professional role, for example, when you advise two competing firms. Although conflicts of interest do not necessarily lead to immoral behavior it is better to avoid them because a conflict can corrupt your professional judgment and diminishes your trustworthiness as an engineer. If a conflict of interest is unavoidable it should at least be disclosed to the interested parties.

Obligations toward clients and employers

Obligations toward clients and employers are mentioned in most professional codes. In many cases, it is stipulated that engineers should serve the interests of their clients and employers and that they must keep secret confidential information passed on by clients or employers.

Engineers shall act for each employer or client as faithful agents or trustees. (NSPE)

Engineers shall not disclose, without consent, confidential information concerning the business affairs or technical processes of any present or former client or employer, or public body on which they serve. (NSPE)

Engineers shall provide impartial analysis and judgement to employer or clients, avoid conflicts of interest, and observe proper duties of confidentiality. (ENGINEERS EUROPE)

Engineers shall be fair and unbiased in all their interactions with the people, their clients, the organizations for which they work, as well as themselves, faithfully and honestly discharging their duties and avoiding any conflicts of interest. (JSCE)

Social responsibility and obligations toward the public

Virtually all professional codes in one way or another emphasize the social responsibility of engineers. Matters frequently referred to are: safety; health; the environment; sustainable development; and the welfare of the public. According to a limited number of professional codes engineers must inform the public about the aspects of the technology in which they are involved and that are relevant to the public, such as the risks and hazards involved.

Engineers shall hold paramount the safety, health, and welfare of the public. (NSPE)

Engineers shall at all times strive to serve the public interest. (NSPE)

Engineers are encouraged to adhere to the principles of sustainable development in order to protect the environment for future generations. (NSPE)

Engineers shall carry out their tasks so as to prevent avoidable danger to health and safety, and prevent avoidable adverse impact on the environment. (ENGINEERS EUROPE)

Engineers shall utilize their expertise and experience to develop and implement comprehensive solutions to issues of public interest, keeping in mind the peace and prosperity of the people and the development of society as their constant concern. (JSCE)

Engineers shall be committed to aiding in protecting the life and property of the people, working with colleagues across a broad range of disciplines, while looking beyond their professional expertise to the concerns of the people, realizing both the capabilities and the limitations of technology with the people. (JSCE)

2.2.2 Corporate codes

Corporate codes are voluntary commitments made by individual companies or associations of companies setting certain values, standards, and principles for the conduct of corporations. Corporate codes are usually more recent than professional codes. They have been formulated since the 1960s and 1970s, particularly in reaction to corporate scandals (Ryan, 1991). Nowadays, most modern corporations have a corporate code of conduct. According to a survey that was carried out in 2013, 95 percent of both Fortune US 100 and Fortune Global 100 companies have a code of conduct (Sharbatoghlie et al., 2013). Next, we will discuss the main elements of the various kinds of corporate codes: the mission and vision, the core values, the responsibilities toward stakeholders, and detailed rules and norms.

Corporate Social Responsibility (CSR) and Sustainable Development Goals (SDGs)

Corporate Social Responsibility The responsibility of companies toward stakeholders and society at large that extends beyond meeting the law and serving shareholders' interests.

The formulation of corporate codes is based on the assumption that companies have a **corporate social responsibility**, that is, a responsibility toward stakeholders and society at large. This assumption has been contested by several authors who maintain that the responsibility of a company is limited to making profit within the limits of the law. This so-called classical view on corporate responsibility can be traced back toward Adam Smith, the founder of modern economics. According to Smith, the invisible hand of the market makes – under certain conditions like the absence of force and fraud – everyone better off if all people, producers and consumers alike, only pursue their own interests (Smith, 1776). An important contemporary defender of the classical view is the economist and Noble Prize laureate Milton Friedman. According to Friedman, companies only have responsibilities toward their shareholders and not to any other stakeholders, society, or the environment (Friedman, 1962). He considers it undesirable that companies take into account other stakeholders' interests and views. He provides two arguments for this statement. First, money spent by a corporation on social responsibility is ultimately the money of the shareholders and this expenditure conflicts with their goal to maximize profits. Second, corporations are not democratically elected. When companies formulate their own ideas about what is morally allowable or desirable they are enforcing their own particular view upon others without any democratic legitimization. If any limits on corporate behavior are desirable, they have to be formulated by the government, not by companies. A similar argument was used by Facebook spokeswoman Dani Lever, when she reacted to the accusation by Frances Haugen (see Chapter 1) that Facebook prioritizes profits over people: "We have no commercial or moral incentive to do anything other than give the maximum number of people as much of a positive experience as possible. Like every platform, we are constantly making difficult decisions between free expression and harmful speech, security, and other issues, and we don't make these decisions inside a vacuum – we rely on the input of our teams, as well as external subject matter experts to navigate them. But drawing these societal lines is always better left to elected leaders, which is why we've spent many years advocating for updated internet regulations."[5]

A number of objections can be raised against Friedman's view. First, although responsibilities to other stakeholders can conflict with shareholders' interests, this is not always the case. First, laws are not always adequate or effective in preventing immoral behavior. Not everything that is morally desirable can be laid down in the law. Laws also tend to lag behind technological development and companies might be in a better position to foretell moral issues raised by new technology than the government. Hence, they have a responsibility that extends beyond what the law requires. Second, companies are the ones that can do something about the huge and fundamental problems we deal with such as sustainability, poverty, and so on. The argument runs as follows:

- There are serious problems in the world, such as poverty, hunger, social exclusion, conflict, environmental degradation, climate change problems, and so on;
- Any agent with the resources, manpower, technology, and knowledge necessary to mitigate these problems has a moral responsibility to do so, assuming the costs they incur on themselves are not great;
- Companies have the resources, manpower, technology, and knowledge necessary to mitigate these problems without incurring great costs.

So, companies should contribute to solutions to these problems.

This is also the reason why, for example, Microsoft is investing in corporate social responsibility initiatives and integrating them into business strategy: "We believe that companies that can do more, should. That's why we focus on four key areas [supporting inclusive economic opportunity; protecting fundamental rights; committing to a sustainable future; and earning trust] in which technology can and must benefit the future of humanity and our planet."[6]

However, CSR initiatives could go beyond moral and philanthropic purposes (Fallah Shayan et al., 2022). They do not necessarily have a negative impact on their bottom line, and they can have an extremely positive impact, which is the third objection against Friedman's view. CSR initiatives enhance company's image and brand, reputation and recognition, public trust and identification, customer satisfaction and loyalty, purchase intention, financial performance, access to capital and markets, and transparency. CSR, therefore, is a win-win value-created strategy, since CSR initiatives create shared value for both companies and their societies (Tundys, 2021). As we have seen in the Dieselgate scandal (see Chapter 1), neglecting CSR may result in large costs, risks, and missed value creation opportunities. The scandal has cost Volkswagen more than 30 billion euros, and has led to 1,200 premature deaths in Europe.

Companies have invested in CSR plans out of moral, philanthropic purposes or financial ones for years, however, not many of these plans are systematically designed: "Each company scanned their society's issues and challenges, picked one or a few of the matters based on their own limited knowledge, and tried to make a small difference in a localized context" (Fallah Shayan et al., 2022, p. 2). Most companies lack a clear vision of how to enhance their CSR practices, and therefore, corporate progress on CSR, in general, is undermined. The United Nations General Assembly offered in 2015 a shared vision: a roadmap by which companies can begin to strategically align their CSR initiatives with the Sustainability Development Goals (SDGs) (ElAlfy et al., 2020). The SDGs consist of 17 goals, as illustrated in Figure 2.3, and 169 associated targets, over the period of 2015–2030. For example, a target for the goal 'no poverty' is "reduce 50% people in poverty by 2030". The SDGs can be considered as a universal set of goals to develop a global vision for sustainable development by balancing economic growth, social development, and environmental protection.[7] Schönherr et al. (2017) state that SDGs may prove beneficial as an underlying framework for companies, especially multinationals, to start mapping their CSR initiatives in order to identify leverage points for enhancing positive impacts and mitigating negative ones. The SDGs contain universally agreed-upon sustainable development priorities broken down in targets of which many are directly relevant to business, and they provide

Figure 2.3 Sustainability Development Goals (UNITED NATIONS/https://www.un.org/en/ sustainable-development-goals/last accessed November 15, 2022).

a common agenda for all stakeholders to rally around and to build partnerships to jointly tackle sustainable development issues beyond the control of a company.

According to the United Nations Sustainability Index Institute, 95 percent of the 500 world's largest companies spent budget on the SDGs and reported their contributions in their annual reports.[8] This illustrates that the SDGs are taken seriously by companies, and that the SDGs are a useful framework to enhance companies' CRS practices. In the Niger Delta Oil Spill case it is obvious that Shell Petroleum Development Company (SPDC) in Nigeria has not aided development in the country as provided by the SDGs, especially Goals 1, 2, and 3. The operations of SPDC have worsen and degenerated the livelihood of people of the host communities into penury, and precarious conditions (see box Shell, Nigeria and the Ogoni: A Study in Unsustainable Development).

Case   Shell, Nigeria and the Ogoni: A Study in Unsustainable Development

"Shell is a global group of energy and petrochemical companies. Our aim is to meet the energy needs of society, in ways that are economically, socially and environmentally viable, now and in the future" (www.shell.com). The company is involved is several voluntary social and environmental initiatives, such as the United Nations Global Compact.

Figure 2.4   Gas flaring (Shell): Woman tending her plot at Shell gas flare site, Rumuekpe, Nigeria. Photo: LIONEL HEALING/AFP/Getty Images.

The decision of the Court of Appeal of The Hague (the Netherlands) on January 29, 2021, that Shell Petroleum Development Company (SPDC) – Shell's Nigerian subsidiary – is responsible for several oil spills in the Niger Delta and liable for the damage the spills cause, offers an opportunity for ending one of the longest running conflicts between a multinational oil company and a local community in the Niger Delta. The Niger Delta was once considered the breadbasket of Nigeria because of its rich ecosystem, a place where people cultivated fertile farmlands and benefited from abundant fisheries. Although SPDC has stopped drilling in 2008, due to the decision of the Nigerian government to replace SPDC as operator of oil concessions in Ogoni area, oil spills continue from leaks of underground pipelines and oil wells, which has rendered local people's fields and fish ponds unusable. In 2008, four Nigerian farmers, represented by Friends of the Earth Netherlands (Milieudefensie) who brought the case that the Court of Appeal of The Hague had decided, sued SPDC for damage caused by the oil spills. They also demanded that SPDC clean up the contamination better and take measures to prevent a recurrence. SPDC denied liability, since the spoils were caused by sabotage and in that case, there is no liability under Nigerian law. In addition, SPDC claimed that the contamination had been cleaned up sufficiently and regardless of the cause of the leaks. In 2013, the District Court of The Hague held SPDC not responsible for the spills, since the court found that the spills were the result of sabotage. This decision was reversed by the Court of Appeal of The Hague. Since SPDS could not provide any evidence of the pipe leaks caused by sabotage, the court found SPDC responsible for most of the oil spills. In addition, the court was not convinced that SPDC has sufficiently cleaned up the oil spills and found that SPDC has violated its "duty of care" with regard to operations oversea. The court ordered SPDC to ensure that a leakage detection system is implemented in the pipelines in Nigeria. This ruling is significant, since establishing a "duty of care" on the parent company (Shell) could break new legal ground: "global environmental case jurisdiction is evolving beyond locality where environmental pollution occur and can be heard in the parent company's country" (Ngwakwe, 2021).

The origins of the conflict between the Ogoni and SPDC date back to the company's discovery of oil in this part of the Niger Delta in 1958. Nigeria was still under British colonial rule, and the Ogoni, like all other minority ethnic groups in the Delta, had no say in the exploitation agreements. Even after independence in 1960, they were not accorded a real stake in oil production.

There were more than 100 oil wells, mostly operated by SPDC. As elsewhere in the Delta, the environmental effects of oil exploration and production in Ogoni territory were severe. Land and water pollution from spills played havoc with the ecosystem. Villagers lived with gas flares burning 24 hours a day (some for over 30 years) and air pollution that produced acid rain and respiratory problems. Above-ground pipelines cut through many villages and former farmland.

SPDC refused to accept responsibility for environmental repercussions and largely denied there was an issue. As late as 1995, for example, an SPDC document insisted that: "Allegations of environmental devastation in Ogoni, and elsewhere in our operating area, are simply not true. We do have environmental problems, but these do not add up to anything like devastation." In response to criticism of its community relations practices, SPDC insisted that most of the Ogoni demands for social benefits and infrastructural development were the responsibility of the government, not an oil company. It maintains that it has responded "promptly, fairly, and completely" to community complaints in Ogoni land but that many, such as those articulated in the Ogoni Bill of Rights, are of a political nature and thus beyond its competence.

In response the Ogoni founded in 1992 the Movement for the Survival of Ogoni People (MOSOP), led by Ken Saro-Wiwa. From the start it adopted a policy of non-violence. MOSOP demanded that SPDC take responsibility for its massive environmental devastation of their homeland and denounced the injustices that Shell has inflicted on the Ogoni and other peoples in the Niger Delta. In 1995, Ken Saro-Wiwa and 13 other MOSOP leaders were subjected to a secret tribunal that, based on unsubstantiated allegations, sentenced nine of the men to death by hanging. They were accused of incitement to murder. All nine were summarily executed without any opportunity for appeal.

Most Ogoni saw SPDC as the architect of the events. The company strongly denied any complicity in the military repression of the Ogoni. However, the impression persisted that it had a hand in the repression. The Ogoni resolved never to allow SPDC to resume operations on their land. Many regarded SPDC's pledge not to use armed escorts and only to resume operations with host communities' consent as mere posturing.

A major issue that has to be dealt with is environmental clean-up. No significant study has been conducted to determine reliably the precise impact of oil industry-induced environmental degradation on human livelihoods in the area, but there are indications of severe damage. In 2011, the United Nations Environmental Program (UNEP) conducted a study and determined that "oil spills continue to occur with alarming regularity."[9] For example, in some areas of the Niger Delta the drinking water containing carcinogenic benzenes at over 900 times the WHO-approval levels, and the life expectancy in the Niger Delta is almost ten years less than in the rest of

Nigeria. According to the UNEP, the devastating oil spills in the Niger delta over the past five decades will cost $1 billion to rectify and take up to 30 years to clean up.

Source: Based on Ngwakwe (2021), International Crisis Group (2008), and Boele et al. (2001).

Mission and vision statement

Many corporate codes contain a mission and vision statement. A mission statement concisely formulates the strategic objectives of the company and answers the question of what the organization stands for. A vision statement is an aspirational statement made by a company that articulates what they would like to achieve in the long-run, generally in a time frame of five to ten years.

Apple mission statement is to bring the best user experience to its customers through its innovative hardware, software, and services, and its' vision statement is to make the best products on earth and to leave the world better than we found it. (Apple mission statement)

Ford's corporate mission is to make people's lives better by making mobility accessible and affordable. Its' vision statement is to become the world's most trusted company, designing smart vehicles for a smart world. (Mission and vision statement of Ford, an automobile manufacturer)

The mission of Merck is to discover, develop and provide innovative products and services that save and improve lives around the world. The vision statement of Merck is to make a difference in the lives of people globally through our innovative medicines, vaccines, and animal health products. We are committed to being the premier, research-intensive biopharmaceutical company, and we are dedicated to providing leading innovations and solutions for today and the future. (Mission and vision statement of Merck, a pharmaceutical company)

Core values

Core values express the qualities that a company considers desirable and which ground business conduct and outcomes. They imply an appeal to the attitudes of employees but do not contain detailed rules of conduct. Often mentioned values include teamwork, responsibility, open communication, and creativity. Also values like customer orientation, flexibility, efficiency, professionalism, and loyalty are regularly mentioned.

The core to do better every day is our magic potion, our culture, which translates into five values:

- Passionate: We dream, we dare, and we do. With the courage to try new things, we can pioneer and continue to grow.
- Together: We only achieve the best results if we work together; with each other and with our partners. We bring out the best in each other and improve every day for our customers.
- Trust: We offer each other freedom and create entrepreneurship by trusting each other's expertise and responsibility, and that of our partners.
- Sincerity: We are open, honest, and straightforward. If we make mistakes we don't ignore them, but learn from them.
- Grounded: Our feet are planted firmly on the ground. We have a sense of humor and offer a clear perspective.
  (bol.com, an online retail tech platform in the Benelux)

At Shell, we share a set of core values – honesty, integrity, and respect for people. By making a commitment to these in our working lives, each of us plays our part in protecting and enhancing Shell's reputation. (Code of Conduct of Shell, a global group of energy and petrochemical companies.)

Responsibility to stakeholders

Most corporate codes also express responsibilities to a variety of stakeholders like consumers, employees, investors, society, and the environment. Competitors and suppliers are also sometimes mentioned as stakeholders. With respect to customers, the supply of qualitatively good products and services is often mentioned as a responsibility. Also enhancing the health and safety of consumers are important topics. With respect to employees, regularly mentioned responsibilities include encouraging personal development, respect, and equal opportunity. With respect to society, the most mentioned responsibility is observing the law. Also being a good corporate citizen and contributing to society are named. Before the adoption of the Sustainability Development Goals (SDGs, see box) in 2015, the responsibilities with regard to enhancing the quality of life, sustainability, and respecting human rights were less often cited in the codes of conduct. More and more, a trend is visible that companies project the SDGs in their codes of conduct.[10] By this projection in their codes of conduct, companies commit themselves to these SDGs and inform the public and their stakeholders about this commitment.

> Nokia continuously seeks to prevent pollution and to reduce the environmental impacts of its products and services during design, procurement, manufacturing, use, and end-of-life. (Code of Conduct of Nokia, a Finnish multinational telecommunications, information technology, and consumer electronics company.)

> Nokia is committed to the principles of the Universal Declaration of Human Rights and the United Nations Global Compact, and we expect our suppliers and business partners to share these values. (Code of Conduct of Nokia.)

> We also expect suppliers to use natural resources responsibly and efficiently, focusing on areas such as carbon and waste reduction. As part of Nike's growth strategy, we seek suppliers who are building agile and resilient management systems which enable them to drive sustainable business growth through developing an engaged and valued workforce, fostering a strong culture of safety, and minimizing their environmental impacts. (Code of Conduct of Nike, an American multinational sportswear company.)

> We are committed to safety, protecting the environment and respecting the communities in which we operate. We are committed to avoiding damage to the environment and related impacts on communities. Our health, safety, security, and environment (HSSE) goals are: no accidents, no harm to people, and no damage to the environment. (Code of Conduct of BP, a British multinational oil and gas company.)

Norms and rules

Norms and rules contain guidelines for employees how to act in specific situations. This may include subjects like the acceptance of gifts, fraud, conflicts of interest, confidentiality, theft, corruption, bribery, discrimination, respect, and sexual harassment.

ASML does not tolerate any form of (sexual) harassment, physical and verbal abuse, mental and physical intimidation, retaliation or any form of aggression or bullying.

We will not make false or misleading claims or statements in any of ASML's financial reports, monitoring reports or other documents submitted to government agencies and investors or published on any media, including advertisements.

Employees should not trade directly or indirectly in any shares or options of ASML while possessing inside information. (Code of Conduct of ASML, the leading supplier to the semiconductor industry)

IBM prohibits bribery and kickbacks of any kind. Never offer or give anyone, or accept from anyone, anything of value that is, or could be viewed as, a bribe or kickback or an attempt to influence that person's or entity's relationship with IBM. And do not do so through others, such as agents, consultants, IBM Business Partners, trade associations or suppliers.

IBM will not tolerate sexual advances or comments, racial or religious slurs or jokes, or any other conduct, such as bullying, that creates or encourages an offensive or intimidating work environment. (IBM Business Conduct Guidelines *Trust Comes First*)

2.3   Possibilities and Limitations of Codes of Conduct

As we have seen, codes of conduct help to express the responsibilities of engineers. They are, therefore, a useful point of departure for discussions about these responsibilities. Still, in the course of time, a number of objections against code of conduct have been leveled. Below, we discuss the main objections. In judging these objections, one should keep in mind that codes of conduct may have different objectives. Especially the difference between aspirational, advisory, and disciplinary codes is relevant here. Objections against disciplinary codes are not always sound objections against advisory codes and vice versa. Although the objections discussed here show some of the limitations of codes of conduct, none of them is strong or convincing enough to conclude that codes of conduct as such are undesirable. Much depends on the actual formulation and implementation of the code.

2.3.1   Codes of conduct and self-interest

Codes of conduct are a form of self-regulation. Sometimes, they are primarily formulated for reasons of self-interest, for example to improve one's image to the outside world, to avoid government regulation or to silence dissident voices. An example in which the latter happened is the case of Jon Tozer (see box).

Case   John Tozer

In 1989 the Australian engineer John Tozer criticized the decision of the Coffs Harbor authorities to pump sewage into the sea. According to him the engineers employed by the local authority had given a misleading impression of the effects upon the environment and they had failed to properly investigate the alternatives. The engineers in question were subsequently successful in removing Tozer from the Association of Consulting Engineers Australia (ACEA). Tozer was accused of having contravened the professional code by openly criticizing the work of other (associated) engineers. Because of his disbarment Tozer, who has his own consulting engineering firm, is no longer able to fulfill any contracts for customers demanding ACEA membership.

Source: Adapted from Beder (1993).

The fact that self-interest plays a role in formulating codes of conduct is not necessarily objectionable as long as the content of the code is moral and serious attempts are made to live by the code of conduct. One way to ensure this is to include a range of stakeholders in the formulation and implementation of the code of conduct to avoid the code becoming one-sided.

Window-dressing Presenting a favorable impression that is not based on the actual facts

A code of conduct serving only the interests of a company or profession may amount to **window-dressing**. We speak of window-dressing if a favorable impression is presented of what the company is doing but that impression does not represent how the company and its employees actually behave. In cases of window-dressing, it may, for example, well be the case that the existence of the code is unknown to members of the organization while in the meantime the code is used in communication with the outside world. The danger of window-dressing is especially present in the case of aspirational codes because they tend to be very general.

The Dieselgate scandal, for example, revealed that Volkswagen engaged in window-dressing. The revelation of the company's efforts to cheat emission tests contradicted Volkswagen's public sustainability messages. For example, their vision statement in 2010, expressed in its *Sustainability Report 2010* was "[b]y 2018 we're aiming to be Number 1 – both economically and ecologically," and in the same report CEO Martin Winterkorn stated that "[s]ustainability is and will remain the foundation of our corporate policy. One clear focus is on 'green' mobility. This is dictated by both social responsibility and sound business thinking."[11]

Case   Google in China: A Case of Window-dressing?

> While removing search results is inconsistent with Google's mission, providing no information … is more inconsistent with our mission. (Google statement.)

Google, the leading Internet search engine company in the world, entered the Chinese market in early 2000 by creating a Chinese-language version of its home page, google.com, that was located in the United States but that could handle search requests from China. In this way, the technology was not subject to Chinese censorship laws as the facilities were not within China's physical boundaries, and Google did not need a license from the Chinese government to operate its business. In 2002, the Chinese version of Google was shut down by the Chinese government for two weeks. When reinstated, it was very slow for all Chinese users and completely inaccessible

Figure 2.5 Google offices in China. Photo: Bloomberg/Getty Images.

for Chinese colleges and universities. By 2005, the Chinese search engine company Baidu emerged as the leading Internet search company in China. To compete with Baidu, Google decided in 2006 to launch a Chinese website (www.google.cn) and agreed to censor its content enforced by means of filters known as "The Great Firewall of China." "Harmful" content included material concerning democracy (e.g., freedom), certain religious groups (e.g., Falun Gong), or antigovernment protests (e.g., Tiananmen Square). Google received much criticism from human rights advocates because it censored information such as human rights.

A moral question is here whether Google's slogan "Don't be Evil" ("It's about providing our users unbiased access to information") and their mission statement "Google's mission is to organize the world's information and make it universally accessible and useful" have been consistently followed. By censoring information, one could argue that Google has strayed from dedication to helping every user get unrestricted access to content on the Internet. Google admitted that the launching of google. cn was problematic with respect to their mission. In the words of Schrage, Google's vice president of Global Communications and Public Affairs: "[Google, Inc., faced a choice to] compromise our mission by failing to serve our users in China or compromise our mission by entering China and complying with Chinese laws that require us to censor search results. … Self-censorship, like which we are now required to perform in China, is something that conflicts deeply with our core principles. … This was not something we did enthusiastically or something we're proud of at all." On March 22, 2010 after a cyber-attack on Google's servers and increased demands for censoring, Google decided no longer to censor its search results. On March 30, 2010, the Chinese government blocked access to Google's search engine from Mainland China.

In 2018, however, an internal leak at Google revealed that Google was attempting for a second time to "launch a censored version of its search engine in China that [would] blacklist websites and search terms about human rights, democracy, religion, and peaceful protest".[12] Google responded stating that the project, known as Dragonfly, was still in an exploration phase and not an official launch. However, from a leaked internal memo, it became clear that Google was aiming to launch Dragonfly at the beginning of 2019. The project came quickly under scrutiny by human rights organizations, investigative reporters, governments, and, most notable, Google employees. In an open letter to Google, signed by over 400 employees, these employees condemned Dragonfly, stating: "Our opposition to Dragonfly is not about China: we object to technologies that aid the powerful in oppressing the vulnerable, wherever they may be. (…) Providing the Chinese government with ready access to user data, as required by Chinese law, would make Google complicit in oppression and human rights abuses."[13] Amid ongoing criticism, Google announced in July 2019 that Project Dragonfly was officially terminated by Google's vice president of public policy, Karan Bhatia, at the Senate Judiciary Committee hearing. However, Bhatia didn't make the promise that Google would not participate in any form of censorship in the future when asked by the senators.[14]

Source: Based on Martin (2008), Dann and Haddow (2008), Congressional Testimony of Schrage (2006), and Ernesto and Xu (2020).

2.3.2   The need for interpretation

In the application of codes of conduct to concrete situations, one is frequently confronted with general concepts and rules that need interpretation.

One relevant notion from codes of conduct that is in need of further clarification and interpretation is "loyalty." Fundamental canon 4 of the NSPE code of ethics, for example, requires that engineers "shall act for each employer or client as faithful agents or trustees." This means that engineers need to be loyal to their company (Harris et al., 2005, p. 191). But what does loyalty exactly amount to? Take, for example the case of the three BART engineers discussed at the beginning of this chapter. Was the engineers' act disloyal because they spoke out against their organization? The answer to this question is yes if one interprets loyalty as **uncritical loyalty**. Harris et al. (2005, p. 191) define uncritical loyalty to an employer as "placing the interests of the employer, as the employer defines those interests, above any other consideration." This we have seen with the engineers who had designed and implemented the cheating software in the Dieselgate scandal (see Chapter 1). Such uncritical loyalty was criticized by Judge Cox who sentenced engineer James Liang who was tasked by Volkswagen with making the defeat device work. Cox said Liang was "arguably a brilliant engineer," but one who had been "arguably too loyal to Volkswagen."[15] Uncritical loyalty can thus be misguided (Martin and Schinzinger, 1996, pp. 193–195). First, it might be doubted whether the interests of the company should always override any other concerns, especially in cases when the public is put at danger, as was the case in the Dieselgate scandal. Second, one might disagree about what the interests of the employer are. In the BART case, it might well be argued that it was not in the interest of the BART organization to keep silent about the technical problems. So conceived, the BART engineers were loyal to the interests of the company. To deal with such objections, Harris, Pritchard, and Rabins propose the notion of **critical loyalty** which they define as "giving due

Uncritical loyalty   Placing the interests of the employer, as the employer defines those interests, above any other considerations.

Critical loyalty   Giving due regard to the interest of the employer, insofar as this is possible within the constraints of the employee's personal and professional ethics.

regard to the interest of the employer, insofar as this is possible within the constraints of the employee's personal and professional ethics."

The example of loyalty shows that rules and terms in codes of conduct need interpretation. In some codes of conduct, the rules and concepts are illustrated by the use of "questions and answers" to help clarify situations that employees may encounter, such as in the code of conduct of Chevron Corporation,[16] an American energy corporation:

Question:

My supervisor told me to destroy documents related to a project that we did last year. Now, the internal auditors are asking questions as though they are concerned. Since my supervisor told me to do this, I should not be in trouble, should I?

Answer:

The auditor is not investigating to get anyone "in trouble." The auditor's role is to ensure that our company follows required policies and processes. You are responsible for understanding our document retention policies. If your supervisor told you to destroy documents that should have been retained, blindly following orders was not the right course of action. The best thing you can do now is to answer the auditor's questions completely and honestly.

Question:

My supervisor asked me to perform a task that I believe violates environmental regulations. What should I do?

Answer:

Never guess about environmental regulations. If you are uncertain, check with your supervisor to be sure you have understood the request. If you still feel the request violates environmental regulations, report the concern to local management or the Chevron Hotline.

Although such questions and answers are helpful, they do not take away the need for interpretation, as always new situations may arise that cannot be fully foreseen and covered beforehand by a code. It would seem no exaggeration to say that following a code of conduct always requires interpretation and judgment. This is not necessarily a bad thing, as we will see below when we discuss the next objection against codes of conduct. Still, the room for interpretation that codes of conduct leave may sometimes result in immoral behavior that follows the code in letter but not in spirit.

Michael Davis (1999) has discussed several undesirable forms of obedience when it comes to codes of conduct. A first is what he calls "**malicious obedience**": in such cases, people may follow the letter of the code but they intentionally leave out certain relevant considerations in their interpretation of the code. In formulating a code of conduct, one may try to avoid malicious obedience by first outlining a number of general "principles," "preambles" or "canons" that set the stage for how more specific rules need to be interpreted. For example, the first fundamental canons of the NSPE code of ethics states that "Engineers, in the fulfillment of their professional duties, shall hold paramount the safety, health, and welfare of the public."[17] This canon, among others, suggests that obligations to employers in the code need to be interpreted in terms of critical loyalty rather than in terms of uncritical loyalty.

> **Malicious obedience** A code of conduct is maliciously obeyed when the code is intentionally followed by the letter but not by the spirit of the code.

In some cases, relevant considerations in interpreting the code of conduct are not left out intentionally, but because the engineer does not pay enough attention to them. In

Negligent obedience A code of conduct is negligently obeyed when an interpretation of the code is followed that does not meet a standard of due care.

such cases, the interpretation of the code may be negligent. We speak of negligence when a certain duty of care is not met. **Negligent obedience** does not need to be intentional, but the engineer did not make enough efforts to properly interpret the code in following it. For example, professional obligation 2b of the NSPE code of ethics states that:

> Engineers shall not complete, sign, or seal plans and/or specifications that are not in conformity with applicable engineering standards. If the client or employer insists on such unprofessional conduct, they shall notify the proper authorities and withdraw from further service on the project.[18]

In this case, compliance may be negligent when the engineer did not make enough efforts to find out what engineering standards might be "applicable" or made an assumption about what the "proper authorities" are without checking.

Stupid obedience A code of conduct is stupidly obeyed when the code is followed by the letter but not by its spirit due to a lack of competence in, for example, moral judgment.

A third form is what Davis calls "**stupid obedience**"; in this case the engineer lacks the competence to properly interpret a code of conduct. One form that Davis mentions is interpreting the rules of a code of conduct in isolation rather than in the context of the entire code. For example, many codes of conduct contain confidentiality duties to employers, but these should be understood in the context of the entire code, not in isolation. Consequently, in most current engineering codes of conduct, confidentiality obligations should be interpreted in the light of obligations to the public and may be overridden by these obligations to the public. Avoiding stupid obedience requires that engineers have at least some basic skills in moral judgment.

2.3.3   Can moral judgments be codified?

Some authors have argued that the idea of drafting a code of conduct is misperceived because moral judgments cannot be codified. In a sense, this objection is the mirror of the previous one. Whereas people who criticize the need for interpretation in codes of conduct are worried that such codes do not uniformly prescribe certain behavior, people who argue that moral judgment cannot be codified are often worried that codes of conduct contain strict prescriptions which conflict with what moral judgments requires according to them (Ladd, 1991). We will consider three different arguments why moral judgment cannot be codified.

One argument is that in the terminology of the philosopher Immanuel Kant, following a code of conduct may be based on heteronymous motives, that is, motives originating outside the acting person like fear for sanctions while moral behavior requires autonomous decisions and behavior (see further Chapter 3). However, even if ethics requires autonomous decision-making, it does not follow that codes of conduct are necessarily objectionable. What is objectionable is a certain uncritical way of using codes of conduct. However, an advisory code need not conflict with the moral autonomy people retain in deciding whether to follow the code or not. Nevertheless, in the case of disciplinary codes the argument may have more strength because disciplinary codes suppose that the code is formally enforced.

A second argument is that codes of conduct are not morally binding (cf. Ladd, 1991). As the box shows, a variety of arguments why codes of conduct are binding can be given. Even if one rejects the view that codes of conduct entail a contract, one might still argue that codes of conduct express already existing moral responsibilities and obligations. In that case, a code of conduct cannot create new moral obligations beyond what was already morally required. From this, however, it does not follow that a code is superfluous. It might still be helpful, for example, to remind people of their moral obligations and responsibilities.

Why are Codes of Conduct Morally Binding?

Three explanations have been offered why codes of conduct are morally binding:

1   One possible explanation is that codes of conduct entail an implicit contract between engineering as a profession and the rest of society (Harris et al., 2005). According to this explanation, professionals serve a moral ideal in exchange for privileges such as status, a monopoly on carrying out the occupation and good salaries. In this explanation, professionals are bound by professional codes because they have implicitly signed a contract with society. This contract creates a moral obligation to follow the code of conduct of a profession.

2   A second explanation is offered by Michael Davis. He defines a profession as follows: "A profession is a number of individuals in the same occupation voluntarily organized to earn a living by openly serving a certain moral ideal in a morally-permissible way beyond what law, market, and morality would otherwise require" (Davis, 1998, p. 417). One important feature of this definition is that being a profession is a voluntary choice. According to Davis, the existence of professional codes for engineers testifies that engineers indeed have made this choice. Such codes are binding because being a member of a profession implies an implicit contract with your colleague professionals. This contract creates a level playing field so that all professionals can pursue the moral ideal.

3   A third explanation is that the codes of conduct as such are not morally binding but that they express moral responsibilities that are grounded otherwise. Michael Pritchard, for example, has argued that engineering codes of conduct are based on common morality (Pritchard, 2009).

Similar arguments may be given for corporate codes. These can also be seen as (1) a contract between a company and society or (2) as a contract between employees of a company or (3) as an expression of the moral responsibilities and obligations a company and its employees have on other grounds.

A third argument against codes of conduct is that they presuppose that morality can be fully expressed in a set of moral rules. One reason why this is questionable is that engineering is too diverse, both in terms of disciplines (civil engineering, mechanical engineering, electrical engineering, aerospace engineering, etc.) and in terms of activities (research, design, testing, maintenance, etc.) for one code to apply. This objection can, however, be dealt with by having a variety of codes of conduct. A more fundamental objection is that sound moral judgment always requires taking into account the particularities of a situation (e.g., Dancy, 1993). According to this line of reasoning, it is not surprising that codes of conduct always require interpretation in particular situations.

Two points are worth noting about these three arguments. First, the arguments are merely directed against disciplinary codes. Such codes are strictly prescriptive and are formally enforced. Enforcement usually requires that the room for interpretation of the code is limited. Moreover, enforcement makes it desirable that the code is morally, or at least legally, binding. The arguments are less, if at all, convincing in the case of advisory and aspirational codes. Second, in as far as especially the first and third argument are sound, they imply that it is neither possible nor desirable to try to avoid all room for interpretation in the formulation of a code of conduct. These are, therefore, more arguments against the second objection against codes of conduct discussed above than against codes of conduct themselves.

2.3.4   Can codes of conduct be lived by?

Codes of conduct sometimes contain provisions that are difficult to follow in practice, or can only be followed against great individual costs. Professional codes can, for example, justify or require actions that go against the interest of the employer. The BART case, which with this chapter started, is an example. More generally, professional codes sometimes require that engineers inform the public timely and completely if the safety, health, or welfare of the public is put at stake in a technological project. This duty to inform the public can conflict with the confidentiality duty that engineers also have according to the law in many countries. If engineers in such situations release information outside the company in which they are working, they are blowing the whistle (see Subsection 1.5.3).

Engineers, and other employees, who blow the whistle are usually in a weak position from a legal point of view (Kenny, 2019; Redman and Caplan 2015). The situation is different from country to country, but the laws that regulate employment contracts in most countries either impose certain **confidentiality duties** on employees or they allow the employer to order the employee to keep silent certain specific information, or they do both. The reason for this is twofold. First, confidentiality may be required to protect the competitive position of one company versus another. Second, such laws are intended to avoid employees disproportionately damaging the company for which they are working by making certain information public. Breaching confidentiality duties may be a ground for dismissal in some countries. In other countries, like the United States, employees can be dismissed at will by the company.[19] However, the employee can hold the company liable for the damage of dismissal on unjust grounds.

> **Confidentiality duties**   Duties on employees to keep silent certain information.

Limits to confidentiality duties

There are limits to the confidentiality duties that companies can impose upon their employees. First, in many countries freedom of speech is legally protected. Historically, freedom of speech is understood to apply to the relation between the state and an individual citizen and not to the relation between a company and an individual employee, which is basically a relation between citizens, according to the law. There is, however, a tendency in law also to apply fundamental rights like the freedom of speech to relations between organizations and individuals. This does not mean that employees have complete freedom of speech, but it might mean that confidentiality duties should be weighed against, or be proportional to the freedom of speech of an employee and the legitimate interests of an employer. Second, in some cases there are legal requirements to make public certain information, or to inform the government or the public prosecutor about certain abuses. These legal requirements may override confidentiality duties. Third, engineers might argue that they have a professional duty, based on their professional code of conduct, to make public certain information. This happened in the BART case and was supported by the professional association of electrical engineers, the IEEE, but to no avail. Fourth, employees can argue that it is in the public interest that certain information is made public. Again, the success of this strategy in court seems limited. In response, several governments have formulated special laws to protect whistleblowers (see box). In the US there has been legislation protecting whistle-blowers for 30 years. In the course of time, this has been adapted to give whistleblowers greater protection. In some cases, large financial rewards have been paid to whistle-blowers who brought to light fraud or tax abuse. Nevertheless, also in these cases whistle-blowers usually only have a limited amount of legal leverage in the first place and they almost always eventually lose their jobs.

Protection of Whistle-blowers

In several countries, attempts have been undertaken to protect whistle-blowers legally. The US and the United Kingdom are at the forefront of enacting whistle-blowing laws that encourage the reporting of fraudulent misconduct. These laws include the protection of whistle-blowers from retaliation and the provision of monetary rewards for whistle-blowing (Hassink et al., 2007; Lee et al., 2020).

In the US, the Sarbanes-Oxley Act (SOX) came into force in 2002. This act requires companies to adopt policies for internal whistle-blowing with respect to accounting and auditing. Companies can also apply such procedures to other kinds of violations covered by their code of conduct. Prior to SOX, federal whistleblower statutes only covered the public sector, or related to more specific areas like safety and the environment. In 2010, the Dodd-Frank Wall Street Reform and Consumer Protection Act was passed, which strengthened and expanded the SOX. Specifically, it includes a provision of monetary rewards by establishing a bounty program where whistle-blowers were entitled to 10% to 30% of the proceeds from successful litigation settlements that they inspire by reporting on bad behavior. It also extends the statute of limitations during which an employee can submit a claim against their employer, doubling it from 90 days to 180 days, and it prohibits retaliation by employers against whistleblowers.

In the United Kingdom, the Public Interest Disclosure Act of 1998 protects both internal and external whistle-blowers from retaliation, but does not have provisions with respect to whistle-blowing policies of companies. The Combined Code on Corporate Governance of 2003, issued by the Financial Services Authority, encourages the institutionalization of whistle-blowing policies by companies. Corporations should follow this code or explain why they did not.

In October 2019, the European Union adopted the Directive on the protection of persons who report breaches of Union law, commonly referred to as the EU Whistleblower Directive. The Directive aims to provide common minimum standards for protection across the EU to whistle-blowers who raise breaches of EU law with their employer. The new rules will require the creation of safe channels for reporting both within an organization – private or public – and to public authorities. It will also provide protection to whistle-blowers against retaliation and require national authorities to adequately inform citizens and train public officials on how to deal with whistle-blowing. Whether the Directive will indeed lead to meaningful protection for whistle-blowers depends "on the transposition of the rules into national laws, the enforcement of the Directive's protections, and the embeddedness of the rules in organizational culture" (Abazi, 2020, p. 641).

In Japan, the Whistle-blower Protection Act (WPA) from 2006 has been amended to broaden the scope of whistle-blower protections and ensure proper whistle-blower systems within companies. This includes a mandatory obligation for companies of a certain size to establish a whistle-blowing system with the aim of ensuring the protection of whistle-blowers by designating personnel who are to be engaged in receiving whistle-blowing reports, investigating allegations and taking corrective measures, and by establishing an internal system to properly respond to whistle-blowing reports.[20] The amended WPA, which has come into force in 2022, does not address retaliation concerns and does not institute a whistle-blower rewards program.

A code of conduct is hardly credible if living by it requires engineers to accept dismissal on a regular base. This is especially a problem for professional codes that require engineers to blow the whistle. Nevertheless, there are a number of initiatives that can be undertaken to improve the degree to which such codes can be lived by. First, the law may be changed to provide better protection for whistle-blowers. Second, companies can include a right to inform the public in certain well-circumscribed cases in their corporate code and can formulate policies so that employees can indeed live by such codes. Some companies, like the chemical concern DSM, have formulated policies or procedures for whistle-blowing (see Appendix V). Also professional associations can undertake initiatives, like providing legal support to individual engineers in cases where adhering to the professional code creates conflict with the employer. The IEEE has done that in the past. Some professional organizations like the NSPE have also published lists of companies that live by the professional code.

2.3.5   Enforcement

Enforcement is only an objective in the case of disciplinary codes. Active enforcement of codes of conduct seems to be an exception, especially for professional codes. We will now elaborate on the reasons for this and discuss what possibilities for enforcement exist.

Professional codes

One obvious reason why professional codes are often not enforced is that they are often advisory and that enforcement is not an objective of advisory codes. An underlying reason for the lack of enforcement, and for the choice to formulate advisory rather than disciplinary codes, is that professional codes do not have a legal status. Moreover, the possibilities for professional associations to enforce professional codes are limited. Enforcement requires sanctions and the most severe sanction that professional societies can exercise with respect to their members is usually loss of membership. The effect of that sanction is limited because in most countries, membership of a professional association is voluntary and is not required to exercise the profession of an engineer. A notable exception is consulting engineering in the United States and Australia. Consulting engineers in these countries have to be registered as engineers in order to carry out their profession if they are not employed by a company but have their own firm. Such registration is also sometimes required for specific groups of engineers in other countries. If registration is required, loss of registration and thus loss of the ability to work as a professional engineer can be the consequence of engineers breaching their professional code. The case of John Tozer, discussed earlier, is an example. In most cases, no attempts are made by professional associations to enforce their code of conduct.

Corporate codes

For corporate codes, enforcement or at least monitoring of the code is more common than in the case of professional codes. Generally speaking, corporate codes offer more possibilities for enforcement than professional codes. The reason for this is that companies usually influence the daily practice of individual engineers to a much larger extent than do professional associations. Companies have more possibilities to stimulate or discourage the behavior of individual engineers than professional associations. Ultimately, they can dismiss engineers if in breach of the code of conduct; a sanction that is much more severe than loss of professional membership.

Corporate codes can also be enforced externally, that is, through an external organization assessing the company in terms of its code of conduct. This is called **external auditing**. An increasing number of companies are audited by accountancy or consultancy firms with respect to, for example, safety, sustainability, social issues, integrity, and CSR policies. An advantage of such external assessment is that

> **External auditing** Assessing a company in terms of its code of conduct by an external organization.

it helps to stop the corporate code of conduct being interpreted and enforced at will. In the absence of external audits, it is conceivable that, for example, those on the work floor are punished severely for not obeying the corporate code of conduct while people at higher levels in the organizations, that is, those persons who also interpret and enforce the code, are judged more mildly. External auditing also increases the credibility, and so the image, of a company. External auditing may also be required for the acquisition of a certificate that guarantees customers of the company that certain standards are met. External auditing or enforcement can also be carried out by branch organizations. This requires a code of conduct on the level of an entire business branch. For example, the International Council of Chemical associations (ICCA) has established for the chemical industry such a code of conduct, the "Responsible Care® Global Charter."[21] Such sectoral codes create a level playing field that all companies in a sector have to meet and, therefore, avoid that companies that live by certain moral standards are commercially worse off than their less scrupulous competitors.

Even if corporate codes are not enforced, they offer better possibilities for stimulating responsible behavior than many professional codes. One reason is that external parties can criticize a company for not living by its own code of conduct (see box). This is of course also the case with professional associations but companies are often more sensitive to external criticism than professional associations.

Case   Brent Spar

According to its code of conduct, Shell is committed to contributing to sustainable development. In 1999 Shell decided to sink the oil platform Brent Spar instead of dismantling it. The British government gave Shell permission to carry out this option, since scientific and technical analysis conducted from 1991 to 1993 had identified deepwater disposal as the "best practicable environmental option." However, subsequently Shell was put under great pressure by environmental organizations, in particular by Greenpeace. Greenpeace argued that dismantling was more environmentally friendly and, moreover, saw the sinking of a platform as an undesirable precedent for the discarding of oil platforms. Because Greenpeace was able to mobilize the public and consumers of Shell products, among others through an occupation of the Brent Spar, Shell eventually felt forced not to sink the Brent Spar.

Source: Adapted from Zyglidopoulos (2002).

2.4   Chapter Summary

Codes of conduct are codes in which organizations lay down rules for responsible behavior of their members. Codes of conduct can be aspirational (mentioning the main values), advisory (assisting individuals in moral judgment) and disciplinary (enforcing rules of behavior). Professional codes are formulated by professional associations of engineers, and corporate codes are formulated by companies in which engineers are employed. Professional codes describe the professional responsibility of engineers, and corporate codes the responsibility of engineers as employees. Most professional codes relate to three domains: (1) conducting a profession with integrity and honesty, and in a competent way; (2) obligations towards employers and clients; and (3) responsibility toward the public and society. Corporate codes usually contain a mission and vision statement (the overall objectives of the company), core values, stakeholder principles and more detailed rules and norms.

A number of objections have been raised against codes of conduct:

1 Codes of conduct sometimes amount to window-dressing;
2 Codes of conduct require interpretation;
3 Moral judgment cannot be codified;
4 Codes of conduct cannot be lived by;
5 Codes of conduct are not enforced.

We have seen that the second and third objection mirror each other. According to the objection that moral judgment cannot be codified, such judgment requires room for interpretation. Objection 3 does not really apply to aspirational and advisory codes, although it may be a problem for disciplinary codes. The same applies to objection 5 because enforcement is only an objective for disciplinary codes and not for advisory and aspirational codes. Objection 4 is serious and may be especially a problem in cases of whistle-blowing, or more generally, tensions between your responsibility as professional engineer and as employee. Partly it can be solved by better attenuating the responsibility of engineers as professionals with the responsibility of engineers as employees, and thus better attenuating professional codes and corporate codes. Some companies have tried to do this.

Study Questions

1 The Software Engineering Code of Ethics and Professional Practice of the Association for Computing Machinery states that "The dynamic and demanding context of software engineering requires a code that is adaptable and relevant to new situations as they occur. However, even in this generality, the Code provides support for software engineers and managers of software engineers who need to take positive action in a specific case by documenting the ethical stance of the profession. The Code provides an ethical foundation to which individuals within teams and the team as a whole can appeal. The Code helps to define those actions that are ethically improper to request of a software engineer or teams of software engineers. The Code is not simply for adjudicating the nature of questionable acts; it also has an important educational function. As this Code expresses the consensus of the profession on ethical issues, it is a means to educate both the public and aspiring professionals about the ethical obligations of all software engineers" (http://www.acm.org/about/se-code, accessed November 2, 2009). Is this code aspirational, advisory, or disciplinary? Explain your answer.

2 Give an example of a situation in which you have a professional responsibility to do something *but* not a legal responsibility.

3 What is meant by "a code is nothing, coding is everything?"

4 What are the most important objectives of professional codes of conduct?

5 Why is enforcement an explicit objective for disciplinary codes? Why is enforcement often difficult to obtain for professional engineering codes of conduct?

6 What are corporate codes? Discuss three objections to and/or shortcomings of corporate codes.

7 What are the two arguments of Milton Friedman's criticism of corporate social responsibility? Give some objections against these arguments.

8 Like engineers, medical doctors and lawyers also have professional codes. Unlike engineering codes, however, these codes typically are accompanied by disciplinary law, so that doctors or lawyers who violate the code can be excluded from practicing the profession. Provide an argument *for* and an argument *against* the adoption of a similar disciplinary law for engineers.

9 What is valuable about loyalty? What is problematic about loyalty? Be careful to indicate what concept of loyalty you are using in answering this question.

10 To gain the protection of the UK's Public Interest Disclosure Act, those who reveal organizational malpractices have to satisfy a number of conditions that witnesses in other criminal investigations do not have to satisfy, for example, deriving no financial gain from the case and not having been involved in the crime at any stage. Critically evaluate the merits of these conditions. Compare them also with the guidelines for whistle-blowing mentioned in Subsection 1.5.3.

11 Look for a professional code of conduct in your own area:
   a. Do you recognize the three general content areas mentioned in the text in this code?
   b. Is the code vague at some points? Where?
   c. Are their potential contradictions between the provisions of the code? Does the code contain provisions to deal with these contradictions?
   d. Are there any provisions in the code that are impossible to live by? Which ones?
   e. Do you agree which the professional responsibility set out in the code? Are you missing anything?

12 Look for a corporate code of an engineering company. In what respects are the responsibilities of engineers that are articulated in this code different from the responsibilities articulated in professional codes (like the code of the NSPE)? Is this code conflicting at certain points with,

for example, the professional code of the NSPE? If there is a conflict what code should, in your view, take precedence and why?

13   Do you agree that engineers have a responsibility for human rights? Is this responsibility restricted to not engaging in violations of human rights or do engineers also have a responsibility to enhance human rights through their engineering projects?

14   Draft a code of conduct to cover e-communications (email, Web use, and so on). Explain and justify your proposed code.

15   One of the two basic moral principles for global engineering related to cross-cultural values, mentioned by Luegenbiehl and Clancy (2017), is: Engineers should endeavor to understand and respect the nonmoral cultural values of those they encounter in fulfilling their engineering duties. Give an example of how respecting a cultural nonmoral value could benefit an engineer and/or the company for which they work.

16   The US government allows employees of aircraft manufacturers like Boeing to serve as inspectors for the Federal Aviation Agency (FAA) which is responsible for regulating the aircraft industry and doing safety and quality inspections. What would be the reasons for the US government to allow this? Is this a conflict of interest? Would it be immoral for an engineer employed by Boeing also to act as inspector for the FAA?

Discussion Questions

1   If you were to give ethical training to engineers, would you stress knowing the law, company rules, and codes of conduct, or would you instead focus on explaining the principles behind these rules? Are there any common principles behind these rules? Which ones?

2   Loyalty or integrity: which should be the most important to engineers?

3   What do you see as the main moral issues arising from globalization?

4   Cases like Shell in Nigeria and Google in China that were discussed in this chapter seem to suggest that codes of conduct are a dead letter when it comes to moral decision-making in practice. Discuss whether codes of conduct are indeed just window-dressing in cases like this or whether they have any positive effect. Can you think of ways to bridge the gap between what companies like Shell and Google say in their codes and what they do in practice? Should multinational companies maybe avoid undemocratic countries like Nigeria and China to avoid tough ethical decisions?

5   Choose any Fortune 500 company. Locate the company's code of ethics published on the company's Web page. Evaluate the code in terms of the Sustainability Development Goals (SDGs).

Notes

1   For a comparable distinction, see Frankel (1989).

2   VDI, "Bekentennis der Ingenieurs" [The Confession of Engineers] (1950), included in Lenk and Ropohl (1987, p. 280).

3   https://www.jspe.org/eng, accessed June 22, 2022.

4   https://www.engineerseurope.com/, accessed February 24, 2023.

5   https://www.theguardian.com/technology/2021/oct/24/frances-haugen-i-never-wanted-to-be-a-whistleblower-but-lives-were-in-danger, accessed June 22, 2022.

6   https://www.microsoft.com/en-us/corporate-responsibility, accessed June 22, 2022.

7   https://sdgs.un.org/2030agenda, accessed June 22, 2022.

8   https://clubofamsterdam.com/contentimages/The%20Future%20Now%20Show/063%20Roland%20Schatz/SCR500_preliminary_top100%202021.pdf, accessed June 22, 2022.

9   https://news.un.org/en/story/2011/08/383512, accessed February 20, 2023.

10 See Balcerzak and MacGregor Pelikánová (2020) for a critical analysis of the projection of SDGs in codes of conduct.

11 https://www.volkswagen.co.uk/assets/common/pdf/annual-reports/2010_sustainability_report.pdf, accessed June 22, 2022.

12 https://theintercept.com/2018/08/01/google-china-search-engine-censorship, accessed June 22, 2022.

13 https://www.cnbc.com/2018/11/27/read-google-employees-open-letter-protesting-project-dragonfly.html, accessed June 22, 2022.

14 https://www.judiciary.senate.gov/meetings/google-and-censorship-though-search-engines, accessed June 22, 2022.

15 https://eu.usatoday.com/story/money/cars/2017/08/25/vw-engineer-gets-40-months-prison-role-diesel-scandal/602584001, accessed June 22, 2022.

16 https://www.chevron.com/-/media/shared-media/documents/chevronbusinessconductethicscode.pdf, accessed June 20, 2022.

17 https://www.nspe.org/resources/ethics/code-ethics, accessed June 17, 2022.

18 Ibid.

19 Convention 158 of the International Labour Organization states that an employee "can't be fired without any legitimate motive" and "before offering him the possibility to defend himself." The US has not ratified this convention.

20 https://www.amt-law.com/asset/pdf/bulletins9_pdf/LELB48.pdf, accessed June 20, 2022.

21 https://icca-chem.org, accessed March 20, 2022.

3

Normative Ethics

Having read this chapter and completed its associated questions, readers should be able to:

- Describe normative judgments, and distinguish them from descriptive judgments;
- Describe norms, values, and virtues;
- Describe the four ethical theories: utilitarianism, Kantian theory, virtue ethics, and care ethics;
- Identify the criticisms of the four ethical theories;
- Apply the ethical theories to moral issues in engineering practice;
- Reflect upon how ethical theories may impact on making moral decisions.

Contents

3.1 **Introduction** 72

3.2 **Ethics and Morality** 75

3.3 **Descriptive Statements and Normative Judgments** 76

3.4 **Points of Departure: Values, Norms, and Virtues** 77

    3.4.1 Values 77
    3.4.2 Norms 78
    3.4.3 Virtues 79

3.5 **Relativism** 80

3.6 **Ethical Theories** 81

3.7 **Utilitarianism** 82

3.7.1 Jeremy Bentham 82

3.7.2 Mill and the freedom principle 87

3.7.3 Criticism of utilitarianism 89

3.7.4 Applying utilitarianism to the Ford Pinto case 90

3.8 **Kantian Theory** 91

3.8.1 Categorical imperative 92

3.8.2 Criticism of Kantian theory 95

3.8.3 Applying Kant's theory to the Ford Pinto case 97

3.9 **Virtue Ethics** 97

Ethics, Technology, and Engineering: An Introduction, Second Edition. Ibo van de Poel and Lambèr Royakkers.
© 2023 John Wiley & Sons Ltd. Published 2023 by John Wiley & Sons Ltd.

3.9.1 Aristotle 98

3.9.2 Criticism of virtue ethics 100

3.9.3 Virtues for morally responsible engineers 100

3.10 Care Ethics 103

3.10.1 The importance of relationships 103

3.10.2 Criticism of care ethics 104

3.10.3 Care ethics in engineering 104

3.11 Capability Approach 106

3.11.1 Applications of the capability approach 109

3.11.2 Criticism of the capability approach 110

3.12 Non-Western Ethical Theories 111

3.12.1 Ubuntu 112

3.12.2 Confucianism 113

3.13 Applied Ethics 115

3.14 Chapter Summary 117

Study Questions 119

Discussion Questions 120

3.1   Introduction

Case   The Ford Pinto

Figure 3.1   Ford Pinto. Photo: Bettmann/Bettmann/Getty Images.

On August 10, 1978, on Highway 33 in the neighborhood of Goshen, Indiana (United States), a tragic accident occurred. A truck rear-ended a five-year-old Ford Pinto carrying three teenagers: sisters Judy and Lynn Ulrich (ages 18 and 16, respectively) and their cousin Donna Ulrich (age 18). The collision caused the gas [petrol] tank to rupture and explode, killing all three teens.

Subsequently an Elkart County grand jury returned a criminal homicide charge against Ford, the first ever against an American company. During the following 20-week trial, the judge advised the jury that Ford should be convicted of reckless homicide if it were shown that the company had engaged in "plain, conscious and unjustifiable disregard of harm that might result (from its actions) and the disregard involved a substantial deviation from acceptable standards of conduct." The key phrase around which the trial hinged was "acceptable standards." Toward the end of the 1960s, Ford Motor Company, one of the world's largest car manufacturers, was gradually losing market share. Ford was losing ground to the smaller and cheaper European cars. In 1968, President Lee Iacocca decided a small cheap car had to be designed quickly. This was to become the Ford Pinto (see Figure 3.1). The decision was made to put it onto the market for less than $2000 in 1970. This was a very competitive price but the time schedule for the car's development was rushed. At the time, car development normally required around 43 months. Only 24 months were reserved for the Ford Pinto. Because the Pinto had to cost a maximum of $2000, a radical design was selected in which styling took precedence over engineering design. The safety aspect of the design did not receive sufficient priority. Among other things, this led to the positioning of the petrol tank just behind the rear axle. Later it was found that the gear construction in the rear axles (the differential) was situated such that it would puncture the petrol tank in the event of a collision. In Ford's tests of the Pinto prototype, this problem occurred at speeds as low as 35 km per hour. The puncture of the tank caused an extremely hazardous situation. These test results were passed on to the highest management level within Ford. From other tests it was shown that there were two simple ways to considerably reduce the risk that the petrol tank would be ruptured. It was possible to alter the design to allow the petrol tank to be situated above the axle. It was estimated that the change in the design would raise the price of the car by $11. A second option was to protect the tank with a rubber layer, which was probably a cheaper option. However, because the design met the safety requirements of the government, the Pinto was taken into production without any alterations.

To justify its actions, Ford made a cost-benefit analysis. In this cost-benefit analysis, which was published under the heading "Fatalities Associated with Crash-Induced Fuel Leakage and Fires," it was asserted that the extra costs of $11 did not weigh against the benefit that society would derive from a smaller number of wounded passengers and fatalities. This statement was argued as follows:

The societal benefits of the riskier design that costs $11 less was estimated at nearly $50 million: 180 lives lost, 180 wounded and 2100 cars burnt out. The calculation for this was 180 lives × $200,000 + 180 seriously wounded × $67,000 + 2100 burnt out cars × $700 = $49.53 million. This was considered to be the total societal benefit.

Against this there was the cost of improving the cars: 11 million cars and 1.5 million trucks had to be called back and retrofitted against an estimated cost per unit of $11, amounting to a total cost of 137 million dollar (12.5 million × $11). A memorandum attached to the report described the costs and benefits as in Table 3.1.

Table 3.1 Benefits and costs

Benefits		
Savings	*Unit cost (US$)*	
180 burn deaths	200,000	36,000,000
180 serious burn injuries	67,000	12,060,000
2,100 burned vehicles	700	1,470,000
Total		*49,530,000*
Costs		
Sales	*Unit cost (US$)*	*Total (US$)*
11 million cars	11 per car	121,000,000
1.5 million light trucks	11 per truck	16,500,000
Total		*137,500,000*

The estimation by Ford of the number of lives lost and wounded incurred was based on statistical data. The estimation Ford made that a human life is worth $200,000 was based on a report of the National Highway Traffic Safety Administration (see Table 3.2).

Table 3.2 Component costs

Component		1971 costs (US$)
Future productivity losses	Direct	132,000
	Indirect	41,300
Medical costs	Hospital	700
	Other	425
Property damage		1,500
Insurance administration		4,700
Legal and Court		3,000
Employer losses		1,000
Victim's pain and suffering		10,000
Funeral		900
Assets (lost consumption)		5,000
Miscellaneous accident cost		200
Total per fatality		*200,725*

The conclusion that Ford drew was clear: a technical improvement costing $11 per car which would have prevented gas tanks from rupturing so easily was not cost-effective for society. The $137 million cost of the safer model clearly outweighed the benefits of $49.53 million. Altering the Pinto for $11 a car would cost society ($137 million – $49.53 million) $87.47 million.

On March 13, 1980, the Elkhart County jury found Ford not guilty of criminal homicide. However, under pressure from various institutions, Ford recalled 1.5 million cars for refitting, and this case and many other similar Pinto accidents cost Ford millions of dollars in legal settlements to accident victims. Ford also suffered a great deal of damage to its reputation.

Source: Adapted from Birch and Fielder (1994).

The Ford Pinto case is a classic example of corporate wrong-doing and is a mainstay of courses in ethics. The argument of Ford is controversial and has evoked a lot of debate. It painfully illustrates that expressing the value of human life in monetary terms involves the danger of neglecting fundamental human rights, such as the right to life. The Ford Pinto case raises many questions of ethical importance. Some people conclude that Ford was definitely wrong in designing and marketing the Pinto, and others believe that Ford was neither legally nor morally blameworthy, and acted right in producing the Pinto. Reflecting on this case, several ethical questions emerge: Was Ford acting wrongly in rushing the production of the Pinto? Even though Ford violated no federal safety standards or laws, should the company have made the Pinto safer in terms of rear-end collisions, especially regarding the placement of the gas tank? Was it acceptable that Ford used cost-benefit analysis to make a decision relating to safety, specifically placing dollar values on human life and suffering? Should companies like Ford play a role in setting safety standards? What were the responsibilities of the Ford design engineers and crash-test engineers?

Different arguments can be used to answer these questions. Despite the apparent differences, some recurrent patterns can be found in the moral arguments that are used in cases like this, and the cases we have already seen: the Dieselgate case (Section 1.1) and the BART case (Section 2.1). These patterns are related to ethical theories that have been developed by various philosophers. Ethical theories help us to sort out our thinking and to develop a coherent and justifiable basis for dealing with moral questions. The role of ethical theories is to provide certain arguments or reasons for a moral judgment. They provide a normative framework for understanding and responding to moral problems, so improving ethical decision-making or, at least, avoiding certain shortcuts, such as neglecting certain relevant features of the problem or just stating an opinion without any justification. Theories each have their own criteria with which they determine whether an action is right or wrong. Before we go into these theories, we shall discuss what we mean by morality and ethics (Section 3.2) and explain the distinction between descriptive and normative judgments (Section 3.3). In Section 3.4, we shall look into the points of departure of ethics: values, norms, and virtues. These points of departure often recur in ethical theories. We will discuss in Section 3.5 an extreme approach to ethics: normative relativism. In Section 3.6, we shall indicate how the three best-known ethical theories – consequentialism, duty ethics, and virtue ethics – are related to each other. These three ethical theories will be discussed at length in the sections that follow (Sections 3.7, 3.8, and 3.9). In Sections 3.10 and 3.11, we will discuss two relatively recent approaches to ethics as an alternative to the familiar moral theories: care ethics and the capability approach. In Section 3.12, we discuss how recently authors try to add the non-Western ethical approaches Ubuntu and Confucianism to the existing literature about engineering ethics. Finally, in Section 3.13, we clarify our position with respect to applied ethics.

3.2   Ethics and Morality

Ethics has had many meanings over the centuries. The term is derived from the Greek word ethos, which can be translated as "custom" or "morals," but also as "conviction." "Ethica" stood for the science that considered what was good or bad, wise or unwise, about people's deeds. The Romans translated "ethos" into the Latin "mos" (plural "mores"), which is the root of the word "moral." Ethics and moral stem from the same source. Over the centuries "moral" has taken on the meaning of the totality of accepted rules of behavior (of a group or culture). In this text, we will distinguish ethics from morality. The term ethics will be reserved for a further consideration of what is morally right.

Ethics is the systematic reflection on what is morally right.

Morality is the whole of opinions, decisions, and actions with which people, individually or collectively, express what they think is good or right.

Ethics   The systematic reflection on what is morally right.

Morality   The totality of opinions, decisions, and actions with which people express, individually or collectively, what they think is good or right.

In this book we define ethics as the systematic reflection on what is morally right. Morality is defined here as the totality of opinions, decisions, and actions with which people express what they think is good or right. This roughly agrees with the often-used definition of morality as the totality of norms and values that actually exist in society.

Systematic reflection on what is morally right increases our ability to cope with moral problems, and thus moral problems that are related to technology as well. Ethics, however, is not a manual with answers; it reflects on questions and arguments concerning the moral choices people can make. Ethics is a process of searching for the right kind of morality.

Descriptive ethics   The branch of ethics that describes existing morality, including customs and habits, opinions about good and evil, responsible and irresponsible behavior, and acceptable and unacceptable action.

Normative ethics   The branch of ethics that judges morality and tries to formulate normative recommendations about how to act or live.

The study of ethics can be both of a descriptive or prescriptive nature. **Descriptive ethics** is involved with the description of the existing morality, including the description of customs and habits, opinions about good and evil, responsible and irresponsible behavior, and acceptable and unacceptable action. It studies the morality found in certain subcultures or during certain periods of history. Prescriptive or **normative ethics** takes matters a step further. Descriptive ethics can discuss the morality of foreign nations or monthly magazines for men without passing judgment. Normative ethics, which is central to this book, moves away from this detachment. By definition normative ethics is not value-free; it judges morality. It considers the following main question: do the norms and values actually used conform to our ideas about how people should behave? Normative ethics does not give an unambiguous answer to this question, but in its moral judgment various arguments are given based on various ethical theories. These ethical theories provide viewpoints from which we can critically discuss moral issues.

3.3   Descriptive Statements and Normative Judgments

One central question in normative ethics is "what is a right opinion, decision, or action?" To answer this question a judgment has to be made about the opinion, decision, or action in question. This is a normative judgment, because it says something about what "correct behavior" or a "right way of living" is. While normative judgments are judgments about

Descriptive statement   A statement that describes what is actually the case (the present), what was the case (the past), or what will be the case (the future).

Normative judgment   Judgment about whether something is good or bad, desirable or undesirable, right or wrong.

what is good or right, **descriptive statements** are related to what is actually the case (the present), what was the case (the past), or what will be the case (the future). Descriptive statements are true or false. The assertion "the Volkswagen diesel cars did not meet the US Environmental Protection Agency's (EPA) standards for nitrogen oxides (NO_x)" is a descriptive statement: the assertion is true or false. Sometimes the truth of a descriptive statement has not yet been determined because testing is impossible. Take for example the statement "God exists." Science plays an important role in determining the truth of descriptive statements. A **normative judgment** is a judgment about whether something is good or bad, desirable or undesirable; they often refer to how the world should be instead of how it is. Such normative judgments often refer to moral norms and values. This can give rise to meaningful discussions, which is not the case for judgments of taste, such as "I do not like Brussels sprouts." Examples of normative judgments are "the Volkswagen engineers should have objected to the cheating software," "Engineers should faithfully provide measurements," and "stealing is bad."

The distinction between descriptive statements and normative judgments is not always that easy. The statement "taking bribes is not allowed" can be both a normative and a descriptive. If the statement means that the law declares that taking bribes is illegal, then it is a descriptive statement. If, however, the statement means that bribery should be forbidden, then it is a normative judgment.

3.4 Points of Departure: Values, Norms, and Virtues

Norms, values, and virtues are the points of departure, respectively, for the three primary normative theories that we will discuss in Sections 3.7, 3.8, and 3.9. We shall discuss them in detail here.

3.4.1 Values

Values help us determine which goals or states of affairs are worth striving for. Moral values are related to a good life and a just society. They have to be distinguished from the preferences or interests of individual people. Preferences or interests are matters people feel they should strive for, for themselves. Moral values are lasting convictions on matters that people feel should be strived for in general and not just for themselves to be able to lead a good life or to realize a just society. A typical example of this is the slogan of the French Revolution: "liberté, égalité, fraternité" (freedom, equality and brotherhood). This slogan did not express a personal preference – such as "I want to be rich" – but expressed values that were felt to be of importance for everyone. Other examples of moral values include justice, health, happiness, and charity. Values are not limited to people; companies have them too. They often formulate their most important moral values (core values) in their mission statement (see Subsection 2.2.2).

> **Values** Lasting convictions on matters that people feel should be strived for in general and not just for themselves to be able to lead a good life or to realize a just society.

A distinction can be made between intrinsic and instrumental values. An **intrinsic value** is an objective in and of itself. An **instrumental value** is a means to realizing an intrinsic value. The value of money for Scrooge McDuck is intrinsic. He values money independently of what you can do with money. For Mother Theresa, however, money was an instrumental value to realize a higher end: helping the poor. A person can consider their work to be both of intrinsic and instrumental value. If work is meant to support the value of becoming rich, it is an instrumental value. If a person has much job satisfaction, then work is an intrinsic value.

> **Intrinsic value** Value in and of itself.
>
> **Instrumental value** Something that is valuable in as far as it is a means to, or contributes to something else that is intrinsically good or valuable.

Case   Biometric Technology and Data Matching
at Super Bowl XXXV

A large spectator event like the Super Bowl presents a prime target for terrorists. Fearing the potential for such an attack or other serious criminal incident,
law enforcement agencies in Florida turned for help to biometrics: the use of
a person's physical characteristics or personal traits for human recognition. At
Super Bowl XXXV in January 2001, a biometric system relying on facial recognition was used. This technology scanned the faces of individuals entering the
stadium. The digitized facial images were then instantly matched against images
in a centralized database of suspected criminals and terrorists. At the time, this
practice was criticized by civil-liberty proponents and privacy advocates. In the
post-September 11 (2001) world, however, practices that employ technologies
such as face-recognition devices have received overwhelming support from the
American public.

Source: Adapted from Tavani (2004).

In the literature on computer ethics, the threat to personal privacy is one of the most
debated ethical problems. The distinction between instrumental and intrinsic values suggests two common ways to attempt to justify privacy. The most common justification is
that privacy has instrumental value. It offers us protection against harm. For example, if
a person is tested HIV+ and this is publicly known, then an employer might be reluctant
to hire them and an insurance company might be reluctant to insure them. The justification of privacy, however, would be more secure if we could show that it has intrinsic
value. While a few authors argue that privacy is only an intrinsic value, others argue that
while privacy is instrumental, it is not merely instrumental. For example, computer ethicist Deborah Johnson proposes that we regard privacy as an essential aspect of autonomy
(Johnson, 2001). Autonomy is fundamental to what it means to be human, to our values
as human beings (see also Section 3.8). So, privacy is a necessary condition for an intrinsic value: autonomy. Johnson argues that the loss of privacy would therefore be a threat
to our most fundamental values. For example, if a person is being watched by constant
surveillance, this has an enormous effect on how the person behaves and how they see
themselves.

3.4.2   Norms

Norms Rules that prescribe what actions
are required, permitted, or forbidden.

Norms are rules that prescribe what concrete actions
are required, permitted or forbidden. These are rules
and agreements about how people are supposed to
treat each other. Values are often translated into rules,
so that it is clear in everyday life how we should act to achieve certain values. One example
of a value within our traffic system is safety. However, the value alone is not enough to
guarantee safety on the road. To this purpose, we need rules of behavior or norms: prescribed actions that indicate what we must do or must not do in a given situation. The
value "safety" in a traffic system is mainly specified by the legal norms from the traffic

Table 3.3 Differences between values and norms

Values	*Norms*
Ends	Means
Global	Specific
Hard to achieve without norms	Ineffective without values

Source: Adapted from Jeurissen and Van de Ven (2007, p. 57).

regulations. In the Dutch regulations, for example, we have the rule that drivers coming from the right always have the right of way.

Moral norms are indications for responsible action. Next to moral norms there are other kinds of norms, such as legal norms (for example, traffic rules), precepts of decorum (for example, "You should not talk when your mouth is full"), and rules of play (for example, in Ludo you can only place a counter on the playing board once you have thrown a six with the die). Some moral norms, like "Thou shalt not kill" and "Thou shalt not steal," have been turned into laws.

The difference between values and norms can be described as follows. Values are abstract or global ideas or objectives that are strived for through certain types of behavior; it is what people eventually wish to achieve. Norms, however, are the means to realize values. They are concrete, specific rules that limit action. Without an interpretation, the objective cannot be achieved. Take for example the need for traffic regulations to guarantee traffic safety. In addition, norms have no meaning or are ineffective if the underlying value is unclear or is lacking. So, one can imagine that the norm "all bicycle bells must be blue" will be largely ignored. The norm has no meaning – there is no underlying value. These differences are summarized in Table 3.3.

3.4.3   Virtues

Next to values and norms we have another moral point of departure: **virtues**. The philosopher Alasdair MacIntyre describes virtues as a certain type of human characteristics or qualities that has the following five features:

> **Virtues**   A certain type of human characteristics or qualities.

1 They are desired characteristics and they express a value that is worth striving for.
2 They are expressed in action.
3 They are lasting and permanent – they form a lasting structural foundation for action.
4 They are always present, but are only used when necessary.
5 They can be influenced by the individual (MacIntyre, 1984a).

The last statement suggests that people can acquire virtues. It is a matter of the shaping of a person's character or personality. This occurs during our upbringing or our learning process within an organization. Examples of virtues are justice, honesty, courage, loyalty, creativity, and humor.

We can distinguish moral virtues (or character virtues) from intellectual virtues. Intellectual virtues focus on knowledge and skills. Moral virtues are the desirable characteristics of people – the characteristics that make people good.

Moral virtues are indispensable in a responsible organization. An organization can formulate nice values like integrity, respect, and responsibility as much as it likes, but without the moral virtues being present in the character of its employees little will be accomplished.

The values indicate which characteristics (virtues) an organization prizes or expects of its employees – what kind of people it expects its employees to be.

> Moral values help us determine which goals or states of affairs are worth striving for in life, to lead a good life or to realize a just society.

> Moral norms are rules that prescribe what action is required, permitted, or forbidden.

> Moral virtues are character traits that make someone a good person or that allow people to lead good lives.

3.5   Relativism

Normative relativism An ethical theory that argues that moral points of view – values, norms and virtues – are only relative to some particular standpoint (such as that of a person, a culture or a historical period) and that no standpoint is uniquely privileged over all others.

Before discussing the three best-known theories in normative ethics, we shall look at an extreme theory that seems to be very tempting at first when it comes to forming a moral judgment: **normative relativism**.

Normative relativism argues that moral points of view – values, norms, and virtues – are relative to some particular standpoint (such as that of a person, a culture, or a historical period) and that no standpoint is uniquely privileged over all others (Westacott, 2012). What is good or responsible from a certain standpoint is not necessarily so for another. A moral judgment or choice is simply an opinion of that particular standpoint: "If in society A it is considered good (or bad) to do A, then it *is* good (or bad) to do A." So, the defense of such a claim is rather subjective and random: there are no guidelines about behavior that are objective and independent of time, place, and culture. Furthermore, normative relativism states that the various values and norm systems for each culture are equally valid, so that it is impossible to say that certain norms and values are better than others. It is often suggested that this means that we have to respect all value and norm systems.

There are three problems with this theory. First, it seems to involve an inherent contradiction. The theory states that there are no universal norms, but at the same time it uses a universal norm: "Everybody has to respect the moral opinions of others." Second, it makes any meaningful moral discussion totally impossible, because you can always appeal to your freedom of opinion, which by definition is neither better nor worse than other opinions. The question is whether this is a valid standpoint. Should the torture of political prisoners be tolerated because this is customary within a given culture? Are there no moral limits to such tolerance? Do we not all object to this kind of relativistic argument to defend the torture of political prisoners? Finally, normative relativism can lead to unworkable or intolerable situations. Engineers work in teams or are employed within a company where there are written – and unwritten – rules to promote cooperation (for example, attending meetings on time). A system that allowed engineers to disregard these rules based on their personal values (which other people have to respect) would create an unworkable situation.

We can state that a choice based only on normative relativism is at the very least *ethically* suspect, since ethics reflects on morality, and calls us to make reasoned judgments about it. The ethical theories we shall discuss in the following section are more rational theories than normative relativism. Two of them originate from the tradition of philosophy of the Enlightenment.

Immanuel Kant summarized the essence of Enlightenment as follows: "Enlightenment is man's release from his self-incurred tutelage. Tutelage is man's inability to make use of his understanding without direction from another. Self-incurred is this tutelage when its cause lies not in lack of reason but in lack of resolution and courage to use it without direction from another. *Sapere aude!* 'Have courage to use your own reason!' – that is the motto of enlightenment." (Kant, 2006 [1784])

3.6 Ethical Theories

We will now discuss three primary ethical theories and attempt to synthesize their applications. These three are consequentialism, deontology, and virtue ethics. We can distinguish these theories from each other by their approach to the structure of human action and the primary focus or point of departure they use to theorize about what is morally good or right (see Table 3.4).

The structure of human action means that an action is carried out by a certain actor (person or institution) with a certain intention, which then leads to certain consequences. So, we can evaluate each moral action from three perspectives: the actor, the action and the consequences.

If we evaluate the action from the perspective of the *action* itself, we make use of deontological ethics or deontology (Greek: δέον *(deon)* meaning *obligation* or *duty*): duty ethics. Here, the point of departure is *norms*. It is your moral obligation to ensure that your actions agree with an applicable norm (rule or principle). One example of such an applicable norm is the "Golden Rule," which can be found in the texts of various religions: "Do unto others as you would have them do unto you."

If we look at the *actor* and their characteristics to pass moral judgment on an action, then we make use of virtue ethics. It is neither the incidental action that counts nor the consequences of the action, but it is the quality of the person acting that makes the action morally right or not. Here, the moral point of departure is *virtues*, which allow people to realize a good life.

If we disregard both the actor and the action in the moral judgment of a certain action, but only consider the consequences, then we apply consequentialism. You ought to choose the action with the best outcomes. The moral point of departure is *values*. Consequentialists focus on realizing certain goals or states of affairs they feel should be strived for, for example, promoting pleasure, avoiding pain, or realizing ambitions.

There are different variants on the ethical theories mentioned earlier. In the following three sections we shall discuss the best-known variant for each theory: utilitarianism as a representative of consequentialism (Section 3.7), Kant's theory as a representative of duty ethics (Section 3.8), and Aristotle's virtues doctrine as a representative of virtue ethics (Section 3.9).

Table 3.4 Differences between the ethical theories

	Actor	*Action*	*Consequences*
Theory	Virtue ethics	Deontology	Utilitarianism
Points of departure	Virtues	Norms	Values

3.7 Utilitarianism

Consequentialism The class of ethical theories which hold that the consequences of actions are central to the moral judgment of those actions.

Utilitarianism A type of consequentialism based on the utility principle. In utilitarianism, actions are judged by the amount of pleasure and pain they bring about. The action that brings the greatest happiness for the greatest number should be chosen.

In **consequentialism**, the *consequences* of actions are central to the moral judgment of those actions. An action in itself is not right or wrong; it is only the consequence of action that is morally relevant. We shall limit ourselves to one type of consequentialism: **utilitarianism**. Utilitarianism is characterized by the fact that it measures the consequences of actions against one value: human pleasure, happiness, or welfare. Utilitarianism therefore is a monistic type of consequentialism. There are pluralistic types of consequentialism too, where various values must be weighed against each other in the assessment of actions.

3.7.1 Jeremy Bentham

Jeremy Bentham was the founder of *utilitarianism*, a word derived from the Latin *utilis* meaning useful. Utilitarianism makes the consequence of an action central to its moral judgment: an action is right if it is useful and wrong if it is damaging. The next question of course is "useful for what?" In other words, what is the purpose for which the action is a means? This purpose has to be something that has *intrinsic* value. So, it has to be good in itself. This means that the utilitarian is primarily concerned with values; he first has a notion of what is intrinsically good and subsequently considers the moral rightness dependent on this notion. The value theory that Bentham connects to his ethics is **hedonism**: the idea that "pleasure" is the only thing that is good in itself and for which all other things are instrumental.

Hedonism The idea that pleasure is the only thing that is good in itself and to which all other things are instrumental.

Jeremy Bentham (1748–1832)

By the principle of utility is meant that principle which approves or disapproves of every action whatsoever, according to the tendency it appears to have to augment or diminish the happiness of the party whose interest is in question: or, what is the same thing in other words to promote or to oppose that happiness. I say of every action whatsoever, and therefore not only of every action of a private individual, but of every measure of government. (Bentham, 1948 [1789])

Jeremy Bentham was born in London on February 15, 1748. At pre-school age his father taught him Latin, Greek and music. A private teacher also taught him French language and literature. His private teacher had him read *Télémaque* by Fénelon. The book had a huge impact on Bentham, who identified strongly

Figure 3.2 Jeremy Bentham. Photo: Classic Vision/AGE Fotostock.

with the hero Telemachus. His dedication to the welfare of humanity was an ideal he held to throughout his life. When he was 12, he was enrolled at Queen's College in Oxford, where he took classical languages and philosophy. As a small and shy but intelligent child, he soon was given the nickname "the philosopher." Bentham looked back on this period in horror. He considered the lectures in Oxford to be useless and a waste of time – the only things he felt had been useful were lessons on logic. As a student he trained to become a lawyer, but after a few years of running a law practice he focused more and more on developing a philosophical and scientific theory of legislation and justice. He fiercely criticized the legal system, because it did nothing to improve the welfare of people. Courts of law could condemn people for "sexual crimes" even if neither party had objections to the sexual act. Bentham thought this was nonsense: if both parties agreed to an act then there could be no crime. As an alternative Bentham wanted to build a new legal system that was rational, clear, and consistent. It was to be based on ethical knowledge and not on tradition or custom. His ethical opinions were set out in *An Introduction to the Principles of Morals and Legislation* (1948 [1789]). Due to the clash between his ethical opinion and conventional Christian thought, Bentham was greatly opposed to Christianity, which he considered to be a form of ascetism where pleasure was condemned. According to him, Christianity was a major obstacle to human happiness and a hindrance in the realization of utilitarianism.

Figure 3.3 Panopticon. Photo: Bettmann/Bettmann/Getty Images.

Bentham was one of the earliest philosophers to argue for a complete equality between sexes, and for decriminalization of homosexuality and equal rights for homosexually inclined people. Furthermore, he is widely recognized as one of the earliest proponents of animal rights. Bentham argued that the ability to suffer, not the ability to reason, must be the benchmark of how to treat other beings. If the ability to reason were the criterion, many human beings, including babies and disabled people, would also have to be treated as though they were things.

Bentham is also known for his design of the "panopticon" (which means all-seeing): it is a dome-shaped prison in which a prison warder can see all prisoners. They are kept in cell rings with windows facing inwards (Figure 3.3). The warder can observe all prisoners, but the prisoners cannot see the warder. The idea behind this is simple: if individuals are checked by an all-seeing eye (without the eye being seen), they will allow themselves to be disciplined and controllable. The panopticon remained an obsession of Bentham's for more than 20 years.

Bentham died on June 6, 1832, in the town he was born aged 85. The real body of Bentham together with a wax head (something went wrong preserving the head) can still be admired in the University College of London in a cabinet with a glass door. During board meetings of the university, he is removed from the cabinet so that he can attend these meetings. Bentham left his fortune to the university with the condition that he would be allowed to attend all meetings of the board.

Bentham calls pleasure and pain the sovereign masters of man. That which provides pleasure or avoids pain is good, and that which provides pain or reduces pleasure is bad. Bentham places experience at the heart of his ethics. According to him, it is an elementary fact of experience that people strive by nature for pleasure and avoid pain. Moreover, people know what provides pleasure and what results in pain, and also how pleasure can be realized. Based on this experience people can form a moral judgment without the intervention of an authority such as a legislator or God.

The only moral criterion for good and bad lies in what Bentham calls the **utility principle**: the greatest happiness for the greatest number. This principle is the only and sufficient ground for any action – both for individuals and collectives (e.g., companies or government). It gives us a reason to act morally. Moral terms like "proper," "responsible," and "correct" only are meaningful if they are used for actions that are in agreement with the utility principle. The greatest happiness can be determined quantitatively according to Bentham. He believed that we can calculate the expected pleasure or pain and can even indicate quite accurately how much will be produced by a given action. Here, pleasure and pain are given in terms of a measurable result, which can be made suitable for calculation. In this context he referred to a **moral balance sheet** and even drew up extensive tables. He made use of a number of circumstances, such as intensity, duration, certainty and extent of an action (see box). Applying this theory to a moral

Utility principle The principle that one should choose those actions that result in the greatest happiness for the greatest number.

Moral balance sheet A balance sheet in which the costs and benefits (pleasures and pains) for each possible action are weighed against each other. Bentham proposed the drawing up of such balance sheets to determine the utility of actions. Cost-benefit analysis is a more modern variety of such balance sheets.

problem means drawing up a moral balance sheet. Here, the costs and benefits for each possible action must be weighed against each other. The action with the best result (providing the most utility) is the one to be preferred. According to Bentham, money can even be used to express quantities of pleasure or pain, because these experiences can (almost) always be bought and sold.

Value of a Lot of Pleasure or Pain, How to be Measured

"Pleasures then, and the avoidance of pains, are the *ends* that the legislator has in view; it behoves him therefore to understand their *value*. Pleasures and pains are the instruments he has to work with: it behoves him therefore to understand their force, which is again, in other words, their value. To a person considered by *himself*, the value of a pleasure or pain considered *by itself*, will be greater or less, according to the four following circumstances:

- its *intensity*;
- its *duration*;
- its *certainty* or *uncertainty*; and.
- its *propinquity* or *remoteness*.

These are the circumstances which are to be considered in estimating a pleasure or a pain considered each of them by itself. But when the value of any pleasure or pain is considered for the purpose of estimating the tendency of any *act* by which it is produced, there are two other circumstances to be taken into the account; these are,

- Its *fecundity*, or the chance it has of being followed by sensations of the *same* kind: that is, pleasures, if it be a pleasure: pains, if it be a pain.
- Its *purity*, or the chance it has of not being followed by sensations of the *opposite* kind: that is, pains, if it be a pleasure: pleasures, if it be a pain.

These two last, however, are in strictness scarcely to be deemed properties of the pleasure or the pain itself; they are not, therefore, in strictness to be taken into the account of the value of that pleasure or that pain. (…) And one other; to wit:

- Its *extent*; that is, the number of persons to whom it *extends*; or (in other words) who are affected by it.

To take an exact account then of the general tendency of any act, by which the interests of a community are affected, proceed as follows. Begin with any one person of those whose interests seem most immediately to be affected by it: and take an account,

1　Of the value of each distinguishable *pleasure* which appears to be produced by it in the *first* instance.
2　Of the value of each *pain* which appears to be produced by it in the *first* instance.
3　Of the value of each pleasure which appears to be produced by it *after* the first. This constitutes the *fecundity* of the first *pleasure* and the *impurity* of the first *pain*.

4   Of the value of each *pain* which appears to be produced by it after the first. This constitutes the *fecundity* of the first *pain*, and the *impurity* of the first pleasure.

5   Sum up all the values of all the *pleasures* on the one side, and those of all the pains on the other. The balance, if it be on the side of pleasure, will give the *good* tendency of the act upon the whole, with respect to the interests of that *individual* person; if on the side of pain, the *bad* tendency of it upon the whole.

6   Take an account of the *number* of persons whose interests appear to be concerned; and repeat the above process with respect to each. *Sum up* the numbers expressive of the degrees of *good* tendency, which the act has, with respect to each individual, in regard to whom the tendency of it is *good* upon the whole: do this again with respect to each individual, in regard to whom the tendency of it is *good* upon the whole: do this again with respect to each individual, in regard to whom the tendency of it is *bad* upon the whole. Take the *balance* which if on the side of *pleasure*, will give the general *good tendency* of the act, with respect to the total number or community of individuals concerned; if on the side of pain, the general *evil tendency*, with respect to the same community.

It is not to be expected that this process should be strictly pursued previously to every moral judgment, or to every legislative or judicial operation. It may, however, be always kept in view: and as near as the process actually pursued on these occasions approaches to it, so near will such process approach to the character of an exact one." (Bentham, 1948 [1789])

The idea behind the calculation above is quite simple: an action is morally right if it results in pleasure, and it is morally wrong is it gives rise to pain. To find out which action leads to the most happiness for the greatest number of people, we need to count the pleasure and pain of all individuals. For example, Google's plans to launch Dragonfly (see Chapter 1) could be considered ethical according to utilitarianism. Google would be complicit in the Chinese government's suppression of information from its citizens, helping it to expand its surveillance powers to flag citizens who have searched for terms that pose a threat to the state. However, the increased access to information, although censored, for a population of 1.39 billion could be justified as the greatest good for the greatest number of people.

The counting of the pleasure and pain of individuals is usually not so simple, because pleasure cannot be measured objectively. First, the pleasure of different people cannot be compared; pleasure is a rather subjective term. A person can enjoy a composition by Mozart, while someone else experiences this quite differently. Second, it is not easy to compare actions: is reading a good book worth more than eating an ice cream? While applying this hedonistic calculus this will often lead to problems, because it is not clear how much pleasure a given experience produces for each person. How much pleasure do social contacts, our health, or our privacy give us? Since this is not clear, making moral judgments about human actions becomes hard. Take, for example, a company that pollutes the environment. If the company were to work in a more environmentally friendly way this would reduce the profits and the numbers of people employed. However, if the company does not become environmentally friendly then the damage to the environment will have repercussions for public health. It seems nearly impossible to draw up a quantitative moral balance sheet for these two options: continuing along the status quo or changing to environmentally friendly production.

3.7.2   Mill and the freedom principle

John Stuart Mill (1806–1873) extended and revised Bentham's thinking. There are two main respects in which Mill's thinking differs from that of his predecessor. According to Mill, qualities must be taken into account when applying the utilitarian calculus: forms of pleasure can be qualitatively compared, in which it is possible that a quantitatively smaller pleasure is preferred over a quantitatively larger one because the former pleasure is by nature more valuable than the latter. According to Mill, "[i]t is better to be a human being dissatisfied than a pig satisfied; better to be Socrates dissatisfied than a fool satisfied" (Mill, 1979 [1863]). Unfortunately, Mill does not answer the question what makes one pleasure more valuable than another. He only gives indications: "higher" desires, like intellectual ones, are to be preferred above "lower" desires, like physical or animal desires. Satisfying the desire to complete a study is more rewarding than watching "As the World Turns" every evening or to be able to eat as much as you want at every meal. The second distinction was a response to the criticism that the position of individuals cannot always be protected if the calculation indicates that the pleasure of the majority outweighs the unhappiness of a few individuals. This could result in the exploitation and abuse of minorities, because Bentham's utilitarianism does not say anything about the division of pleasure and pain among people. Therefore, Mill introduces the **freedom principle**: everyone is free to strive for their own pleasure, as long as they do not deny or hinder the pleasure of others. Mill illustrates this principle using the example of drunkenness. The right to interfere with someone who is drunk only arises when the person who is

> **Freedom principle**   The moral principle that everyone is free to strive for their own pleasure, as long as they do not deny or hinder the pleasure of others.

drunk starts to do harm to others. Mill's principle also provides a foundation for the discussion nowadays about legalizing soft drugs (or even heroin). According to Mill, the sale and use of soft drugs should not be a matter for penal law, as this would be a violation of freedom. The fact using soft drugs is bad for your health cannot be a consideration for the legislator to intervene, because the legislator has no right to be involved with personal decisions in Mill's view. Mill illustrates this principle on the basis of drunkenness.

> Drunkenness, for example, in ordinary cases, is not a fit subject for legislative interference; but I should deem it perfectly legitimate that a person, who had once been convicted of any act of violence to others under the influence of drink, should be placed under a special legal restriction, personal to himself; that if he were afterwards found drunk, he should be liable to a penalty, and that if when in that state he committed another offence, the punishment to which he would be liable for that other offence should be increased in severity. (Mill, 1859, Chapter 5)

The freedom principle is also known as the **no harm principle**: "one is free to do what one wishes, but only to the extent that no harm is done to others." Howevjer, the principle can hardly ever be applied in full, since any moral problem involves possible harm to others, or at least the risk of harm.

> **No harm principle**   The principle that one is free to do what one wishes, as long as no harm is done to others. Also known as the freedom principle.

John Stuart Mill (1806–1873)

Figure 3.4 John Stuart Mill. Photo: Everett Collection/Shutterstock.

The only freedom which deserves the name, is that of pursuing our own good in our own way, so long as we do not attempt to deprive others of theirs, or impede their efforts to obtain it. Each is the proper guardian of his own health, whether bodily, or mental and spiritual. Mankind are greater gainers by suffering each other to live as seems good to themselves, than by compelling each to live as seems good to the rest. (Mill, 1859)

John Stuart Mill was born in 1806; he was the oldest son of James Mill and proved to be a prodigy. James Mill had special ideas about raising children. At the age of three he taught his son Greek, at age four he taught him Latin, and shortly after he taught him mathematics. At age twelve John Stuart Mill wrote a book about Roman history. Mill was a proponent of utilitarianism as proposed by his godfather Jeremy Bentham. When he was 18, Mill founded a utilitarian society for youths, where lectures and discussions were held about the utility principle. When he was 20, Mill had a nervous breakdown and he suffered from severe depressions. He found that the utilitarianism of Bentham was not making him happy. Following this, he distanced himself from Bentham's ideas.

In 1823 he started to work for the East India Company under his father's authority. This work provided him with much opportunity to study and write. In 1830 he met the 23 year-old Harriet Taylor. They were highly impressed by each other. However, Harriet was married to the businessman John Taylor and she decided not to sacrifice her family because of her feelings. Her husband eventually allowed her to meet with Mill on a regular basis. According to Mill's testimony their love for each other was purely platonic. After John Taylor's death in 1849 there was no more reason not to marry, which they did in 1851. In Mill's view, Harriet's opinions had a major influence on him and especially his socio-philosophical work. Together with her, Mill called for the emancipation of women and also argued for women's right to vote. In 1869 he published *The Subjection of Women*, which is now the classical theoretical statement of the case for woman suffrage. Harriet died in 1858 in Avignon. Between 1866 and 1868, Mill was a Member of Parliament. He was considered a radical, because he supported the public ownership of natural resources, the development of labor organizations, compulsory education, birth control, an end to slavery, and equality of women. His advocacy of women's suffrage in the Reform Bill of 1867 led to the creation of the suffrage movement. He died in 1873.

John Stuart Mill was the most influential British thinker of the nineteenth century. Mill's essay *On Liberty* (1859) remains his major contribution to political thought. He proposed that self-protection is the only reason an individual or the government can interfere with a person's liberty of action. Outside of preventing harm to others, the state has no legitimate reason to compel a person to act in the way the government wishes.

3.7.3   Criticism of utilitarianism

Although utilitarianism has a strong intuitive attraction because of its simplicity, it has nevertheless received much criticism. Two important points of criticism were discussed above: happiness cannot be measured objectively and utilitarianism can lead to exploitation. Four other points of criticism are discussed below. In many cases the criticism was incorporated by utilitarians to improve utilitarianism.

The first criticism is that the consequences cannot be foreseen objectively and often are unpredictable, unknown, or uncertain. An obvious solution is to work with expected consequences and the accompanying pleasure. In the twentieth century this notion was even given a mathematical foundation using statistics.

Next to this, there is the problem of distributive justice. **Distributive justice** refers to the value of having a just distribution of certain important goods, like income, happiness, and career. Utilitarianism can lead to an unjust division of costs and benefits. According to the political philosopher John Rawls utilitarianism suffers from this problem because it does not recognize the fundamental separateness of persons (Rawls, 1971). Instead of that utilitarianism treats society as a whole in which pleasure must be increased via the criterion "the greatest happiness for the greatest number." The question concerning the distribution of happiness is neglected, even under Mill's formulation of utilitarianism. It is a tricky question because numerous issues in technology are concerned with this problem, such as how the risks and benefits of technology should be justly distributed (see Chapter 8). Despite Rawls criticism, utilitarians have tried different ways to pay attention to justice and the distribution of welfare. Henry Sidgwick, for example, believed that although the total amount of societal happiness should be considered in the first place, it should be the situation with the most equitable distribution of happiness that must be selected from various situations with equal happiness (Sidgwick, 1877). Other utilitarians argue that the classical utilitarianism – with the emphasis on the greatest happiness for the greatest number of people – does not require such a clause, because it leads to a just and balanced distribution of welfare. The modern utilitarian Richard Hare mentions two reasons for this. First, a rich person experiences less added pleasure on average from an increase in income of 100 Euros than a poor person. This phenomenon is known in economics as decreasing **marginal utility** (the term marginal utility refers to the increase in utility with an increase in income for example). An improvement in income for poor people will sooner lead to maximization of happiness than an increase in income for people who already are rich. Second, inequality of income leads to jealousy and thus to pain and is thus to be avoided (Hare, 1982).

> **Distributive justice** The value of having a just distribution of certain important goods, like income, happiness, and career.

> **Marginal utility** The additional utility that is generated by an increase in a good or service (income for example).

A third point of criticism is that utilitarianism ignores the personal relationships between people. In the hedonistic balance of Bentham each individual counts as an anonymous unit. Who receives the pleasure is irrelevant; it is only to total amount of pleasure that counts. In other words, the total happiness counts and not the individual happiness of specific persons. For this reason, Mill called Bentham's followers reasoning machines. In daily life, some people's happiness has a greater impact on us than the happiness of others. If you were to be shipwrecked and had to make a choice between saving a friend or a famous surgeon, utilitarian theory dictates that saving the surgeon is the right thing to do, because they are more useful to society. This choice ignores the fact that it is *specific individuals* that want to be happy and that it really depends on *who* is made happier. The question, therefore, is whether we have special moral obligations to the people that we have a personal relationship with, and whom we want to make happy.

Finally, certain actions are morally acceptable even though they do not create pleasure and some actions that maximize pleasure are morally unacceptable. In the next section we will see that Kant always considers lying to be morally wrong, even if it results in more or maximal pleasure in certain situations. According to utilitarianism, even the most fundamental rules, such as the human rights formulated in the *Universal Declaration of Human Rights* (proclaimed and adopted by the United Nations General Assembly in 1948), can be broken if the positive consequences are greater than the negative ones: "the end justifies the means." On utilitarian grounds, an engineer could be asked to bend a fundamental rule of professional conduct because of the positive consequence it would have. Say, for example, that an engineer is asked to falsify the measurements they gave in a report by the party commissioning the work, because the correct measurement results would have major negative consequences, such as the payment of damages or bankruptcy. According to the traditional utilitarian view, this behavior would be justified in a certain situation. This traditional view is known as **act utilitarianism** because it judges the consequences of individual acts. A solution to this problem is proposed by one variant of utilitarianism: **rule utilitarianism**. Rule utilitarianism recognizes the existence of moral rules, if only because life would be very complicated without them. For each situation we would have to judge whether it was morally correct or not, because each situation is slightly different from another. Rule utilitarianism looks at the consequences of rules (in contrast with actions) to increase happiness. Though the falsifying of measurements may

> **Act utilitarianism**   The traditional approach to utilitarianism in which the rightness of actions is judged by the (expected) consequences of those actions.
>
> **Rule utilitarianism**   A variant of utilitarianism that judges actions by judging the consequences of the rules on which these actions are based. These rules, rather than the actions themselves, should maximize utility.

increase societal utility in a specific situation, a rule utilitarian will not allow it because the rule "measurement data should be presented correctly" generally promotes happiness within society. If such a rule withstands the test of promoting happiness, then it is turned into a moral rule. Within rule utilitarianism there are a number of variants. There is a variant where the moral rules are viewed as conditional rules (they are more like rules of thumb), and a variant that views the rules as unconditional ones (they apply to all people in all circumstances without exception). Rule utilitarianism is close to duty ethics, which is the subject of the next section, although their conceptual foundations are very different.

3.7.4   Applying utilitarianism to the Ford Pinto case

In the Ford Pinto case, the Ford company provided an act-utilitarian argument by making a cost-benefit analysis to justify that the defective vehicle model was not recalled and retrofitted by Ford. This cost-benefit analysis, according to Ford showed that the total social costs of retrofitting all the cars were higher than the social costs of the expected accidents. It is important to note that the cost-benefit analysis refers to social cost rather than to costs for Ford. For this reason, Ford's argument was utilitarian rather than egoistic.

The Ford Pinto case clearly illustrates some of the objections against utilitarianism. First, the amounts of money that Ford attached to different kinds of pain (dead, injuries) seem rather arbitrary, even if some of the amounts were based on government documents. Second, one might wonder how reliable the estimates of, for example, number of fatalities are. A change in these estimates may change the conclusion of the cost-benefit analysis. Apart from such more practical objections, the case also illustrates some of the more fundamental objections to utilitarianism. In making a decision solely based on considerations of overall welfare or happiness, Ford adopted a policy of allowing a certain number of people to die or be injured even though they could have prevented it. One could also argue that the Ford Pinto case reveals exploitation or abuse because the victims were sacrificed to optimize overall welfare (the ends justify the means). Moreover, the case shows how a utilitarian argument may lead to abandoning inherent principles, like "you cannot put a value on human life" or the freedom principle of Mill. According to the latter principle, Ford should have recalled and repaired the car.

Some of these objections might be overcome by applying rule utilitarianism to the case. Then, one should ask whether or not following rules like "companies must recall a car if it is unsafe" or "companies should produce safe cars" maximizes overall happiness. Since this seems to be the case, Ford was ethically obliged to recall the car, because this is required by rules from which everyone in the society would benefit most in the long run. So, in the case of rule utilitarianism, the fact that an action maximizes utility on a *particular* occasion does not show that it is right from an ethical point of view.

3.8 Kantian Theory

According to **duty ethics** (also known as deontological ethics), an action is morally right if it is in agreement with a moral rule (law, norm, or principle) that is applicable in itself, independent of the actual or probable consequences of that action. There are two important points of difference between the various duty ethics theories. First, some theories rely on one

> **Duty ethics** Also known as deontological ethics. The class of approaches in ethics in which an action is considered morally right if it is in agreement with a certain moral rule (law, norm, or principle).

main principle from which all moral norms can be derived (monistic duty ethics). Other theories, the pluralistic theories, are based on several principles that apply as norms for moral action. A second important difference concerns the foundation or origin of the moral rules. These rules can be given by God, such as in the Bible or the Koran, or they make an appeal to a social contract that the involved parties have implicitly agreed to (e.g., a company code), or they are based on reasonable arguments.

The best-known Western system of duty ethics has been developed by Immanuel Kant. Since Aristotle, the basis for ethics had been sought in striving for happiness or welfare (e.g., Bentham and Mill). According to Kant, moral laws or normative ethics cannot be based on happiness. Happiness is an individual matter and changes for each person during their lifetime. Moreover, it is hard to determine what increases happiness, so striving for happiness can even lead to immorality. Thus, Kant argued that duty was a better guide for ethics.

A core notion in Kantian ethics is *autonomy*. In Kant's opinion man *himself* should be able to determine what is morally correct through reasoning. This should be possible independent of external norms, such as religious norms. The idea behind this is that we should place a moral norm upon ourselves and should obey it: it is our *duty*. We should obey this norm out of a *sense of duty* – out of respect for the moral norm. It is only then that we are

Good will   A central notion in Kantian ethics. According to Kant, we can speak of good will if our actions are led by the categorical imperative. Kant believes that the good will is the only thing that is unconditionally good.

Hypothetical norm   A condition norm, that is, a norm which only applies under certain circumstances, usually of the form "If you want X do Y."

acting with **good will**. According to Kant, we can speak of good will if our actions are led by the moral norm. Thus, the notion of good will is different from having good intentions.

Since a moral norm has validity independent of time and place, it means that a moral norm is unconditionally applicable (or categorically applicable) to everyone in all circumstances in Kant's view. Often a norm follows the form of "thou shalt …," such as "thou shalt not kill," or "thou shalt not lie." In contrast to a categorical norm, a hypothetical (conditional) norm only applies under certain circumstances. A **hypothetical norm** usually has the following shape: "if you wish to achieve this goal, then you will have to act in this way." An example of such a norm is "if you do not wish to betray your friend, then you may not lie," in which the rule of behavior ("you may not lie") is not unconditional but can only be applied under certain conditions ("you do not wish to betray your friend").

3.8.1   Categorical imperative

According to Kant there is one universal principle from which all moral norms can be derived, which makes his ethics a monistic duty ethics. This principle, which is the foundation of all moral judgments in Kant's view, is referred to as the **categorical imperative**. An imperative is a prescribed action or an obligatory rule. By arguing reasonably, any rational person should be capable of judging whether an optional action is morally right. The categorical imperative was formulated by Kant in different ways.

Categorical imperative   A universal principle of the form "Do A" which is the foundation of all moral judgments in Kant's view.

Universality principle   First formulation of the categorical imperative: Act only on that maxim which you can at the same time will that it should become a universal law.

The first formulation of the categorical imperative, the **universality principle**, is as follows:

"Act only on that maxim which you can at the same time will that it should become a universal law."

A maxim is a practical principle or proposition that prescribes some action. Kant states that the maxim should be unconditionally good, and should be able to serve as a general law for everyone without this giving rise to contradiction. We must oblige ourselves to follow generally applicable laws. Perhaps a woman decides to recycle her bottles and cans to help the environment. She should ask herself whether the maxim or rule behind her action – that one should recycle containers to help the environment – could be applied to all people. In this case, there is no apparent problem. She could consistently wish that everyone follows the rule or maxim behind her action. However, when you break a promise, this is different. Sometimes people are in a situation where it would be more convenient to break a promise. Say that one wonders whether it is morally acceptable to break one's promise. The maxim of the action to be undertaken is "I may break my promises when doing so is convenient for me." The categorical imperative states that it is morally acceptable if I can wish everyone to break their promise without *contradiction*. Breaking a promise is only possible if people trust in the custom of making (and keeping) promises. If breaking a promise when convenient becomes a general law, no one would trust anybody to keep a promise. The contradiction now is that you cannot wish to break a promise and want the breaking of promises to become a general law. If the latter were to become true, then promises would lose their meaning and it would be no use to make a promise.

According to Kant, the categorical imperative also implies a postulate of equal and universal human worth. His reflections on autonomy and self-legislation lead him to argue that the free will of all rational beings is the fundamental ground of human rights. The **equality postulate** is defined as the prescription to treat persons as equals, that is, with equal concern and respect (Dworkin, 1977, p. 370). To recognize that human beings are all

> **Equality postulate**   The prescription to treat persons as equals, that is, with equal concern and respect.

equal does not mean having to treat them identically in any respects other than those in which they clearly have a moral claim to be treated alike. Opinions diverge concerning the question what these claims amount to and how they have to be balanced with competing claims (based on, for example, the principle of freedom). For example, how should goods be distributed if we set out to treat people as equals?

The second formulation of the categorical imperative is, according to Kant, equivalent to the first.

> The second formulation of the categorical imperative, the **reciprocity principle**, is as follows:
>
> Act as to treat humanity, whether in your own person or in that of any other, in every case as an end, never as means only.

> **Reciprocity principle** Second formulation of the categorical imperative: Act as to treat humanity, whether in your own person or in that of any other, in every case as an end, never as means only.

Humanity in this version of the imperative is presented as equivalent to "reason" or "rationality," for humans differ from things without reason (objects and animals) because humans can think. This imperative states that each human must have respect for the rationality of another and that we must not misguide the rationality of another. In other words, Kant here stresses the rational nature of humans as free, intelligent, self-directing beings. In saying they must never be treated as a means only, he means that we must not merely "use" them as means to our ends. They are not objects or instruments to be used. To use people is to disrespect their humanity. Say I borrow money from someone and promise to pay him back although I know that I will not do so. In this case, I am using the person I made a promise to as a means and not as a goal. I am misleading them, or I am misleading their rationality. I have provided insufficient information about the fact that I will not keep my promise, so that they cannot make a rational choice. Probably they would not have lent me money if they had known that I did not intend to pay them back. I use their rationality as a means to achieve my own aim. The reciprocity principle is strongly anti-paternalistic by nature (on paternalism see Chapter 1), since, a person – as a rational being – should have the right to make up their own mind.

The reciprocity principle tells us that we should respect people *as* people, and not "use" them. However, we need to be careful in interpreting the idea of using people, or treating people merely as a means. The difference between treating someone as a means versus treating someone as a *mere* means is not always clear-cut. Suppose someone has religious objections to taking medication (a Christian Scientist, for example), and yet the doctor forces the person to be medicated for the person's own good. Now the doctor is treating the patient as a mere means to the patient's own welfare – paradoxical as it might sound – and that is unacceptable, according to the reciprocity principle. Note that to treat someone as an end does not simply mean doing what they want. If a consumer argues about the purchase price of a car, and the salesperson does not want to bargain about the price, this does not mean that the salesperson treats the consumer not as an end. If the salesman informs the consumer about the price of the car and the condition of the car, the salesman treats the consumer as an end.

Figure 3.5   Immanuel Kant. Photo: Unknown author/Wikimedia Commons/Public Domain.

Immanuel Kant (1724–1804)

Nothing can possibly be conceived in the world, or even out of it, which can be called good, without qualification, except a good will. Intelligence, wit, judgment, and the other talents of the mind, however they may be named, or courage, resolution, perseverance, as qualities of temperament, are undoubtedly good and desirable in many respects; but these gifts of nature may also become extremely bad and mischievous if the will which is to make use of them, and which, therefore, constitutes what is called character, is not good. (Kant, 2002 [1785])

Immanuel Kant, one of the most influential philosophers in history, was the fourth of nine children born to a poor saddle maker. He was born in 1724 in the university city of Königsberg in East Prussia, which was a rich trading place at the time. He was brought up in a tradition of devout Christianity that he strongly rejected in later life. After completing pre-university education, he first studied theology and then philosophy, mathematics, and physics in Königsberg. After completing his studies in 1746, he became a teacher for various families. From 1755, when he attained the title of Magister, he became a private teacher at the University of Königsberg. At 46 he accepted a professorship in logic and metaphysics. He had great admiration for the enlightened king Frederick the Great of Prussia, but in 1794 he came into conflict with the King's successor due to his theological philosophy. He valiantly defended the right of scientists to think in freedom and to publish for fellow scholars. Kant died at the age of 80 (1804). His life was known to be highly disciplined – he had a great fervor for work and a strict daily routine. The inhabitants of Königsberg could set the clock by the time, when Kant passed by for his daily walk. The reason for this way of life was his poor physical health, which he tried to improve through his strict routine.

Kant's theory of mind represents a turning point in the history of philosophy, since it radically revised the way that we all think about human knowledge of the world. He built his systematic theoretical philosophy around the idea that the world as we experience it does not exist independently of us. Our own minds are responsible for its form and structure. This introduced the human mind as an active originator of experience rather than just a passive recipient of perception. As Kant puts it, it is the representation that makes the object possible rather than the object that makes the representation possible. This idea, in his words, effected a Copernican revolution. Before Nicolaus Copernicus (1473–1543) – the founder of modern astronomy – astronomical data were explained by assuming that the sun revolves around the earth. Reversing this, Copernicus explained the data by taking the earth to revolve around the sun. In moral philosophy, Kant proposed an equally revolutionary idea. In morality we are not required to obey laws imposed by God or eternal moral principles; instead, we must understand morality as resting on a law that springs from our own practical rationality. Kant's ethics, which he expounded in the *Critique of Practical Reason* (1788) and the earlier *Groundwork of the Metaphysics of Morals* (1785]), was based on the principle known as the "categorical imperative," an unconditional obligation derived from the concept of duty.

3.8.2   Criticism of Kantian theory

There are two primary criticisms of Kantian theory. According to Kant all moral norms can be derived from the categorical imperative. The question arises whether all these norms form an unambiguous and consistent system. Often there are several contradictory norms, as we saw earlier in the case of the whistle-blower. Another example is the situation in which one can only save one's friend from an emergency situation by lying. It means breaking a norm: either you break the norm that you must always speak the truth or you break the norm about helping people when they need it. In Kant's theory there is no such thing as bending a rule. Kant does not allow for any exceptions in his theory.

To cope with this problem, William David Ross developed a pluralistic theory of moral obligations (Ross, 1930). Ross states that right is often situated on two levels: what seems to be right at first and that which is good once we take everything into consideration. The norms of the first level are called **prima facie norms** and those of the second level are called *self-evident norms* ("duties sans phrase"). Usually, the *prima facie* norms are our self-evident norms, but this is not necessarily the case. An example can illustrate this. Say you promise your students that you will check their work by the end of next

> **Prima facie norms**   Prima facie norms are the applicable norms, unless they are overruled by other more important norms that become evident when we take everything into consideration.

week. Later on, a good friend of yours gets into trouble and needs aid. The fact that you have promised to check the work does not disappear. Both norms are prima facie norms, but upon closer inspection only the norm "you must help your friend" is a self-evident norm while the other norm ("you must keep your promise") is not.

Note that here too we have to weigh the different norms: the norm to keep one's promise and the norm to help a friend. We are never certain that the norm we identify as the self-evident one truly is the self-evident norm. How we should weigh the norms remains unclear here too. Ross states that our choices are never more than considered judgments. Though this is perhaps not very satisfactory, it does pay respect to the complexity of our moral world. Examples of regular *prima facie* norms that are common in duty ethics include the following:

- Norms concerning faithfulness: freely given promises should be kept.
- Norms concerning reciprocity: this can refer to things like the Golden Rule in a positive or a negative sense (treat others as you would like to be treated/do not treat others as you would not like to be treated yourself).
- Norms of solidarity: help people in need regardless of their achievements or usefulness to society or to you as an individual.

Second problem is that duty ethics, and thus Kantian theory, often elicits the objection that a rigid adherence to moral rules can make people blind to the potentially very negative consequences of their actions, as becomes clear in the child labor case.

Case   Child Labor

The International Labour Organization and UNICEF (2021) estimated in 2020 that 79 million children are engaged in hazardous child labor, which is defined as "work that that, by its nature or circumstances, is likely to harm children's health, safety or morals" (International Labour Organization and UNICEF 2021, p. 19). Child labor is particularly an issue in the fashion industry because much of the supply chain requires low-skilled labor and some tasks are better suited to children than adults.[1] In cotton picking, for example, employers prefer to hire children for their small fingers, which do not damage the crop. Child labor is often seen as opposed to the rights and interests of children, but many actions taken against child labor in the past have ended up doing more harm than good, because they take away a relatively good opportunity for children to provide themselves a living, which will decrease their chances of survival. As a result of losing a job, many of these children may end up in slavery or prostitution. Moreover, trade and industry can contribute to the improvement of the working conditions of the children, such as working times, medical care, training, etc. (see, e.g., Radfar et al., 2018).

This case demonstrates, on the one hand, the value of adhering to a strict moral principle: that child labor should not be condoned, regardless of the consequences, which befits Kantian theory. We could argue with Kant that child labor is unethical because it violates the autonomy of the children. To make a rational choice, a child needs to understand the situation and the consequences of their choice. Children of young age, who still lack full rational and moral capacities, cannot rationally make such a decision. Moreover, they may be manipulated by their parents or other adults. On the other hand, utilitarians would emphasize the negative consequences of such strict adherence to principle. If child labor is the only means for a family to survive, breaching the principle counts for nothing in the face of the consequences of abolishing child labor. As this example shows, both theories appeal to our moral intuitions, but they can come to diametrically opposed conclusions concerning the moral correctness of an action. Ross' approach could offer a solution to the rigidity of Kantian theory. According to Ross the reason why a norm is a self-evident norm depends on the situation in which one finds oneself. We must do what is more of a duty in a given situation. The *prima facie norm* "child labor is not permitted" is not a self-evident norm in this situation, because the situation calls for us to help the children in need and prevent them from becoming slaves or prostitutes. The norm "children should not be forced into slavery or prostitution" would be the self-evident norm instead of "child labor is not permitted."

3.8.3 Applying Kant's theory to the Ford Pinto case

To apply Kant's first categorical imperative, the universality principle, to the Ford Pinto case, we must examine whether the maxim of Ford: "Ford will market the Ford Pinto, knowing that the car is unsafe in certain conditions and without informing the consumers" can be universalized. To do this we have to explain whether this maxim can become a universal law, and can be willed without contradiction. The universal law would read as follows: "Marketing unsafe cars without informing the consumers is allowable." If this were to be a universal law marketing a car would become impossible because no rational person would buy a car anymore, because they could not trust that the car would be safe. It may be clear then that the maxim cannot be universalized and should, therefore, not be followed by Ford.

The second categorical imperative, the reciprocity principle, tells us that people should not be treated as mere means. As we have seen this principle implies respect for people's **moral autonomy** in making their own choices. From this, it follows that Ford should have informed its consumers because otherwise they cannot make an autonomous rational decision to buy

> **Moral autonomy** The view that persons themselves should (be able to) determine what is morally right through reasoning.

the car or not. If consumers had known what Ford knew about the safety of the Ford Pinto, they would probably have thought twice before buying the car. By failing to inform them, the rational agency of the consumer was thus undermined, and they were used as *merely* a means (and thus not as an end) to achieve Ford's aim: increasing Ford's profit. It is not just that Ford endangered people's lives; rather, it is that Ford did so without informing car drivers about the risks.

3.9 Virtue Ethics

Utilitarianism and Kantian theory both are theories about criteria concerning action. Rather than taking action as point of departure for moral judgment, **virtue ethics** focuses on the nature of the acting person. This theory indicates which good or desirable characteristics people should have or develop and how people can achieve this. Virtue ethics is not

> **Virtue ethics** An ethical theory that focuses on the nature of the acting person. This theory indicates which good or desirable characteristics people should have or develop to be moral.

exclusively aimed at reason, as the previous two theories were, but is more a mixture of ethics and psychology with an emphasis on developing character traits.

Virtue ethics is based on a notion of humankind in which people's characters can be shaped by proper nurture and education, and by following good examples. The central theme is the development of persons into morally good and responsible individuals so that they can lead good lives. To this purpose, developing good character traits, both intellectual and personal character traits, is essential. These characteristics are called virtues. They not only indicate how to lead a good life but also what a good life is. Examples of virtues are reliability, honesty, responsibility, solidarity, courage, humor, and being just.

3.9.1 Aristotle

Virtue ethics stems from a long tradition and was already popular in ancient Greece with philosophers like Socrates, Plato, and Aristotle. Aristotle was the first to define virtue ethics as a field of inquiry in itself. According to Aristotle, the final goal of human action is to strive for the highest good: *eudaimonia*. This can be translated as **"the good life"** (or as "welfare" or "happiness"). This does not refer to a happy circumstance that brings pleasure (the goal of classical utilitarians), but *human flourishing*, a state of being a good person. It means leading a life as humans are meant to lead it; one should excel in the things that are part of being human. As only humans can reason, this is where happiness lies. If we wish to become happy as humans, we must use our reasoning to its fullest extent. The good life is not only determined by activities related to reasoning, but is also realized by virtuous activities according to Aristotle. The good life therefore is an active life in agreement with the virtues necessary to realize one's uniquely human potential.

> **The good life**   The highest good or *eudaimonia*: a state of being in which one realizes one's uniquely human potential. According to Aristotle, the good life is the final goal of human action.

Figure 3.6   Aristotle. Photo: Argus/Adobe Stock.

Aristotle (384–322 BC)

Virtue, then, is a state of character concerned with choice, lying in a mean, that is, the mean relative to us, this being determined by a rational principle, and by that principle by which the man of practical wisdom would determine it. Now it is a mean between two vices, that which depends on excess and that which depends on defect; and again, it is a mean because the vices respectively fall short of or exceed what is right in both passions and actions, while virtue both finds and chooses that which is intermediate. Hence in respect of its substance and the definition which states its essence virtue is a mean, with regard to what is best and right and extreme. (Aristotle, 1980 [350 BC])

Aristotle was born in Stageira, Macedonia in 384 BC. His father was the personal physician to the Macedonian King Amyntas II. As a result, he was sometimes referred to as the Stagerite. He came from a family of doctors, which probably explains his interest in physics and biology. In 367 BC Aristotle entered Plato's academy in Athens. He took lessons there for 20 years and taught there himself. Political circumstances made him leave Athens in 347 BC. He first moved to Assos (the north coast of Asia Minor) and then to Mitulene on the island of Lesbos. There he became fascinated with aquatic animals. Aristotle had a far greater interest in biological questions than his predecessors. He realized that the biology of humans could never be understood without studying the biology of lower animals. Up until the nineteenth century, Aristotle's research on water animals was unsurpassed in biological literature.

In 343 BC, Aristotle went to Pella in Macedonia to take up the duty of raising the 13-year-old Alexander the Great. Despite his election as head of the Academy in 339, he was only able to return to Athens in 334. Up to 323 BC he had his own philosophical school in the Lyceum, which was situated in the north-east of the city. Its name, Peripatos, is taken from Aristotle's habit of teaching while he was walking, so that his pupils were often referred to as walkers (peripatetic).

The news of the death of Alexander the Great in 323 BC resulted in a strong anti-Macedonian response in Athens, forcing Aristotle to flee "to stop the people of Athens committing a second atrocity against philosophy." Aristotle was referring to Socrates trial in 399 BC. A year later, in 322 BC, Aristotle died in Chalcis aged 63.

Aristotle is one of the most important founding figures in Western philosophy. He was the first to create a comprehensive system of Western philosophy, encompassing morality and aesthetics, logic and science, politics and metaphysics. For example, he is credited with the earliest study of formal logic, and his conception of it was the dominant form of Western logic until nineteenth-century advances in mathematical logic. His work *Ethica Nicomachea* is one of his most accessible texts. It is also the first systematic approach to ethics in Western philosophy. Though Christian Europe ignored him in favor of Plato until Thomas Aquinas reconciled Aristotle's work with Christian doctrine, this work was the origin of certain types of philosophical ethics: the so-called happiness ethics, which was a dominant philosophy until the time of Immanuel Kant.

Each moral virtue (also referred to as a character virtue by Aristotle) holds a position of equilibrium according to Aristotle. A moral virtue is the middle course between two extremes of evil; courage is balanced between cowardice and recklessness for example, generosity between stinginess and being a spendthrift, and pride between subservience and arrogance. This is an expression of an old Greek notion: there is a certain ratio that is essential to humans that must be kept in balance and should not lean to the left or right if one wishes to achieve an optimal human state. A courageous person will not act as a coward in a dangerous situation, but they will also not be reckless and ignore the danger. According to Aristotle, moral virtues are not given to us at birth nor are they supernatural; they can be developed by deeds. In other words, they can be practiced just like all arts: "For the things we have to learn before we can do them, we learn by doing them, for example, men become builders by building and lyre players by playing the lyre; so too we become just by doing just acts, temperate by doing temperate acts, brave by doing brave acts" (Aristotle, 1980 [350 BC], 1130a).

People must seek a middle course, but this is not a simple matter. Aristotle believed that people know what they want instinctively, but not what they should do. Moreover, the middle course depends on the circumstances in a given situation. In other words, what is good in one case is not necessarily so in another. Unlike Plato (and, later, Kant), Aristotle argues that the good is sometimes ambiguous. However, people are not powerless in

Practical wisdom The intellectual virtue that enables one to make the right choice for action. It consists in the ability to choose the right mean between two vices.

finding the middle course. The intellectual virtue sagacity or **practical wisdom** is aimed at making the right choices for action concerning what is good and useful for a successful life. According to Aristotle, a wise person can see what they have to do in the specific and often complex circumstances of life. Sagacity implies a capacity for moral judgment, which is the middle course. Moral virtues and the intellectual virtue go hand in hand.

The influence of Aristotle spread across Syria and through the Islamic world. From the thirteenth century on Aristotle's work started to influence Europe too, because the Christian philosopher Thomas Aquinas (1225–1274) reconciled the heathen virtue ethics with Christian doctrine. Thomas Aquinas distinguished seven virtues. These include the four cardinal virtues of prudence, temperance, justice, and fortitude. These virtues are natural and revealed in nature, and they are binding on everyone. There are also the three theological virtues of faith, hope, and charity. These are supernatural and are distinct from other virtues in their object, namely, God. From 1600 on virtue ethics was falling into oblivion because a new ethics was arising that was focused on rules and paid less attention to virtues. In recent years there is growing interest in the origins of virtue ethics; this is particularly due to the influence of the philosopher Alasdair MacIntyre.

3.9.2   Criticism of virtue ethics

William Frankena argues that virtue ethics is not essentially different from duty ethics (Frankena, 1973). According to him each virtue is accompanied by a moral rule for action and there is a virtue for each moral rule. However, it appears that not all obligations to act can be reduced to virtues and vice versa. Virtues characterize the person and provide insight into the background to action. A person's good character traits do raise expectations, but they do not provide a measure for judging an action. For example, the argument that the actions of an engineer are moral by definition because they are upstanding and reliable will not readily be accepted in a moral discussion. Moreover, it is hard to check whether the engineer acted with proper intentions. So, virtue ethics does not give concrete clues about how to act while solving a case, in contrast with utilitarianism and Kantian ethics. Opposite this we can argue that having the right virtues does facilitate responsible action, as will become evident in the LeMessurier case that is discussed later.

Finally, we can join Kant in wondering whether we can simply declare a moral virtue to be good in itself without any reservation. Kant's example for this is a cold psychopath whose virtues moderation of conscience and passion, self-control and cool deliberation make them, much more terrible than they would have been without those virtues.

3.9.3   Virtues for morally responsible engineers

Virtues like reliability, honesty, responsibility, and solidarity, are quite general and most are virtues that morally responsible engineers need to possess too. If we look more specifically at the virtues engineers need, then we must focus on engineering practice. Michael Pritchard lists a number of virtues that are more specific than those mentioned above and that are required for morally responsible engineers (see box).

Virtues for Morally Responsible Engineers

- expertise/professionalism;
- clear and informative communication;
- cooperation;
- willingness to make compromises;
- objectivity;
- being open to criticism;
- stamina;
- creativity;
- striving for quality;
- having an eye for detail; and
- being in the habit of reporting on your work carefully. (Pritchard, 2001)

Stipulations in professional codes of conduct often refer to some of these virtues. The professional code of conduct of FEANI (Fédération Européenne d'Associations Nationales d'Ingénieurs or European Federation of National Engineering Associations) recognizes such virtues as integrity and impartiality. A list of virtues, however, does not say exactly how they are expressed in engineering practice, but the presence of certain virtues can have an important influence on the quality and ethical integrity of the work (see the LeMessurier case).

Case   LeMessurier

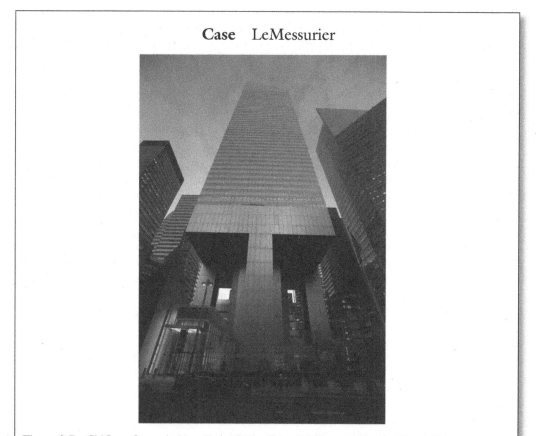

Figure 3.7   CitiCorp Center in New York. Photo: Orjan F. Ellingvag/Corbis Historical/Getty Images.

William LeMessurier designed the Citicorp Center (see Figure 3.7) in Manhattan, which was built in 1977. It was an innovative design because the skyscraper had to be built on pillars nine floors high placed at the middle of the sides of the construction site. This unusual structure was designed to accommodate a church that was being resurrected under one corner of the Citicorp edifice.

During construction, the contractors decided to attach the supports with bolts rather than welding them due to the high costs of welding – this was done without LeMessurier being informed to avoid potential delays. The fact that this would result in much weaker connections was not viewed as a problem because the choice of using bolts was technically correct. In the original design welding was chosen because this would mean that there would be less movement in the skyscraper, which would improve the comfort of its future inhabitants. When LeMessurier heard about the alteration later, he did not worry about the safety risks because a connection using bolts met the safety requirements.

This changed when a month later, in 1978, he received a telephone call from a student whose professor had informed him that the construction was dangerous. The pillars should have been placed in the corners of the sites according to him. LeMessurier explained why the pillars were positioned in the middle of each side to allow the skyscraper to be better able to cope with storms than other standard constructions. Following this, he decided to deal with the technical aspects of his design and safety in his own lectures for building engineers. Because LeMessurier took the remarks of the student and the professor seriously, he decided to carry out other wind-resistance tests beyond the standard ones. From this intellectual game, LeMessurier came to the conclusion that the building was not as safe as he had thought. A 16-year storm (one that passes every 16 years) could possibly rip loose one of the connections and the whole building could collapse. LeMessurier knew how to solve this problem however. As soon as he made his finding, he considered his options, including keeping silent and committing suicide. These two options he rejected, because these options may endanger human lives and struck him as a coward's way out. Instead, he first informed his company and its liability insurers and lawyers, the chief architect, and, thereafter, the chief executive at Citicorp Center and the city hall. All parties (against all expectations) were highly cooperative and the corrections that LeMessurier advised were carried out. The building is now safer than the way it was originally planned when its construction began.

The media too had to be informed, because so much activity around a brand-new building could not go unnoticed. LeMessurier was highly dubious of being involved with the media, as the press could turn it into quite a story. After the news was published that the building was being altered to withstand more powerful storms, it received no further attention because the press happened to go on strike.

Source: Adapted from Morgenstern (1995).

For many engineers, LeMessurier's actions with regard to the Citicorp building exemplify the highest virtues of the engineering field. Nevertheless, many will wonder why LeMessurier deserves so much praise, since it was his professional duty to report mistakes to the authorities. Michael Pritchard, however, indicates that the way in which LeMessurier acted was exemplary – and therefore praiseworthy – for the following two reasons (Pritchard, 2001). First, much courage was needed to report the error, even though not reporting it would have been highly reprehensible. The report could have damaged his reputation considerably. Second, LeMessurier not only reported the problem, he also proposed a solution to it, which is characteristic of a virtuous engineer in Pritchard's opinion.

By taking seriously the objections of the student and their professor, LeMessurier also demonstrated another virtue: openness to criticism. Instead of ignoring the criticism because the construction met the safety requirements, LeMessurier decided to check everything and recalculate it. This demonstrates his dedication to safety of the general public.

3.10   Care Ethics

Using utilitarianism and Kantian theory, we try to form a balanced moral judgment about the right way to act in a given situation. In these ethical theories, an appeal is made to abstract and general principles, such as the utility principle and the reciprocity principle (or the universality principle). No attention is paid to the specific social context of the moral situation in question. These theories presuppose an independent and rational actor who makes decisions in a vacuum.

Care ethics – initially inspired by the work of Carol Gilligan (Gilligan, 1982) – emphasizes that the development of morals does not come about by learning general moral principles. Its basis is that people learn norms and values within specific contexts and by encountering concrete people with emotions. By recognizing the vulnerability of the

> **Care ethics**   An ethical theory that emphasizes the importance of relationships, and which holds that the development of morals does not come about by learning general moral principles.

other and by placing yourself in their shoes to understand their emotions, you can learn what is good or bad at that particular time. Care ethics focuses attention on the living and experienced reality of people in which mutual relationships can be viewed from different perspectives, and where people's abilities and limitations impact moral decision-making.

3.10.1   The importance of relationships

Philosophers of care ethics argue that there is moral significance to the specific details of our lives, details that tend to be disregarded when we formulate "the good" in terms of general principles. We have seen that utilitarianism tends to ignore personal relationships, but that it does make a difference with whom we have a personal relationship. We have special relationships with our children, relatives, friends, and colleagues. These relationships are coupled to special responsibilities and moral obligations. Moral problems are first and foremost understood in terms of responsibility of the individual with respect to the group. The solution of moral problems must always be focused on the maintenance of relationships the people have with each other. Besides people, companies can have special relationships too, such as the employees, suppliers, and people living close to a factory.

In care ethics the connectedness of people is key; the mutual responsibility and care for each other. People are connected to each other and through this connection there is attention for your fellow human being. People feel responsible for each other. Care arises from this involvement. Care encompasses all typically human activities that we carry out to maintain, continue, and repair our world, so that we can live in it as best as we can (cf. Tronto, 1993). Though care in this description is primarily described as an action, it is also important to look at care as a certain attitude or motivation.

Care ethics has some grounds in common with virtue ethics (see, e.g., Nair and Bulleit, 2020). Care ethics places the relationship central together with the acquired attitude of the person who can provide care. A proper attitude that one has acquired involves compassion, attention, and being caring. These virtues stimulate people to become emotionally involved and responsible. Moreover, both approaches are aimed at the good life; virtue ethics is based on good character traits and care ethics is based on care.

Care ethics specifically focuses the attention on the relationships that people have with each other. In relationships the recognition of vulnerability and dependence play an important role, especially if the relationships are asymmetrical, such as the relationship between parent and child, between employer and employee, or between doctor and patient. Thus, it is important to be conscious of the types of relationships we have and the role we have within them. Roles determine to what degree we can expect care from each other and they also determine whether we should take the other into account in our actions. Besides that, it is important to know how people respond to each other's vulnerabilities. The degree to which we respond appropriately and the way we shape our responsibility cannot be indicated in advance using rules, but has to be answered in the context in which the need for care arises.

3.10.2 Criticism of care ethics

A frequently voiced criticism of care ethics is that it is philosophically vague. This is mainly due to the fact that it is unclear what "care" exactly entails. The term is used in numerous contexts and is also used to indicate more than one attitude or action. As a result, it is not very normative. Care ethics assumes that caring is good in itself, thus it can tell us neither what makes a particular attitude or action right, nor what constitutes the right way to pursue them. Care ethics judges a situation by means of "good care" and not according to principles. But the question is what turns "care" into "good care"? Finally, care ethics like virtue ethics does not give concrete indications how one has to act in a particular situation, in contrast with utilitarianism or Kantian ethics. Care ethics is more focused on the attitude of the person who can provide care than on indications for ways to solve a concrete moral problem.

3.10.3 Care ethics in engineering

Although the attempt to develop a care ethics approach to engineering ethics is still in its infancy, care ethics' emphasis on care (e.g., for safety and sustainability) responsibility, and other concerns shared by engineers, suggests that is has a contribution to make to engineering practice. One of the possible applications is a social ethics of engineering (see box). The ideas of care ethics can also work for companies: they can contribute to the vision and mission of a company and can be a major influence on the practice of corporate social responsibility. One essential characteristic of a business situation is that one is working within an intersection of different relationships. A company has dealings with various parties and institutions, which have diverse and sometimes contradictory expectations. Employees have relationships with clients or contractors, with their employer, with consumers, with suppliers and sometimes even with the natural environment. The point is that an employer or employee has to ask themselves how they as part of the enterprise can best deal with the interests and rights of others. This has to be achieved through an attitude of compassion, attention, and care.

Social Ethics of Engineering

Most of the approaches to ethics in engineering focus on the individual. Such approaches tend to neglect the relationships with others in which engineers enter in their work and that are morally relevant. A social ethics approach would pay more attention to such relationships and would inquire into the social arrangements in which engineering

> **Social ethics of engineering** An approach to the ethics of engineering that focuses on the social arrangements in engineering rather than on individual decisions. If these social arrangements meet certain procedural norms the resulting decisions are considered acceptable.

decisions are made. Relevant social arrangements include, for example, the way a design team or the engineering company is organized and the way that relations with stakeholders are structured. In the case of the Boeing MAX 737 case (see Section 6.1), for example, it was striking that the pilots were not involved in the design of the automatic correction system MCAS for its new 737 MAX aircraft, and were even not informed about its operation and impact.

Richard Devon has proposed a number of values for the participation of engineers in group processes, involving both engineers and non-engineers. These values are:

- competency;
- cognizance, requiring interdisciplinary skills and breadth built into the group;
- democratic information flows;
- democratic teams;
- service-orientation;
- diversity;
- cooperativeness;
- creativity; and
- project management skills

Some of these values are rather similar to the virtues for morally responsible engineers mentioned by Michael Pritchard (see 3.9.3). The main difference is that whereas virtues are usually seen as individual character traits, these values are understood at the level of group processes and social arrangements.

A social ethics approach emphasizes procedural values for dealing with moral problems in a group process rather than substantial moral norms that are to be applied by individuals. It leaves open the possibility "that the individual may be unhappy with the outcome but be able to accept it because the process was perceived as the most acceptable way for a group" (Devon, 1999, p. 91) to reach a decision.

Source: Based on Devon (1999), Devon (2004), and Devon and Van de Poel (2004).

In the case of child labor, in Section 3.7, the following reasoning would be applied from the perspective of care ethics. Since the children are involved in the production of a company's articles, for example, say a fast fashion company. The fast fashion company has a relationship with those children. The children are extremely vulnerable and dependent on this company: if the company stops with child labor, then the consequences could be that the children would end up performing slave labor or becoming prostitutes. The company has the responsibility to care for these children. In this specific case, this care could involve improving work conditions, providing medical care, and developing a schooling program. In this example of an application of care ethics, we see that the question whether child labor should be abolished fades into the background and that care ethics makes the fast fashion industry's involvement with the children the central issue. Another example is the care for employees by the employer in cases of mass unemployment, or mergers, takeovers, down-sizing or relocation of the enterprise. The employer has a relationship with the employees and from this relationship follow certain obligations of care between the employer and the employees. This care can consist of active involvement in transfer of employees within the company or finding places of work outside the company. From these two examples it becomes clear that a care ethics approach places high demands on an enterprise concerning responsible entrepreneurship.

In the Ford Pinto case, Ford had a relationship with the consumers. This relationship was asymmetrical since the consumers had in general no clear idea of all the relevant technical aspects of the Ford Pinto, and the consumers were dependent on the information Ford gave. Ford should recognize this vulnerability of the consumer, and therefore Ford had the responsibility to inform the consumer about the (un)safety of the car, or Ford should not have marketed the car.

3.11   Capability Approach

The **capability approach** was pioneered by the Nobel-prize winning economist Amartya Sen and the philosopher Martha Nussbaum, whose writings over the last decade have become influential in a number of fields with an ethical or polical dimension, such as welfare economics and international development. It is used, for example, as a guiding framework to improve public health and for the attainment of the Sustainable Development Goals (see Chapter 2). In this approach, the prosperity or poverty of a country is no longer assessed solely on the basis of its Gross Domestic Product and economic growth, but on basis of whether people have certain valuable human capabilities, or the real and effective freedoms to do and be certain things in life.

Capability approach A normative framework which emphasizes the importance of people's freedom to choose between different life paths and to enhance their capabilities in order to do so.

While the capability approach is relatively recent, the underlying ideas have a much longer history. The capability approach has been influenced, for example, by the virtue ethical conception of "flourishing" as the basis of the good human life, Kantian notions of dignity and autonomy, and the Rawlsian conception of distributive justice of important goods, like income, happiness, and career. The capability approach is generally understood as a conceptual framework for a range of normative exercises, including most prominently the following: (1) the assessment of individual well-being; (2) the evaluation and assessment of social arrangements; and (3) the design of policies and proposals about social change in society (Robeyns, 2020). There are numerous ways to turn this general conceptual framework into a more specific theory of justice or of well-being. A well-known example is Nussbaum's capability theory of well-being and human development (more details later).

The key concepts of the capability approach are *capabilities* and *functionings*. Capabilities are the real freedoms (or effective opportunities) to choose and obtain those things that one has reason to value. They refer to both what we are able to do (activities), as well as the kind of person we can be (dimensions of our being). These "beings and doings," that is, the various states of human beings and activities that a person can undertake, are called "functionings." Examples of "beings," are, for example, being well-nourished, being educated, being literate, being part of a social network, and being happy. Examples of "doings," are, for example, travelling, caring for the elderly, taking part in a debate, playing football, eating animals, and donating money to charity. Instead of focusing on the achievement of certain functionings, the capability approach concentrates on our freedom and ability to achieve them.

A focus on just functionings would be problematic because it does not value free choice and human agency, that is, the ability to pursue goals that one values and has reason to value. Sen (1992) illustrates this with an example. A person who is fasting is in a state of under-nutrition may seem very similar to a person who is starving. If one concentrates on the functionings alone, then both person's functionings are being equally impeded (e.g., to be well-nourished or free from hunger). Without analysis of their capabilities, it would give a misleading view of their well-being. The fasting person *could* eat, although chooses not to, whereas the starving person has no choice at all, and would do it if they could. So, the fasting person has the capabilities at their disposal to be healthy, whereas the starving person does not.

For decades, Amartya Sen has been offering fundamental criticisms of traditional welfare economics, which typically conflates well-being with either resources (income, wealth, basic goods) or utility (happiness, desire fulfillment). Utility, or in general the utilitarian theories, are criticized on the grounds that "utility can be easily swayed by mental conditioning or adaptive expectations" (Sen, 1999, p. 62). For example, factors such as oppression and deprivation can blind people to the reality of their situation and interests, one possible consequence being that they take disproportionate pleasure in what by another standard would count as a very minor improvement in their well-being.

The reason that the capability approach prefers capabilities over resources is the misconception that resources are the ends of economic and social policies, while they are only a means of human flourishing and well-being. The focus should be placed on the capabilities that are generated with resources, since what is important is one's ability to convert resources into freedoms and wellbeing. The conversion of resources depends on personal, social, and environmental factors. For example, having a computer may allow a person to play games, browse the web, send emails, work from home, and increase freedoms. To realize these capabilities requires that the person has the cognitive and physical abilities to use the computer (personal conversion factor), that there is education on computer literacy (social conversion factor), and that there is access to, for example, Wi-Fi (environmental conversion factor). If one is unable to convert the resources into useable and practical benefits for their wellbeing, then the equal distribution of resources does not equate to equal levels of freedom or well-being. As a consequence, focusing on the just distribution of resources may lead to grave inequalities because of the differences in abilities to use and convert those goods into benefits.

The capability approach emphasizes that each and every person needs sufficient capabilities to lead a flourishing life. Hence, policies should aim at expanding, strengthening and safeguarding human capabilities. But what are the relevant capabilities? Here, Nussbaum's proposal of a list of ten "central human capabilities" offers a starting point (see box). These ten central human capabilities should, according to Nussbaum (2011), be the basis of constitutional guarantees, human rights, and development policy. Influenced by Aristotle's work, Nussbaum puts human flourishing at the center of her capability theory, and sees these capabilities as the moral entitlements of every human being on earth. Achieving some threshold level of all these capabilities is held to be necessary for a flourishing life.

Figure 3.8   Martha Nussbaum.
Photo: Ullstein bild / Getty Images.

Nussbaum's ten central human capabilities

"My claim is that a life that lacks one of these capabilities, no matter what else it has, will fall short of being a good human life" (Nussbaum, 2011).

1 *Life.* Being able to live to the end of a human life of normal length; not dying prematurely, or before one's life is so reduced as to be not worth living.
2 *Bodily Health.* Being able to have good health, including reproductive health; to be adequately nourished; to have adequate shelter.
3 *Bodily Integrity.* Being able to move freely from place to place; to be secure against violent assault, including sexual assault and domestic violence; having opportunities for sexual satisfaction and for choice in matters of reproduction.
4 *Senses, Imagination, and Thought.* Being able to use the senses, to imagine, think, and to reason – and to do these things in a "truly human" way, a way informed and cultivated by an adequate education, including, but by no means limited to, literacy and basic mathematical and scientific training. Being able to use imagination and thought in connection with experiencing and producing works and events of one's own choice, religious, literary, musical, and so forth. Being able to use one's mind in ways protected by guarantees of freedom of expression with respect to both political and artistic speech, and freedom of religious exercise. Being able to have pleasurable experiences and to avoid non-beneficial pain.

5 *Emotions.* Being able to have attachments to things and people outside our-selves; to love those who love and care for us, to grieve at their absence; in gen-eral, to love, to grieve, to experience longing, gratitude, and justified anger. Not having one's emotional development blighted by fear and anxiety. (Supporting this capability means supporting forms of human association that can be shown to be crucial in their development.)

6 *Practical Reason.* Being able to form a conception of the good and to engage in critical reflection about the planning of one's life. (This entails protection for the liberty of conscience and religious observance.)

7 *Affiliation.* A. Being able to live with and toward others, to recognize and show concern for other human beings, to engage in various forms of social interaction; to be able to imagine the situation of another. (Protecting this capability means protecting institutions that constitute and nourish such forms of affiliation, and also protecting the freedom of assembly and political speech.) B. Having the social bases of self-respect and non-humiliation; being able to be treated as a dignified being whose worth is equal to that of others. This entails provisions of non-discrimination on the basis of race, sex, sexual orientation, ethnicity, caste, religion, national origin.

8 *Other Species.* Being able to live with concern for and in relation to animals, plants, and the world of nature.

9 *Play.* Being able to laugh, to play, to enjoy recreational activities.

10 *Control over One's Environment.* A. Political. Being able to participate effectively in political choices that govern one's life; having the right of political participa-tion and protections of free speech and association. B. Material. Being able to hold property (both land and movable goods), and having property rights on an equal basis with others; having the right to seek employment on an equal basis with others; having the freedom from unwarranted search and seizure. In work, being able to work as a human being, exercising practical reason and entering into meaningful relationships of mutual recognition with other workers.

3.11.1 Applications of the capability approach

The capability approach is not an ethical theory that specifies which actions are right and which are wrong, but it offers a normative framework that focuses on one's ability to con-vert resources and goods into freedoms and well-being, or on one's capacity to realize well-being. Nussbaum's version of the capability approach can, for example, be used for the normative assessment of technological options, since technology has a huge impact on human capabilities. Oosterlaken (2013) proposes the idea of "capability sensitive design," analogue to the idea of "value sensitive design" (see Chapter 5). In the approach of capa-bility-sensitive design, the key is to design technology in such a way that it increases human capabilities, as understood by the capability approach. Special attention may be given to (vulnerable) user groups of which the capabilities may be limited by technological designs rather than enhanced. For example, an architectural design of a library, which is a public space, that is designed with a three-step stairway in front of the entry, is inaccessible for people in a wheelchair. Such an architectural design may be considered morally unjust. because it fails to bring people in a wheelchair to the threshold level of Nussbaum's capa-bility "bodily integrity," which requires that people can move freely from place to place.

Figure 3.9   Paro. Photo: Bart van Overbeeke Photography.

Amanda Sharkey (2014) uses the capability approach as a framework for the assessment of the possible effects on human dignity of robots in elderly care. She suggests that robot pets, such as robot seal Paro (See Figure 3.9), can expand the set of capabilities available to vulnerable elderly. For example, these robots by acting as social facilitators encourage social interaction, which enhances the capability "affiliation" (see Nussbaum's ten central human capabilities) by creating more opportunities to engage in social interaction. There is evidence that interacting with Paro reduces stress and anxiety in the elderly, and as a consequence enhances the capability "emotions." On the other hand, some authors (e.g., Sparrow and Sparrow, 2006) describe these companion robots for the elderly as "simulacra" replacing real social interaction, and depriving and deceiving elderly with dementia. This would humiliate the elderly and limit their capability for "affiliation." According to Sharkey, the reduction in non-humiliation may be justified by the health and social benefits that they gained from these companion robots. We return to the question of possible trade-offs among capabilities in the next subsection.

The capability approach can also be used for assessing new technologies. Hillerbrand et al. (2021) provide the example of a mobility system based on shared autonomous vehicles. One promising feature of autonomous cars is the reduction of accidents, which relates to the capabilities of "life" and "bodily health." In a scenario of sharing autonomous vehicles, there will be a significant reduction in the number of cars, so that parking spaces can be converted to areas for recreation, sports, socializing, or commercial development. This will increase the range of options for social engagement related to the capability "affiliation." On the other hand, autonomous cars can be hacked, which is a serious threat since the entire mobility system is organized by software. This threatens the capabilities "life" and "bodily health." Such future-oriented technology assessment provides insight into the positive and negative capability impacts, and reveals some pressing challenges. Since the introduction of a mobility system based on shared autonomous cars is a decades-long process, there is time to anticipate and to avoid or reduce negative capability impacts, while expanding positive capabilities.

In the case of child labor, according to the capability approach, the prohibition of child labor depends on whether the child's capabilities will be expanded or limited. Children in very deprived countries work to prevent themselves and their families from starving, or feel that the world of work gives them more relevant knowledge, desirable skills and attitudes, the social basis for acceptance and self-respect in society, than, for example, going to school (Githitho-Muriithi, 2010). In certain circumstances, such as when families are very poor, "being well-nourished" is a more valued functioning for children than "being-well educated." These children, however, have no real freedom to choose which function they value, since the extreme deprivation of poverty and hunger limits freedom substantially. So, according to the capability approach, poverty and hunger should be eradicated, which also corresponds with the first two goals of the Sustainable Development Goals adopted by the United Nations (see subsection 2.2.2), so that children's capabilities will be expanded.

3.11.2   Criticism of the capability approach

The capability approach is sometimes criticized for being too individualistic (see, e.g., Stewart and Deneulin, 2002). According to Sen, however, this critic is a misunderstanding of what he means: "The options that a person has depends greatly on relations with others and on what the state and other institutions do. We shall be particularly concerned with those opportunities that are strongly influenced by social circumstances and public policy" (Drèze and Sen, 2002, p. 6). For example, women in sexist societies accepting their inferior position are not independent of social conditions, and expanding their capabilities is linked to the advancement of the broader affairs of society (see, Sen, 2009, p. 244).

Another criticism is that it is not clear which capabilities matter and who should determine this. For example, Nussbaum's list of the ten capabilities, which according to her is an outcome of "years of cross-cultural discussion" (Nussbaum, 2000, p. 76), has been criticized because of its lack of democratic legitimacy (Robeyns, 2020). And even if there would be a consensus about the relevant capabilities, important questions still remain on how to rank or weigh these capabilities when they conflict. For example, according to Nussbaum, the ten capabilities on her list cannot be traded off against another, because they are incommensurable. In design contexts, however, it is often the case that a capability conflict occurs in the sense that some capabilities are enhanced and others are diminished by a product. For example, a care robot monitoring the activities of elderly with mild dementia to warn caregivers when they leave the house or even prevent them from leaving reduces the capability "bodily integrity" since it limits the ability "to move freely from place to place," but at the same time, it gives the elderly the opportunity to live independently for longer, increasing their well-being and self-respect (related, e.g., to the capabilities "bodily health" and "affiliation"). Nussbaum's approach would require making sure that all relevant capabilities are met to a minimal degree, but how do we determine what is minimally required? And is this level the same for all humans irrespective of the phase of their life? In Chapter 5, we will discuss different approaches for dealing with conflicting values in design, which may also be relevant to deal with capability conflicts. The approach that Nussbaum proposes amounts to setting a minimal threshold for each capability, but as we will see in Chapter 5 such an approach also has disadvantages and there may be other approaches to deal with capability conflicts.

3.12 Non-Western Ethical Theories

The ethical theories we have discussed above by and large originate from the West. Of course, the validity of an ethical theory does not depend on its geographical origins. Moreover, the Western tradition is diverse and many elements of the discussed theories can also be found outside the Western hemisphere. Still, it is worthwhile to look at some non-Western ethical theories. First of all, we live in a globalized interconnected world, and intercultural debates have therefore become an indispensable means to generate ethical guidance. Secondly, we can always benefit and learn by being more receptive to different ideas and perspectives. Finally, failing to do so can perpetrate ethnocentric biases in academia, where works in Western philosophical traditions dominate the global academic debate. We focus here on the African ethics of Ubuntu and the East Asian ethics of Confucianism, which both can help to make engineering ethics education itself more inclusive (Jing and Doorn, 2020). Further research is still needed to develop a comprehensive account of these non-Western ethical theories for engineering ethics, since these theories are not full-blown normative theories, such as utilitarianism and Kantian ethics. There are numerous interpretations of these theories from their early history to the present and various insights into how these theories can be applied to society. First attempts, however, are already made by, for example, Jing and Doorn (2020), Mhlambi, (2020), Wong (2012, 2020), and Zhu (2020).

In this section, we restrict ourselves to some main concepts of two non-Western ethical theories, Ubuntu and Confucianism, and some examples of how these concepts can be related to engineering ethics.

3.12.1   Ubuntu

Ubuntu is an old African term for "humanness," and can be considered the root of all African philosophy. It is a way of life and emphasizes the importance of community, solidarity, sharing, and caring. It is about establishing relations with others (Ujomudike, 2016). One is considered human by accepting and respecting the humanity of others and vice versa. This means that, as a community, we must care for the well-being of others and support each other. We must recognize and accept each other's rights and take responsibility for others as individuals and as a community (Mugumbate and Nyanguru, 2013).

> **Ubuntu** An African concept of humanness which emphasizes the importance of community, solidarity, sharing, and caring.

From an Ubuntu perspective, Western ethics is too individualistic. Whereas Western ethics begin with the individual (expressed by the Cartesian cognito ergo sum: "I think therefore I am"), Ubuntu begins with the group (expressed by the Zulu aphorism *umuntu ngumuntu ngabantu*: "a human is human being only through their relationships with other humans" (Ewuoso, 2021) or roughly translated as "I am because we are").

Some authors have recently explored how Ubuntu could ameliorate existing discussions in engineering ethics, especially with respect to robotics and artificial intelligence (Langat et al., 2020; Ewuoso, 2021; Van Norren and Verbeek, 2021; Coeckelbergh, 2022; Friedman, 2022). For example, Ubuntu can be very helpful to shed some light on the ethical design and deployment of care robots. From an individualistic perspective, care robots are designed to function as companions for the elderly who feel lonely, or as caregivers. Community and solidarity in this perspective is lacking without "a truly relational perspective, the 'social' in social robots means that it is designed to interact with individuals in order to respond to unique, individual interests, needs, wishes and commands and to support their human dignity understood as individual autonomy" (Coeckelbergh, 2022, p. 8). From an Ubuntu perspective, it is ethically unacceptable if care robots replace human caregivers, since this will deprive caretakers of human relationships. Although care robots can interact with humans in a reciprocal way, this reciprocity is not indicative of genuine human sentiments (Friedman, 2022), since these robots will function as "simulacra" replacing real social interaction (Sparrow and Sparrow, 2006). Robots, at least current robots, cannot build a meaningful relationship with humans. Characteristic of our relationship with another human being is that the other party has its own needs and wishes, regardless of our own needs and desires. "The demands that our friends … make on us are therefore unpredictable, sometimes unexpected and often inconvenient. This is an essential part of what makes relationships with other people, or animals, interesting, involving and rewarding" (Sparrow and Sparrow, 2006, p. 149).

This does not mean that we may not develop care robots from an Ubuntu perspective. "Rather, it is a case of thinking carefully about how we design these robots, and where in society we advocate for their use. The point here is that, given the moral concern that arises in the context of human–robot relations, we should at least be aware of how important it is to also maintain and cultivate relations with other humans. Thus, we must be careful not to allow robot relations to crowd out human ones" (Friedman, 2022, p. 10), and try to develop robots that stimulate or encourage human relationships. For example, Paro, the seal robot, can encourage relationship seeking. In a study by Hung et al. (2019), they show that the use of Paro in nursing homes significantly improves communication between senior citizens and strengthens their social bonds.

3.12.2 Confucianism[2]

Confucius (551–479 BC) was a Chinese philosopher, politician, and teacher. His thoughts on ethics were written down by his disciples in several books. The most authoritative source of his thoughts is the Lunyu or the "Analects," a collection of sayings by Confucius and dialogues with his disciples. Some of the morals Confucius taught, are easily recognizable, such as his version of the Golden Rule: "Do not do unto others what you don't want done to yourself." But some of them also seem very strange, or old-fashioned, especially for people from Western cultures, such as that we shouldn't travel far away while our parents are alive, and should cover for them if they steal a sheep. Confucius was not the first person to discuss many of the important concepts in Confucianism. While most people today considered Confucius an innovator, he always portrayed himself as someone who was concerned with preserving traditional Chinese knowledge from earlier thinkers. After Confucius' death, several of his disciples, such as Mencius (372–289 BC) and Xunzi (c. 310 – c. 220 BC), developed Confucius' thoughts further.

Confucianism is a philosophy from ancient China, which laid the foundation for much of East Asian culture and societies. **Confucian ethics** is characterized by the promotion of personal moral cultivation through virtues such as benevolence, righteousness, loyalty, wisdom, filial piety, and sincerity. In Confucianism, persons are viewed as partly constituted by their personal social relations; Confucian moral cultivation entails the development of harmonious social relationships (Herr, 2003). Each one has a role to play in a web of relationships that create community, and when these roles are fulfilled and played well, better human relationships will ensure, and peace and social harmony will exist. The core relationship of the family provides an idealized model for all other relationships: "When these family relationships are harmonized, it does not only foster people's internal harmony of their selves, but also extends to the harmony in society" (Li, 2008, p. 430).

> **Confucian ethics** An ethical approach that is characterized by the promotion of personal moral cultivation through virtues such as benevolence, righteousness, loyalty, wisdom, filial piety, and sincerity.

We will discuss some key concepts of Confucianism, and relate these to two relevant issues in engineering ethics: whistle-blowing and robot design.

A key notion of Confucianism is "filial piety" whose basic connotation is to treat parents well (Feng, 2020, p. 288). We must look after our parents, be obedient to our parents, respect our parents, and be loyal to our parents. Confucius' idea of filial piety is expressed by the following rather controversial passage:

> The Governor of She said to Confucius, 'In our village we have an example of a straight person. When the father stole a sheep, the son gave evidence against him.' Confucius answered, 'In our village those who are straight are quite different. Fathers cover up for their sons, and sons cover up for their fathers. In such behaviour is straightness to be found as a matter of course.' (Confucian Analects 13:18)

Essential is to maintain a harmonious family, which is the basis for a harmonious society, and therefore the son should not disclose this wrong-doing of his father. However, sons are not necessarily obedient followers and uncritically loyal. They have a duty to remonstrate in case of unrighteous conduct by their fathers:

[W]hen a case of unrighteous conduct is concerned, a son must by no means keep from remonstrating with his father, nor a minister from remonstrating with his ruler. Hence, since remonstrance is required in the case of unrighteous conduct, how can (simple) obedience to the orders of a father be accounted filial piety? (A text from Confucian School, translated by Legge (1879).)

From this quote, the notion of filial piety can be expanded to other relations, such as the relationship between employer and employee (see also Low and Ang, 2013; Jing and Doorn, 2020). With respect to abuses in an organization, Confucianism may discourage whistle-blowing (see Subsection 1.5.3), and encourage remonstration. An assumption of remonstration is that, under certain circumstances, it is better to try reforming a "corrupt" organization, than to quit the organization to maintain "clean hands" or to create disruptions in the social benefits that are produced by that company, for example, the employment of thousands of people. To maximize the chance that the employer treats the criticism not as an attack but as a sincere concern from someone who is loyal, Confucius teaches us to remonstrate gently and "if they are disinclined to listen, remain reverent but do not abandon your purpose" (Analects 4:18).

Remonstration relates to critical loyalty (see Subsection 2.3.2) in which the employee gives due regard to the interest of the employer. However, if there is not a culture of trust in the organization, or no harmonious relationship between employees and employer, then remonstration takes a great deal of courage. This is also expressed by a nice quote from Peter Wei in which he proposes a shift from a culture of continuous improvement to a culture of trust to avoid severe abuses in organizations:

While whistleblowing requires courage in extraordinary situations, remonstrance teaches us to practice the courage to speak up about everyday problems. In the risk-averse professional culture that predominates today, there is a tendency for workers to keep their heads down and not raise problems until they become unignorable. (...) Until the crisis comes, such a workplace may seem nice. But it's a niceness born of cowardice and mistrust. Remonstrance ethics naturally prizes both the courage to speak up and the culture of trust that makes it possible. It is the moral counterpart of the culture of continuous improvement that has become standard in advanced manufacturing. (Wei, 2022)

Confucian ethics can be considered a role-based approach. The action we perform depends on the specific roles we have in a certain situation. The way you speak to your parents may differ from how you communicate, for example, with your employee. Becoming a good person depends, according to Confucianism, on the extent to which we live our social roles in a way that is conducive to the well-being of all parties involved in the relationship and practice the moral responsibilities prescribed by these social roles (Zhu et al., 2019). Concerning engineering ethics, we have to investigate whether technological innovations "are conducive or detrimental to our performance of the social roles" (Wong, 2012, p. 81). The moral quality of a technological innovation is thus assessed by the extent to which the innovation helps us exercise our role-based responsibilities in order to cultivate virtues and personal social relationships. For example, if a household robot can free us from routine operations of the home, so that we can spend more time caring for our children, then such technology is supported by Confucianism. It becomes more difficult when a robot is designed to play a social role, since then the robot must fulfill its assigned role, and need "to render assistance to other human beings in their pursuit of moral improvement" (Liu, 2017). For example, a companion robot is supposed to play the social role of a friend, which entails that such a robot must remonstrate when the robot observes that its human friend commits wrong-doing, since remonstration is a requirement for a harmonious healthy friendship. A companion robot that does not remonstrate or demands anything from its human friend is problematic from a Confucian perspective, since people will not learn to deal with real-life people to exercise their moral responsibilities, and may limit "the human dimension of interpersonal relations" (Tan, 2022).

3.13 Applied Ethics

Some philosophers believe that applied ethics is essentially the application of general moral principles or theories to particular situations (cf. Smart, 1973; Gert, 1984; and Hare, 1988). This view is, however, problematic for a number of reasons (cf. Beauchamp, 1984; and MacIntyre, 1984b). One is that no moral theory is generally accepted. Different theories might yield different judgments about a particular case. But even if there were one generally accepted theory, framework, or set of principles, it is doubtful whether it could be straightforwardly applied to particular cases. Take a principle such as distributive justice. In many concrete situations, it is not clear what distributive justice exactly amounts to. What does, for example, a just distribution of technological risks mean? Should everybody be equally safe? Should everybody have the same minimum level of safety? Or does someone's right to safety depend on the amount of taxes they pay? All these can be considered as an application of the principle of distributive justice to the distribution of risks, but clearly these answers reveal different moral outlooks. Without doubt, part of this confusion could be solved on the theoretical level, that is, by further elaborating the notion "distributive justice" and developing an ethical theory about it. It seems doubtful, however, whether this would solve all applications issues. This brings us to a third point. Theory development in ethics in general does not take place independent of particular cases. Rather, theory development is an attempt to systematize judgments over particular cases and to provide a rational justification for these judgments. So, if we encounter a new case, we can of course try to apply the ethical theory we have developed until then to that case, but we should also be open to the possibility that the new case might sometimes reveal a flaw in the theory we have developed so far. For example, human contact is often seen as crucial for the provision of "quality of care," and some authors argue that care robots cannot meet the emotional and social needs that the elderly have in relation to almost all aspects of caring (e.g., Royakkers and Van Est, 2016). However, care robots are likely to perform more and more sophisticated caring activities that some will consider comforting and valuable. Should we reconsider the concept of "quality of care" in which human contact is essential if robots can assist and ease people's suffering?

If ethical theories do not provide moral principles that can be straightforwardly applied to get the right answer, what then is their role, if any, in applied ethics? Their role is, first, instrumental in discovering the ethical aspects of a problem or situation. Different ethical theories stress different aspects of a situation; consequentialism, for example, draws attention to how consequences of actions may be morally relevant; deontological theories might draw attention to the moral importance of promises, rights and obligations. And virtue ethics may remind us that certain character traits can be morally relevant. Ethical theories also suggest certain arguments or reasons that can play a role in moral judgments or decisions, for example, in the design of algorithms (see box).

Ethical issues in algorithm-design

Fairness metrics: COMPAS

Perhaps, the most famous case study of algorithmic fairness concerns the COMPAS algorithm used to predict recidivism. COMPAS (Correctional Offender Management Profiling for Alternative Sanctions) is a proprietary tool that assigns risk scores to individuals on the basis of a questionnaire consisting of 137 questions that are either answered by defendants or pulled from criminal records. The survey asks defendants about prior arrests, criminal behavior among friends and family, employment, housing, substance use, and personality traits. COMPAS is used in jurisdictions across the United States to make decisions on whether to grant the benefit of parole/probation.

ProPublica, an independent, nonprofit newsroom that aims to produce investigative journalism in the public interest, raised several criticisms of COMPAS, such as that the algorithm was "biased against blacks" (Angwin et al., 2016). ProPublica compared COMPAS's risk assessments for 7,000 people arrested in Broward County, Florida, with how often they reoffended. They took the 1–10 COMPAS scores and binned them into two categories: "low risk" (1–4) and "high risk" (5–10). Using these two categories, they defined a *false positive* prediction as a "high-risk" defendant who did not reoffend, and a *false negative* prediction as a "low-risk" defendant who reoffended.

ProPublica found that when the COMPAS algorithm was wrong in its prediction, the results were differently for black and white offenders. Through COMPAS, black offenders were seen almost twice as likely as white offenders (45% as opposed to 23%) to be labeled a higher risk while not actually reoffending. While the COMPAS software produced the opposite results with white offenders: they were mistakenly labeled as low risk almost twice as much as black ones (48% as opposed to 28%). So, it seemed that COMPAS's assessment was affected by racial bias. On the one hand, black non-reoffenders were more likely than white ones to be mistakenly subjected to the adverse consequences linked to a recidivism prediction. On the other hand, white reoffenders were more likely than black ones to mistakenly obtain the more favorable decision, that is, granting parole, for expected non-reoffenders.

Equivant (formerly Northpointe), the company behind COMPAS, responded that the algorithm was not racially biased, because its predictions were equally correlated for the two groups (Dietrich et al., 2016): black defendants who were predicted to reoffend (high-risk score) actually did reoffend at (approximately) the same rate as the white ones (63% as opposed to 59%), and similarly, white defendants who were predicted not to reoffend (low-risk score) actually did not reoffend at (approximately) the same rate as black ones (71% as opposed to 65%). So, a defendant who has, for example, a high-risk score is equally likely to reoffend regardless of the race of the defendant.

In this case, both organizations used different fairness metrics to justify their claims. ProPublica focused on equalized odds, meaning that the false positive rate (i.e., the fraction of negative cases incorrectly predicted to be in the positive class out of all actual negative cases: $FP/(FP+TN)$) as well as the false negative rate (i.e., the fraction of positive cases incorrectly predicted to be in the negative class out of all actual positive cases: $FN/(TP+FN)$) is the same for both groups, while Equivant focused on calibration: the proportion of correct predictions should be equal for each score within each group.[3] In an ideal world, we want to have both fairness metrics, but research has shown that these two fairness metrics are not jointly satisfiable except in marginal cases (see, e.g., Kleinberg et al., 2017). From this, we could conclude that at best in these kinds of situations we should try to make an optimal trade-off between different kinds of relevant fairness metrics depending on the context. This decision is clearly an ethical one, and ethical theories on justice and fairness can provide arguments that can guide designers and policymakers as they make such a decision. For example, Lee et al. (2021) show how lessons from moral philosophy can be learned on what constitute acceptable or unacceptable discrimination or inequalities in algorithms.

The next case (adjusted from Kraemer et al. (2011)) illustrates how deontology and utilitarianism provide arguments for a certain threshold in a disease detection algorithm.

The choice of a certain threshold in a disease detection algorithm
In this scenario, it is assumed that an algorithm predicts whether somebody has a serious disease. We use false positives to refer to patients who are mistakenly diagnosed with the disease, and false negatives to refer to patients who are not diagnosed with the disease, but have the disease. The algorithm is used to determine whether a patient will receive treatment.

The algorithm makes use of a threshold or a cutoff value to determine when someone is considered to have the serious disease or not. The positioning of a threshold constitutes a compromise between sensitivity (i.e., picking up everyone who has the disease) and specificity (i.e., avoiding healthy people being diagnosed with the disease). The higher the threshold, the fewer false positives. Ethical theories provide arguments from which we can critically discuss this issue. From a deontological point of view, we would probably protect the individual patients, and ensure that they will be treated when needed. This would not be possible if the disease remained undetected. As a consequence, deontologists would accept more false positives to avoid that a severe disease remains undetected and therefore untreated. Utilitarianism, however, would prefer fewer false positives. The statistical data used by the algorithm will be more biased if we accept too many false positives, since this could "contaminate" the statistical data. Data that are more accurate will lead in the long run to better consequences, that is, to larger overall well-being in society, in spite of the fact that some individual patients who have the disease have to suffer along the road because they are not diagnosed with the disease.

3.14   Chapter Summary

While morality is the totality of opinions about what is good and right, ethics is the critical reflection on morality. Normative ethics not just describes what morality is but it judges morality and tries to formulate answers to questions like: "what kind of person should I be?," and "how should I act?" Normative ethics, therefore, tries to come to certain normative judgments. However, it is not a manual or an unambiguous code in which you can look up the answer how to act in a difficult situation. Rather it is an area that is characterized by a variety of partly conflicting ethical theories about how to act. The three best-known ethical theories in Western philosophy are consequentialism, deontology, and virtue ethics. Whereas virtue ethics focuses on the acting person and their character traits, deontology focuses on the actions themselves and consequentialism focuses on the consequences of actions.

Utilitarianism is a main variety of consequentialism. It measures consequences by their effect on one value: pleasure or human happiness. It is based on the so-called utility principle: the greatest happiness for the greatest number. Utilitarianism requires drawing up a moral balance sheet or a cost-benefit analysis to determine what the action with the best consequences is. Despite its intuitive attractiveness, utilitarianism has been heavily criticized. Typically, these criticisms have led to adaptations in the original theory. One criticism is that utilitarianism can lead to exploitation. To deal with this problem, John Stuart Mill has formulated the freedom principle: everyone is free to strive for their own pleasure, as long as they do not deny or hinder the pleasure of others. Another criticism is that actions are sometimes right or wrong independent of their consequences. Lying is an example. Rule utilitarianism is an attempt to deal with this criticism: it focuses on the utility of rules of action rather than on the utility of individual acts. Other criticisms of utilitarianism are that happiness is difficult to measure, that consequences are hard to predict, that it ignores the distribution of pleasures and pains and that it ignores personal relationships.

Immanuel Kant is the main representative of deontology. He formulated a principle for judging the rightness of actions that is independent of the actual consequences of those actions, the universality principle: Act only on that maxim which you can at the same time will that it should become a universal law. According to Kant, this principle is basically the same as his reciprocity principle: Act as to treat humanity, whether in your own person or in that of any other, in every case as an end, never as a means only. Two main criticisms of Kant's theory are that it ignores conflicts between norms and that it is too rigid. A way of dealing with such criticisms may be to conceive of norms as prima facie norms rather than as universal norms that apply to each and every situation.

Virtue ethics focuses on the character of the acting person rather than their actions or the consequences of those actions. It goes back to Aristotle but is still relevant today for engineers. Relevant virtues for engineers include professionalism, objectivity, being open to criticism, stamina, creativity, and having an eye for detail. The main criticisms of virtue ethics include that it does not tell you how to act and that virtues are not unconditionally good. To the first, virtue ethicists might reply that an engineer who possesses the right virtues acts differently from one who is lacking them. To the second, they would probably say that virtue ethics does not involve just isolated virtues but also practical wisdom: the ability to make ethical judgments in complex situations.

In addition, we have briefly discussed two relatively recent ethical theories: care ethics and the capability approach. These theories do not focus on abstract principles but rather on the relations between people (care ethics) and on capabilities that contribute to one's well-being (capability approach). Care ethics is relevant to engineering ethics because engineers are often involved in complex projects with many stakeholders who partly rely on them. One way to apply care ethics to engineering is to look for norms that the social arrangements in engineering should meet to express due care to all relevant stakeholders.

The capability approach is relevant to engineering ethics because it has added value to the design of technical artifacts, and to the assessment of technology from the perspective of human dignity. We also briefly paid attention to two non-Western ethical approaches: Ubuntu and Confucianism.

As the diversity of ethical theories testifies there is not a single answer to the question of what is right or wrong. However, one should not conclude from this that anything goes. Some things are morally good or morally bad according to all theories. Moreover, if theories disagree, they are still helpful in distinguishing the ethical questions in a concrete situation, in more precisely analyzing the situation, and in suggesting possible reasons and arguments for acting in one way rather than the other. Even if using one or more of the theories is no guarantee for making the right decision, a moral decision that just ignores the ethical theories, and the underlying ethical concerns, is usually plainly unethical.

Study Questions

1 Mention a number of differences between values and norms, and between values and virtues.
2 Which of the following statements are descriptive, and which are normative?
   a. People should accept the risks of nuclear energy.
   b. The majority of your colleagues finds this proposal unacceptable.
   c. There is life on Mars.
   d. Engineers who blow the whistle are usually in a weak position from a legal point of view.
3 Describe the main ideas of "normative relativism." What are the criticisms of normative relativism?
4 John Stuart Mill has argued that Kant's ethics is really a masked version of consequentialism because the consequences of actions do play an important role in his ethics – in spite of what Kant himself says. Why do you think that Mill is arguing this despite the fact that Kant himself denies that consequences are relevant to his theory? Do you agree with Mill? Explain why or why not.
5 Describe the main ideas of Bentham's utilitarianism. What criticism of Bentham's theory did Mill articulate?
6 What is rule utilitarianism? Describe how a rule utilitarian would go about determining whether I may "copy and paste" my essay from the Internet for a course. How would this differ from how an act utilitarian would reach a decision on this matter?
7 Describe the main idea of care ethics. What criticism do care ethicists have of utilitarianism and Kantian theory?
8 Define and contrast *functionings* and *capabilities* of the capability approach.
9 An engineer helped a colleague with her work; and this colleague happened to be rather pretty. The engineer thinks "Why not have an affair with her?" And since she is Kantian, the engineer argues as follows: "If you really want to express respect for me, then you should join me for a drink in my room. Otherwise, you would treat me only as a means and – as a good Kantian – that is not something you can possibly want!"
   a. What is it to treat someone only as a *means* according to Kant?
   b. What is it to treat someone as an *end* according to Kant?
   c. What is her answer if she is a good Kantian?
10 Suppose that it is possible to download copyrighted music through the Internet without paying for it. Suppose that the makers of such music (the artists) do not want their music freely copied in this way.
   a. How would Kant address the question of whether it is morally permissible to download such music by such artists without paying for it? Explain in detail what a Kantian would say.
   b. Would a utilitarian give a different answer? Why or why not?

11   Sarah is an engineer working for the company AERO that produces aero-engines. The company is developing a new type of aero-engine called the FANX. Sarah is responsible for the testing of the FANX. She is in the middle of conducting a range of crucial tests for the reliability of the new aero-engine. Yesterday, Bill – who is Shara's boss – has asked Shara to finish her test reports within a week because an important potential customer will visit AERO next week and wants to have a look at the first test reports. Sarah first reaction is to refuse Bill's request: she is not able to finish the test report within a week; she first needs to do more tests. Sarah considers these additional tests crucial for gaining good insight in the reliability of the FANX. Bill tells Sarah to abandon the planned other tests and to start writing her report immediately. Later, there will be more time to do the other tests. Bill also tells Sarah that if she refuses, he will ask Eric to write the report. Sarah says that she really needs more time. Moreover, she objects, Eric is not knowledgeable of the tests and will not be able to write a sound report. After the meeting, Sarah contacts Eric who says that he agrees with Bill and that he will finish the test reports if Bill asks him to do so.

Suppose that Sarah the next day decides to follow Bill's order and to finish the reports immediately, abandoning the other tests.

a.   Can this choice of Sarah be justified in utilitarian terms? Explain why or why not.
b.   What should Sarah do if she would try to apply Kant's categorical imperative to this situation? Argue your answer.
c.   What virtues are relevant for an engineer doing tests like Sarah? Mention at least four.
d.   What action is supported by these virtues? Argue your answer.
e.   Which normative theory is in your opinion best able to deal with this moral problem? Argue why.

Discussion Questions

1   Are there any absolute rules that should never be broken, whatever the circumstances? Defend your view.
2   Choose an event in your life where you believe you acted ethically. Discuss the event in terms of virtue ethics, Kantian ethics, utilitarianism, Ubuntu, and Confucianism.
3   What makes a decision an ethical one according to the capability approach?
4   What does Virginia Held (2005), a feminist philosopher and expert in care ethics, mean when she states, "our relations are part of what constitute our identity"?
5   How much should we take potential consequences into account when making an ethical choice? How much work should we put into making sure that the assessment of outcomes is correct?
6   Should we always do what is morally best? Is there a difference between morally decent and heroic behavior?
7   Discuss whether the robot seal Paro (see Subsection 3.11.1) will have a negative or positive impact on the essential capabilities of people with Alzheimer's disease.

Notes

1   https://labs.theguardian.com/unicef-child-labour, accessed June 14, 2022.
2   Confucianism is a complex system of moral, social, political, and religious thought, and we are aware of the fact that the brief discussion of some concepts may oversimplify Confucian ethics.
3   Formally, calibration means that the ratio between true positives and the total positive prediction (true positives plus false positives) should be same for both groups, and similarly, the ratio between true negatives and total negative predictions should be equal in both groups.

4

The Ethical Cycle

Having read this chapter and completed its associated questions, readers should be able to:

- Explain why moral problems are ill-structured;
- Explain the analysis steps of the ethical cycle;
- Explain the role of the ethical cycle in moral decision-making;
- Apply the ethical cycle to concrete moral problems in engineering;
- Analyze and evaluate the complex consequences and motives that typically attend moral issues in engineering practice;
- Describe wide reflective equilibrium and its relation with the analysis step reflection;
- Deliberate and discuss moral issues with other people.

Contents

4.1 **Introduction** 122

4.2 **Ill-Structured Problems** 124

4.3 **The Ethical Cycle** 125

    4.3.1 Moral problem statement 127
    4.3.2 Problem analysis 128
    4.3.3 Options for actions 130
    4.3.4 Ethical evaluation 131
    4.3.5 Reflection 134

4.4 **An Example** 135

    4.4.1 Moral problem statement 135

    4.4.2 Problem analysis 135
    4.4.3 Options for actions 136
    4.4.4 Ethical evaluation 137
    4.4.5 Reflection 138

4.5 **Collective Moral Deliberation and Social Arrangements** 140

4.6 **Chapter Summary** 142

Study Questions 143

Discussion Questions 143

Ethics, Technology, and Engineering: An Introduction, Second Edition. Ibo van de Poel and Lambèr Royakkers.
© 2023 John Wiley & Sons Ltd. Published 2023 by John Wiley & Sons Ltd.

4.1   Introduction

Case   Highway Safety

David Weber, age 23, is a civil engineer in charge of safety improvements for District 7 (an eight-county area within a US Midwestern state). Near the end of the fiscal year, the district engineer informs David that delivery of a new snow plow has been delayed, and as a consequence the district has $50 000 in uncommitted funds. He asks David to suggest a safety project (or projects) that can be put under contract within the current fiscal year.

After a careful consideration of potential projects, David narrows his choice to two possible safety improvements. Site A is the intersection of Main and Oak Streets in the major city within the district. Site B is the intersection of Grape and Fir Roads in a rural area.

Pertinent data for the two intersections are shown in Table 4.1.

Table 4.1   Pertinent traffic data for both intersections

	Site A	*Site B*
Main road traffic (vehicles/day)	20,000	5,000
Minor road traffic (vehicles/day)	4,000	1,000
Fatalities per year (3 year average)	2	1
Injuries per year (3 year average)	6	2
PD* (3 year average)	40	12
Proposed Improvement	New signals	New signals
Improvement Cost	$50,000	$50,000

* PD refers to property damage only accidents.

A highway engineering textbook includes a table of average reductions in accidents resulting from the installation of the types of signal improvements David proposes. The tables are based on studies of intersections in urban and rural areas throughout the United States, over the past 20 years (see Table 4.2).

Table 4.2   Average reductions in traffic accidents resulting from improvements that David proposes

	Site A	*Site B*
% reduction in fatalities	50	50
% reduction in injuries	50	60
% reduction in PD	25	−25*

* Property damage only accidents are expected to increase because of the increase in rear-ends collisions due to the stopping of high-speed traffic in rural areas.

David recognizes that these reduction factors represent averages from intersections with a wide range of physical characteristics (number of approach lanes, angle of intersection, etc.); in all climates; with various mixes of trucks and passenger vehicles;

various approach speeds; various driving habits; and so on. However, he has no special data about Sites A and B that suggest relying on these tables is likely to misrepresent the circumstances at these sites.

Finally, here is some additional information that David knows about.

1 In 1975, the National Safety Council (NSC) and the National Highway Traffic Safety Administration (NHTSA) both published dollar scales for comparing accident outcomes, as shown in Table 4.3.

Table 4.3 Estimated accident costs for traffic safety analysis

	NSC ($)	NHSTA ($)
Fatality	52,000	235,000
Injury	3,000	11,200
PD	440	500

A neighboring state uses the following weighting scheme:

Fatality 9.5 PD
Injury 3.5 PD

2 Individuals within the two groups pay roughly the same transportation taxes (licenses, gasoline taxes, etc.).

Which of the two site improvements do you think David should recommend? What is your rationale for this recommendation?

Source: Harris (2005, pp. 325–326) / Reproduced with permission from Cengage Learning

Highway safety is a fictional case presented by Harris, Pritchard, and Rabins in their book *Engineering Ethics: Concepts and Cases* (2005) in which a young engineer has to decide how to act in a difficult situation. It is the kind of situation you may also find yourself in after starting working as an engineer. Such situations call for moral judgment, using the tools we have introduced in the preceding chapters. However, moral judgment is not a straightforward or linear process in which you simply apply ethical theories to find out what to do. Instead, it is a process in which the formulation of the moral problem, the formulation of possible "solutions," and the ethical judging of these solutions go hand in hand. This messy character of moral problems, however, does not rule out a systematic approach. In this chapter, we describe a systematic approach to problem-solving that does justice to the complex nature of moral problems and moral judgment: the ethical cycle. Our goal is to provide a structured method of addressing moral problems which helps to guide a sound analysis of these problems. In Section 5.4, we will apply the whole ethical cycle to this Highway Safety case. In Section 5.3 we will describe the ethical cycle, and illustrate the steps of the ethical cycle on the basis of the Dieselgate case from the introduction in Chapter 1. Section 5.5, we will discuss how the ethical cycle, which is mainly part of individual moral judgment, can be integrated into collective deliberations on moral issues. But, first, we will pay attention to the fact that moral problems are ill-structured, which explains their messy and complex character.

4.2   Ill-Structured Problems

Moral problem-solving is a messy and complex process, as a design process. The design analogy has been introduced by engineering ethicist Caroline Whitbeck.[1] Mainstream ethics, Whitbeck argues, has been dominated by rational foundationalist approaches (Whitbeck, 1998). As a result, ethics has focused primarily on the analysis of moral issues and on a quest for the ultimate rational foundations of morality. This means that the field of ethics as we know it now is typically searching for one, or a limited number of, basic moral principle(s), and tends to build on unrealistic decision-making problems. The rational foundationalist approach, according to Whitbeck, is unnecessarily reductive and therefore misleading. She holds that moral philosophy should be tolerant toward different approaches, and should overcome the idea that dealing with moral problems is only about analyzing preset moral problems, and selecting the best option through justified principles.

One need not agree completely with Whitbeck's criticism of moral philosophy in general to appreciate the alternative she seeks to offer with her design analogy. This analogy can best be understood by considering the central notion of ill-structured problems. Whereas well-structured problems (such as basic arithmetical calculations), usually have clear goals, fixed alternatives to choose from, usually maximally one correct answer, and rules or methods that will generate more or less straightforward answers, **ill-structured problems** have no definitive formulation of the problem, may embody an inconsistent problem formulation, and can only be defined during the process of solving the problem. In cases of ill-structured design problems, thinking about possible solutions will further clarify the problem and possibly lead to a reformulation

> **Ill-structured problem**   A problem that has no definitive formulation of the problem, may embody inconsistent problem formulations, and can only be defined during the process of solving the problem.

of the problem (Cross, 1989). Moreover, ill-structured problems may have several alternative (good, satisfying) solutions, which are not easily compared with each other (cf. Cross, 1989; Van de Poel, 2001; Dringenberg and Purzer, 2018). This is due to the fact that for ill-structured problems, no single criterion exists to order uniformly the possible solutions from best to worst (Simon, 1973). Another characteristic is that it is usually not possible to make a definitive list of all possible alternative options for action (Simon, 1973). This means that solutions are in some sense always provisional.

For Whitbeck, the fundamental mistake rational foundationalists make is that they fail to see that moral problems are *ill-structured*. By framing moral problems as "multiple-choice" problems (where we have a fixed number of possible alternatives to choose from, of which only one is right), moral philosophers implicitly suggest that moral problems are well-structured. As an alternative, Whitbeck proposes to take the ill-structured nature of moral problems as a starting point for considering moral problem-solving. Given the fact that designers have to deal with ill-structured problems all the time, Whitbeck holds we can learn a lot from designers and engineers when dealing with moral problems in domains that are not traditionally associated with "design."

The most important lesson to be learned from designing is that practical problem-solving is not only about analyzing the problem and choosing and defending a certain solution, but also about finding (new) solutions. Whitbeck calls this "synthetic reasoning." Designers engage in a design *process*, during which new information may arise, uncertainties and unknowns are taken to be defining characteristics of the problem situation, and several possible solutions are pursued simultaneously. Another lesson from designing is that designers seem well able to satisfy apparently conflicting demands at once. Whitbeck maintains that even though some moral problems may be irresolvable, it is misleading to present moral problems as such from the start, "because it defeats any attempt to do what design engineers often do so well, namely, to satisfy potentially conflicting considerations simultaneously" (Whitbeck, 1998, p. 56).

Apart from these characteristics, which moral problems share with design problems (and other ill-structured problems), moral problems have their own peculiarities which make them even more messy and complex. One of them is that in identifying a moral problem one needs a conception of what morality and ethics are. Such a conception is partly theory-dependent as different ethical theories emphasize different parts of reality as morally relevant. Nevertheless, despite such differences, there is much common ground in ethical theories on what are moral concerns or problems. As a first approximation it will often be possible to define a problem based on common sense and one's own theoretical commitments. This formulation may later be refined during the process of moral problem-solving.

A second peculiarity of moral problems is related to the first one. The different ethical theories are not only relevant in identifying and formulating moral problems but also in judging them. The diversity of theories also reveals a diversity of reasonable moral opinions among different people on moral issues. This does, however, not mean that any solution to a moral issue will do. Solutions are better if they are based on systematic reasoning about the moral problem, on taking into account different viewpoints and theories, and on the exercise of a critical and reflective attitude.

4.3   The Ethical Cycle

Moral problem-solving is thus a messy and complex process. This does, however, not preclude the possibility of a systematic approach to the identification, analysis, and solution of moral problems. A systematic approach might even be required to avoid the reduction of moral judgment to mere gut-feeling without any attempt to understand the moral problem or to justify one's actions. The approach we propose, the ethical cycle, aims at an improvement of moral decision-making or at least it tries to avoid certain shortcuts. Such shortcuts for example consist in neglecting certain relevant features of the problem or in just stating an opinion without any justification.

The **ethical cycle** is a helpful tool in structuring and improving moral decisions. The cycle helps you to make a systematic and thorough analysis of the moral problem and to justify your final decisions in moral terms. Ultimately, moral problem-solving is directed at finding the morally best, or at least a morally acceptable, action in a given situation in which a moral problem arises. It is, however, hard to guarantee that the ethical cycle

> **Ethical cycle**   A tool in structuring and improving moral decisions by making a systematic and thorough analysis of the moral problem, which helps to come to a moral judgment and to justify the final decision in moral terms.

indeed delivers such a solution, because people may reasonably disagree about what is the morally best, or a morally acceptable, solution. We will discuss this further in Section 4.5.

The ethical cycle consists of a number of "steps" (Figure 4.1). It is important to stress that by distinguishing these steps we do not want to suggest that moral problem-solving is a linear

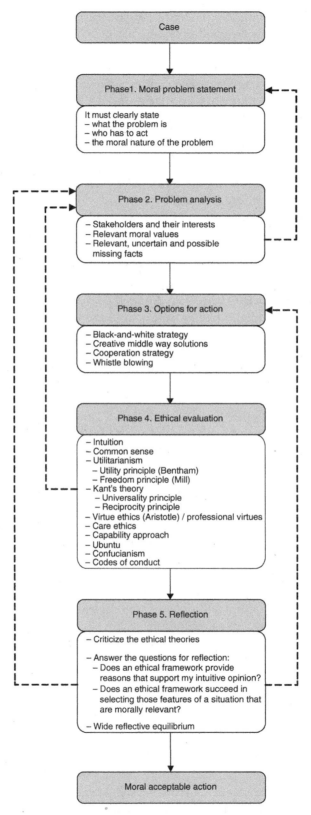

Figure 4.1 The ethical cycle.

process. Rather, it is an iterative process, as the feedback loops in Figure 4.1 already suggest. The cycle, for example, starts with formulating a moral problem. In many actual cases, the moral problem only becomes clear after further delving into the facts of the situation, by distinguishing stakeholders, looking at ethical theories, et cetera. In other words, formulating a good problem statement is an iterative process that continues during the other steps. Nevertheless, it is important to start with formulating a moral problem to get the process going.

4.3.1 Moral problem statement

The start of the ethical cycle is the formulation of a **moral problem**. A characteristic of a moral problem is that there are two or more positive moral values or norms that cannot be fully realized at the same time. Ethicists often call situations like these **moral dilemmas** instead of moral problems. Originally "dilemma" means "double proposition" implying that there are only two options for action. The crucial feature of a moral dilemma is, however, not the number of actions that is available but the fact that all possible actions are

> **Moral problem** Problem in which two or more positive moral values or norms cannot be fully realized at the same time.
>
> **Moral dilemma** A moral problem with the crucial feature that the agent has only two (or a limited number of) options for action and that whatever they choose they will commit a moral wrong.

morally unsatisfactory. The agent seems condemned to moral failure; no matter what she does, she will do something wrong (or fail to do something that she ought to do). A well-known example of a moral dilemma is taken from William Styron's *Sophie's Choice* (Styron, 1979). Sophie and her two children are at a Nazi concentration camp. On arrival, she is "honored" for not being a Jew by being allowed a choice: One of her children will be allowed to live and one will be killed. But it is Sophie who must decide which child will be killed. Sophie can prevent the death of either of her children, but only by condemning the other to be killed. The guard makes the situation even more excruciating by informing Sophie that if she chooses neither, then both will be killed.

Although some moral problems are real dilemmas, many moral problems are not. Often the problem is not an impossible choice between two or more evils. Therefore, we will use the term "moral problem" instead of moral dilemma. In order to apply the ethical cycle successfully, it is important that the moral problem is stated as precisely and clearly as possible. This can best be done by formulating a moral question. A good moral question meets three conditions: (1) it must clearly state what the problem is; (2) it must state who has to act; and (3) the moral nature of the problem needs to be articulated. Sometimes, the second condition is not relevant; for example, when we ask a general question about the moral acceptability of a particular course of action or a technology. An example of such a question is: Is cloning morally acceptable? or, more precisely: Under what conditions – if any – is cloning morally acceptable?

Dieselgate: Moral Problem Formulation

One possible problem formulation is:

> What is the effect on public health of the software manipulation device in the 482,000 2009–2015 vehicle models with 2.0-L diesel engines?

Although it is essential to ask and answer questions like this one in dealing with the moral problem, so that one knows what one is talking about, the question is not a good moral problem formulation because it is a factual question rather than a moral question (condition 3 in the text).

Another possible formulation is:

How can Volkswagen pass the stringent nitrogen oxide standards set by the Environmental Protection Agency (EPA)?

Again this is not a sound moral problem formulation, because this is a practical question about how to achieve a given goal (fulfilling legal norms) rather than a moral question about what to do given a range of (potentially conflicting) moral considerations.

One possible problem formulation that meets all three criteria in this case would be:

Should a Volkswagen engineer who knows about the company's decision to install secret software to cheat emissions tests inform the public about this or keep quiet about this for, e.g., loyalty reasons or for fear of reprisals?

Often it will not be possible to formulate a definitive formulation of the moral problem at this stage. The reason is that at later stages analyses will be made, like the identification of relevant values, which are crucial for a good problem formulation. Nevertheless, one can start with a somewhat vaguer notion of the moral problem and try to make the formulation of the moral problem clearer and more precise once some of the other steps have been carried out. We will illustrate this below in the boxes.

4.3.2   Problem analysis

During the problem analysis step, the relevant elements of the moral problem are described. Three important elements can be distinguished: the stakeholders and their interests, the moral values that are relevant in the situation, and the relevant facts. These elements are to be described during this step because they give a good sketch of the current situation with respect to the moral problem; moreover, they are indispensable for the carrying out of the later steps of the ethical cycle.

Dieselgate: Relevant Values

- Public health
- Environmental care
- Honesty (speaking the truth)
- Loyalty to the company
- Integrity (i.e., living by one's own moral standards and commitments)
- Profit

These values can also be used to reformulate the moral problem a bit, for example:

> What should a VW engineer who knows about the company's decision to install secret software to cheat emissions tests do given on the one hand moral considerations of public health, environmental care, honesty, and integrity and on the other hand their loyalty to the company and the importance for Volkswagen to increase its market share?

This problem formulation places more emphasis on the relevant moral values than the one before and it does not directly focus on one possible solution, so leaving more room for creatively looking for solutions that meet the various moral concerns.

Stakeholders are both the people who can influence the options for action being chosen and the eventual consequences of this action as well for the people suffering or profiting from those consequences. Stakeholders can be individuals, like colleagues, groups, like the design team, organizations, like a company or society, as far as it concerns the common interest. For each of the stakeholders, it is to be indicated what interests they have.

> **Stakeholders**   Actors that have an interest ("a stake") in the development of a technology.

Dieselgate: Main Stakeholders and their Interests

- *Management of Volkswagen*: increasing market share, meeting legal requirements, good reputation
- *American Environmental Agency (EPA)*: protecting human health from pollution and waste, reducing the impact of climate
- *Consumers*: safe and reliable car
- *General public*: health, environmental protection
- *VW engineers who know about the cheating software*: being a reliable and honest engineer, keeping their job, meeting the law

Stakeholders may disagree about the facts. Usually, not all facts are undisputed in a moral problem situation. Facts can also be uncertain or unknown. Disputed, uncertain, or unknown facts are certainly not irrelevant for the analysis of the moral problem. In later steps, they can make a distinct difference. One way to deal with such facts is to make explicit assumptions about them. Naturally, different people will often make different assumptions. Since the final option chosen at the end of the ethical cycle can depend on the assumptions made with respect to facts, it is advisable to formulate the moral standpoint sometimes in a hypothetical form: "If x is the case, than option for action A is morally acceptable; but if it turns out that y is the case then option B is morally acceptable."

Dieselgate: Some Unknown or Disputed Facts

- The estimation of the human costs of the scandal, and the estimation of the impact on human health due to the cheating software.
- The legal and effective solution for emission reduction technology Selective Catalytic Reduction (SCR), which was rejected, would threaten the profits and might mean job losses.
- VW engineers who know about the cheating software might lose their job if they go public.

4.3.3   Options for actions

After the analytic step in which the moral problem is formulated, a synthetic step follows in which possible solutions for action are generated in the light of the formulated problem analysis. Often a moral problem is formulated in terms of whether it is acceptable to engage in a certain action or not. In this **black-and-white-strategy** only two options for actions are considered, doing the action or not, other actions are simply not considered.

Black-and-white-strategy A strategy for action in which only two options for actions are considered: doing the action or not.

While this strategy may be helpful in better understanding and formulating the moral problem, in many more complex situations it is too simplistic. In real life, options are usually not given but have to be thought out or "invented" by the agent. In fact, by thinking out new options for action a seemingly irresolvable moral dilemma can sometimes be resolved or made less dramatic. During this step creativity is therefore of major importance. It can invite us to find options for actions that bridge seemingly conflicting moral obligations playing a role in the moral problem.

Strategy of cooperation The action strategy that is directed at finding alternatives that can help to solve a moral problem by consulting other stakeholders.

Also, the **strategy of cooperation** can be helpful in thinking out possible options for action. This strategy is directed at finding alternatives that can help to solve the moral problem by consulting other stakeholders. Sometimes, such cooperation and consultation can lead to win-win situations – solutions which make nobody worse off. Often such win-win situations are not self-evident and one should creatively look for new options for action.

Whistle-blowing (speaking to the media or the public on an undesirable situation against the desire of the employer, see Subsection 1.5.3), is a last resort strategy because it usually brings large costs both to the individual employee and to the organization. Nevertheless, some situations may require whistle-blowing, for example, if human safety or health is at stake and there are no other options of actions available.

Dieselgate: Options for Action

Our original problem formulation was:

Should a Volkswagen engineer who knows about the company's decision to install secret software to cheat emissions tests inform the public about this or keep quiet about this for, e.g., loyalty reasons or for fear of reprisals?

This formulation suggests a black-and-white-strategy: either the VW engineer should inform the public or they should not. In this black-and-white strategy one of the options is whistle-blowing because it is clear that Volkswagen is against making the information public.

Now consider the reformulated problem:

What should a VW engineer who knows about the company's decision to install secret software to cheat emissions tests do given on the one hand moral considerations of public health, environmental care, honesty, and integrity and on the other hand their loyalty to the company and the importance of Volkswagen to increase its market share?

This formulation suggests a range of other options, including:

1 Develop or adapt, in cooperation with other engineers, an emission technology that fulfills the legal requirements;
2 Make use of the internal procedures within the organization to discuss the illegal use of the cheating software (*internal whistle-blowing*);
3 Contact the engineering society for advice and help;
4 Contact EPA to inform them about this problem and ask them to take action (*whistle-blowing*); or
5 Investigate how serious the problems with the cheating software are for public health and the environment.

Most of these additional options employ the strategy of cooperation and pay more attention to the relationship of the engineer, and their company Volkswagen, with relevant stakeholders. The fifth option is especially relevant in the light of the disputed or unknown facts we have identified.

4.3.4 Ethical evaluation

In this step, the moral acceptability of the various options for action is evaluated. This can be done on the basis of both formal and informal moral frameworks. Formal moral frameworks are based on professional ethics discussed in Chapter 2: the codes of conduct, and the ethical theories discussed in Chapter 3.

Ethical evaluation also can be based on more informal ethical frameworks. We distinguish two such frameworks here: intuitions and common sense. The **intuitivist framework** is rather straightforward: indicate which option for action in your view is intuitively most acceptable and formulate arguments for this statement. The **common sense method** asks to weigh the available options for actions in the light of the relevant values. In a specific case, it might, for example, be possible to argue that although making a profit is important, the value that is really at stake (or dominant) is public safety. In determining which value is dominant, certain guidelines can be followed, such as, "dominant values are

Intuitivist framework The ethical framework in which options for action are evaluated on basis of one's view about what is intuitively most acceptable and that formulates arguments for this statement.

Common sense method The method that weighs the available options for actions in the light of the relevant values.

usually intrinsic values and not merely instrumental values," and "if more people find a value important, it is more likely that it is a dominant value." Once the dominant value has been selected, the option can be chosen that best meets that dominant value (Brady, 1990).

The fourth step results in the moral evaluations of the various options for action. In the next box, we will illustrate the application of some ethical frameworks with regard to the option "keep quiet about the deception" by an VW engineer who knows about the company's decision to install secret software to cheat emissions tests.

The application of some formal ethical frameworks with regard to the option "keep quiet about the deception" in the Dieselgate scandal

If we look at the ethical framework of "code of conduct," then the question of whether the option "keep quiet about the deception" is morally acceptable or not is not so difficult. For example, the NSPE Code of conduct states that "[e]ngineers shall hold paramount the safety, health, and welfare of the public," that "[e]ngineers shall at all times strive to serve the public interest," and that "[e]ngineers are encouraged to adhere to the principles of sustainable development in order to protect the environment for future generations." From this, it follows that the VW engineer must speak out and take action to discuss this issue with the management of Volkswagen, and if necessary, report the violation to the code enforcement official EPA.

The application of act utilitarianism is not so unequivocal. The consequences of the action "to keep quiet about the deception" depends on whether the fraud comes to light. Given the uncertainty about this, we must make use of probability. Let us assume, for the sake of simplicity, that the exposure of the company's deliberate cheating in emissions testing has a very high probability in our calculation to determine the greatest happiness for the greatest number. Then, almost all the stakeholders are adversely affected, ranging from job losses, pay cuts, stock devaluation, sales drop, damaged company reputation, public health damage, and environmental pollution. So, the action does not lead to greater overall good. Even if we assume that the fraud will never be exposed, then it is plausible to say that the "speaking-out" is the most acceptable action according to act utilitarianism, because this action will avoid the very high costs of public health damage and environmental pollution in contrast to the action "keep quiet about the deception."

The VW engineer also violated rule utilitarianism by keeping quiet about the deception. A general rule as "you should not speak out when companies violate legal rules that affect public health and environment" will not withstand the test of generally promoting happiness within society.

It also contradicts the reciprocity principle: "Act as to treat humanity, whether in your own person or in that of any other, in every case as an end, never as means only." By keeping quiet about the deception, customers, for example, are treated only as a means, since they cannot make a rational choice. If they knew that their purchased car is not reliable, they would not have considered buying the car. So, according to this principle of Kant, the VW engineer should inform the public about the deception.

Applying care ethics leads to the same conclusion. Care starts with a relationship and the responsibilities it entails. Although the relationship of engineers to society is mainly an indirect relationship, with industry, in its various forms, as the intermediary, engineers play a major role in the development and application of technology, and so have a special responsibility in serving as guardians of various public health, safety, and welfare issues. In this case, the VW engineer should therefore address the problem with the cheating software to ensure that this problem is meaningfully addressed, or to disclose this problem to the general public.

In the Dieselgate case, all the ethical frameworks seem to point in the same direction. However, this need not always be the case, because different ethical frameworks may point to different preferred options for action in a given situation.

An Illustration of Conflicting Ethical Frameworks

To see how different ethical frameworks may lead to conflicting recommendations, consider the following case:[2]

> Jasmine is director of a building department in a big city. Due to budget constraints, the city has been unable to hire a sufficient number of qualified individuals to perform building inspections. This makes it difficult for the inspectors to do a good and thorough job. At the same time, a new and tougher building code was adopted by the city. While this code promotes greater public safety than the last one, it also contributes to the difficulty inspectors have in doing a good and thorough job.
>
> Jasmine sets up an appointment with the chairman of the city to discuss her concerns. The chairman agrees to hire additional code officials for the building department on the condition that Jasmine agrees to permit certain specified buildings under construction to be inspected under the older, less rigid enforcement requirements. Should Jasmine agree to concur with the chairman's proposal?

Applying Kant's universalization test to this case would yield an argument like this: If Jasmine complied with the older, less rigid requirements, she acts from the maxim "apply less rigid requirements if this improves the situation in the long run." One may be doubt whether this maxim can be universalized. That is, if everybody would act like that, rules would seem to become meaningless. Apart from that, the profession of building inspectors might be rendered meaningless. If rules are altered in order to improve the situation in the long run, what would a building inspector inspect?

Applying an act utilitarian framework, the main question is how Jasmine can achieve the best results. In this case, it seems obvious that the best results are achieved if she agrees with the chairman's proposal. Otherwise she will have too few inspectors and will not even be able to inspect all building according to the old less strict code.

It is worth noting that according to the Kantian framework, the actual consequences of her action are irrelevant, while in the utilitarian framework these consequences are crucial for the ethical judgment.

4.3.5   Reflection

Since the different ethical frameworks, including the informal frameworks, do not necessarily lead to the same conclusion, a further reflection on the outcomes of the previous step is usually required. The goal of this reflection is to come to a well-argued choice among the various options for actions, using the outcomes of the earlier steps.

> **Wide reflective equilibrium**
> Approach that aims at making coherent three types of moral beliefs: (1) considered moral judgments; (2) moral principles; and (3) background theories. Also the resulting coherent set of moral beliefs is often called a wide reflective equilibrium.

The approach to reflection we want to advocate here is known as the method of **wide reflective equilibrium** (Daniels, 1979, 1996). This approach aims at making coherent three types of moral beliefs: (1) considered moral judgments; (2) moral principles; and (3) background theories. The background theories include ethical theories, but also other relevant theories such as psychological and sociological theories about the person. The inclusion of theories is important because they block the possibility of simply choosing those principles that fit our considered judgments. Achieving wide reflective equilibrium forces us to bring our judgments not only into coherence with principles but also with background theories. Because such theories also apply to other cases, our various considered moral judgments become connected, so that we are forced to examine critically our various considered judgments and eventually have to achieve coherence between the different layers of our moral beliefs.

The basic idea is that in a process of reflection different ethical judgments on a case are weighed against each other and brought into equilibrium. As we see it, this process is not so much about achieving equilibrium as such, but about arguing for and against different frameworks and so achieving a conclusion that might not be covered by one of the frameworks in isolation.

Central to the reflection step is thus argumentation. Arguments for or against ethical frameworks can be positioned at two levels. One level is the general criticism of the ethical frameworks. Utilitarianism can, for example, be criticized for neglecting duties or moral rights, while deontological theories might be criticized for not taking into account the consequences of actions. Such criticisms are well-known in moral philosophy and might be helpful for the reflection in this step. The second level of criticism is the concrete situation in which a certain option for action has to be chosen. It might for example be the case that a certain general objection to an ethical theory is not so relevant in a particular case. For example, a general objection against the utility principle ("the greatest happiness for the greatest number") of classical utilitarianism is that it neglects distributional issues (see Subsection 3.7.3), but it might be that in the particular situation different options for actions hardly have distributional effects, so that in that situation this objection is not so relevant. In general, we suggest two types of questions for reflection on this second level:

- Does an ethical framework provide reasons that support my intuitive opinion? If not, do I have other reasons that support my intuitive opinion? If I have other reasons are they strong enough to override the reasons within the ethical framework? If not, do I have to revise my intuitive opinion and in what way?
- Does an ethical framework succeed in selecting those features of a situation that are morally relevant? Are there any other moral relevant features that are not covered? Why are these relevant and how could they be accounted for?

The result of the fifth step is the choice for one of the options of action; a choice that can be argued in relation to the different ethical frameworks.

4.4 An Example

Earlier we already applied parts of the ethical cycle to the Dieselgate case. We will now apply the whole ethical cycle to the Highway Safety case described in the Introduction. This case description is, we must admit, rather stylized. We have chosen, however, to leave out certain complexities and uncertainties as to be able to show more clearly and straightforwardly how the ethical cycle would proceed in a case like this. In particular, we show that the ethical cycle by including a reflection step moves beyond the simple opposition between a consequentialist and a deontological ethical approach for which this case description was originally devised.

4.4.1 Moral problem statement

In the original case, the moral problem statement is already given: "Which of the two improvements do you think David should recommend?" This is not the only possible moral problem statement in this case. One might for example wonder whether making this decision is actually David's responsibility. The case concerns spending of public funds and it might be argued that such a decision is to be made by the relevant city council or state council. One might formulate as problem statement: "Is it David's (moral) responsibility to make this decision?" We will, however, restrict ourselves here to the problem statement formulated by Harris, Pritchard, and Rabins. This problem statement meets two of the three earlier mentioned conditions for a good problem statement: it is clear what the problem is (which option to choose) and it is clear who has to act (David). It is not clear from the statement itself, however, why it is a *moral* problem. Maybe this is simply a practical decision about what to do or an economic decision about how to spend public fund most efficiently. In fact, many people respond to this problem by stating that it is obvious that option A should be chosen because it results in the highest level of reductions in fatalities and injuries. However, from a deontological point of view it might be argued that site B is the best option because it is fairer to reduce the risk for people who are now subject to the highest risk factor (see Subsection 4.4.4 for more details). So, considering the case from a deontological point of view helps to realize that there is a potential moral problem here. Ethical theories thus help in recognizing the moral relevant characteristics of a situation and in formulating the moral problem. This also underlines the iterative character of the ethical cycle: it might well be that someone only recognizes the fairness considerations in step 4 of the ethical cycle. They might then go back to step 1 and reformulate the moral problem and redo steps 2 and 3.

4.4.2 Problem analysis

Now, we have to state the relevant facts, stakeholders, and interests and values. The main facts are already listed in the detailed case description. Some facts are uncertain. It is, for example, not known whether the general reduction factors for municipal and rural intersections apply to the specific case. There are no indications to the contrary, but this does not guarantee that these factors do apply. Such uncertainties could make a difference for the final judgment on the case.

Apart from David, drivers and their passengers, taxpayers, and the relevant city or state council could be distinguished as relevant stakeholders. As a first approximation, one could say that the main interests of these stakeholders are safety (drivers and their passengers), minimal costs (taxpayers) and highest safety for the lowest cost (city or state council). On closer examination, these stakeholders are not really uniform. Some drivers will use only the city intersection, some – but probably less – only the rural, some will use both, and some will use neither; which might result in different preferences about where to place traffic lights. Moreover, some drivers will prefer speeding above safety and will maybe prefer that no traffic lights are placed at all! Most drivers will, as taxpayers, prefer minimal costs, which may conflict in this case with increasing safety. To determine which option of action is the "best," it is necessary to make compromises concerning the various interests: you trade off a certain level of safety for a certain level of costs.

Although it is difficult to draw up a definitive list of stakeholders and interests, the given analysis is helpful in distinguishing relevant values. In the formulation of the problem, we have already distinguished two relevant values: safety and fairness. We might now add a third one which is related to the interest of keeping costs low. Low cost is, however, hardly a moral value as such. The moral value at stake here seems to be something like "public utility," which in this particular case implies that higher costs, ultimately resulting in higher taxes, may pay themselves back in higher public utility through higher safety.

4.4.3   Options for actions

In this case, the options for action are already given in the problem formulation. One might, however, wonder whether these two options are really the only ones. Whitbeck, for example, comments on this case:

> Notice ... that the problem is presented as a forced choice between spending all the remaining resources on one intersection and spending it all on the other. In fact, there would likely be many other choices. For example, putting up traffic signs at both intersections may be an alternative to installing traffic lights at either one. (Whitbeck, 1998, p. 65)

So, it might be useful to think of other options in the light of the relevant values. In Subsection 4.3.3, we suggested a number of strategies that could be helpful in devising options. The black-and-white-strategy has been chosen in the original formulation of the problem. This has probably been done for didactical considerations, that is, illustrating the difference between a consequentialist – in particular a utilitarian ethical – framework and a deontological framework. While this may be illuminating, it might also give the wrong impression that the actual problem is best solved by a black-and-white strategy, which is usually not the case. Another strategy, for example, could be the cooperation strategy, which is directed at finding alternatives that can help solve the moral problem by consulting other stakeholders. In this case, it might be useful, for example, to consult drivers and people who live in the neighborhood of the intersections because they may have more specific knowledge about why and what accidents occur at the intersections, or may have creative solutions. Whistle-blowing is not really relevant here because there is not a hidden abuse that needs to be uncovered.

4.4.4   Ethical evaluation

Common sense

According to this approach we first look at the relevant values. In this case the values at stake are safety, fairness, and public utility. You might argue that the dominant value in this case is safety and, consequently, you could argue that the best option for action is the action that reduces the most fatalities and injuries in *absolute* numbers. In this interpretation of safety – using the data that are given in the case description – you should recommend site A. However, other interpretations are possible as well, which can lead to other recommendations. For example, the best option is the action that reduces the most fatalities and injuries in *relative* numbers: a reduction of 0.5 fatalities per 6,000 vehicles/day for site B (which corresponds with a reduction of 2 fatalities per 24,000 vehicles/day) is "better" than a reduction of 1 fatality per 24,000 vehicles/day for location A. In this case, the recommendation will be site B. The common sense approach gives no clear-cut answer, but it stresses the importance of the interpretation of safety (assuming that this is the dominant value). So, you have to look for arguments concerning relative versus absolute numbers to motivate and justify your choice.

Utilitarianism

The utilitarian framework selects the option that brings "the greatest happiness for the greatest number." The expected social utility can be calculated with a cost-benefit analysis using the different "pricing schemes" suggested in the case description where money is used to express quantities of pleasure (benefits) or pain (costs). For the sake of simplicity, we leave out the effect of uncertainty in making these calculations, but it is important to recognize that such uncertainties might affect your final judgment.

As Table 4.4 shows, the available data suggest that site A in the city area is to be chosen. In all calculations site A has the largest gross benefit, and also the largest net benefit, since the costs of $ 50,000 is the same in all calculations. The data in the calculations according to the pricing schemes of NSC and NHTSTA, moreover, suggest that the costs of $ 50,000 are recovered within one year for both choices.

Kantian theory

The application of Kantian theory in this case is based on fairness considerations. Kant's first categorical imperative "Act only on that maxim which you can at the same time will that it should become a universal law" implies the equality postulate, that is, the duty to treat persons as equals, that is, with equal concern and respect (see Subsection 3.8). One

Table 4.4   Gross benefit per year of placing traffic lights at the two sites using different "pricing schemes"[3]

	Site A	Site B
NSC	$ 65,400	$ 58,400
NHTSTA	$ 273,600	$ 129,440
Neighboring state	30 PD	5.95 PD

PD refers to property damage only. The numbers in the table indicate the expected reduction expressed in the unity "property damage only" according to the pricing scheme of the neighboring state mentioned in the case description.

Table 4.5 Current risk of fatality and of injury for individuals approaching the intersection per year (under the assumption that there is one person in each vehicle)[4]

	Site A	Site B
Fatalities	2.3 E-07	4.6 E-07
Injuries	6.9 E-07	9.1 E-07

could argue that, as a consequence of this postulate, everybody has a right to the same level of protection, so that the same maximum risk factor applies to everyone. In this case, individuals approaching intersection B face a higher risk than individuals approaching intersection A (see Table 4.5). A choice for site B would therefore be fairer, since this decreases the current inequality in risk factors.

Kant's second categorical imperative (the reciprocity principle): "Act as to treat humanity, whether in your own person or in that of any other, in every case as an end, never merely as a means" is difficult to apply to this case. This imperative states that each human must have respect for the rationality of others and that we must not misguide the rationality of others, but in this case the rationality of others is not an issue.

Virtue ethics

From a virtue ethics point of view, one might try to formulate a list of virtues that are relevant for engineers (see Subsection 3.9.3). One may then ask how a virtuous engineer, employing the relevant virtues, would act in this situation. For example, how can a virtuous engineer make the decision *objectively?* This might reveal new relevant moral considerations, or might even lead to a reformulation of the moral problem (step 1 of the ethical cycle). One might, for example, begin to wonder whether it is desirable that David makes this choice himself or whether he should merely inform the public authorities who then make the decision.

Professional ethics

If we look at the code of conduct of the *National Society of Professional Engineers*, the following article is relevant to David:

> 2. Engineers shall perform services only in the areas of their competence.

David has competence in determining the nature and the magnitude of the safety improvements, but it may be argued that this choice requires not only engineering knowledge, but is political in nature, and can therefore only be legitimately be made by public authorities like the city council. You could argue, therefore, that David should not make the choice himself, but that he should inform completely the public authority so that they can make a conscious choice.

4.4.5   Reflection

Since the applied ethical frameworks provide different outcomes, further reflection is required. First of all, in this case, one could reflect internally on the frameworks. For convenience, we will focus on the utilitarian and Kantian framework, and we will leave aside the other ethical frameworks. With respect to the utilitarian framework, one could for example question whether the provided data on the monetary value of a human life, injuries, and property-damage only accidents are adequate. Nevertheless, the various monetary schemes and the weighing scheme of the neighboring state all suggest the choice of site A over site B. In fact, it is not possible to devise a monetary scheme in which site B would score better unless one's weights human lives negatively and/or injuries and property damage positively. So the outcome that the utilitarian test selects site A is rather robust.

This is less so for Kant's approach or the fairness test. The rural intersection is more dangerous in terms of the probability of a fatality or injury per vehicle approaching the intersection. However, we do not know the average number of people in a car and whether this number is the same for the rural and urban intersection. The data therefore do not rule out that the individual risk of a car driver or passenger in expected fatalities *per year* is actually higher on site A than on site B, contrary to what Table 4.5 suggests.

There are also other reasons to doubt whether fairness considerations necessarily suggest the choice for section B. If fairness is understood in terms of a right to protection, this is perhaps best understood in terms of an equal level of minimal safety for everyone. It might well be that that level is already met at both intersections. Alternatively, one could understand fairness in terms of equal absolute safety. This would mean that everybody has a right to the same absolute level of risk. This would have rather absurd consequences, however. It would, for example, imply that if someone would be very safe off, for example due to chance, everybody would have the right to that level of safety, even if that would be very hard, if not impossible, to realize. It would even imply that it would be desirable to make the safest person less safe, even if that would increase the safety of nobody else, because in this way a more equal distribution of risks is achieved.

The last remarks already make clear that applying only the Kantian framework without considerations of overall safety or public utility does not make much sense in this case. Conversely, one might argue that public utility or overall safety considerations alone are also not enough, which would mean that the utilitarian framework alone is too narrow to judge this case. What seems required then is a certain balancing of the various moral frameworks or considerations, including possibly also one's intuitive opinion and common-sense considerations.

The approach that we advocate here is that of wide reflective equilibrium. Suppose that someone has the considered judgment that location A is best (belief a). They might defend this choice by referring to the principle "the greatest happiness for the greatest number" (belief b). This principle, in turn, might be justified on basis of the ethical theory of utilitarianism (belief c). Utilitarianism is not only a theory about where to place traffic lights but a much broader theory that is related to a whole range of moral judgments, including the judgment that – for the sake of comparison – we can express human lives in a common value like money (belief d). The same person judging that location A is best (belief a) might reject the moral judgment that we can express the value of human live one way or the other in money (belief d). In this case, the set of beliefs a, b, c, and not-d is incoherent.

There are several ways you can solve the incoherence between a, b, c, and not-d. We mention some:

- You could give up the belief not-d. After all, you might come to the conclusion that human life is not priceless, even if you intuitively thought so. So you might choose to adopt the belief d.
- You could also look for another ethical theory (c) or another ethical theory with moral principles (b and c), which would still justify a, but would not imply other moral judgments, like d, that you consider dubious.

- You might also try to look for a theory that better fits your judgments about valuing human life. You might, for example, have the considered moral judgment that since we cannot put a price on human lives, you should treat humans equally and respect their freedom. On the basis of such a belief you might embrace – at least for the moment – a deontological ethical theory and some principle of fairness. On that basis, you might revise your initial belief a about the case, and now choose site B.

This list does not exhaust the possibilities. One could also try to combine utility and fairness considerations in several ways. One could for example argue that fairness considerations imply that all drivers and passengers have a right to the same minimal level of safety. One might then argue that this level is actually met at both intersections, so that one can choose without scruples the option with the highest public utility – location A.

The important point about this example, however, is not how you solve the incoherence between your different beliefs. The important thing is that by trying to achieve a wide reflective equilibrium you are forced to engage in a broader and more systematic theoretical consideration of the case, including a range of arguments and reasons.

4.5   Collective Moral Deliberation and Social Arrangements

The emphasis in the ethical cycle is on individual judgment. However, in many, if not most, situations in real life, other people will be involved in and affected by your choices. You might doubt whether in such situations, individually achieving a conclusion on how to act is justified. In particular, you might wonder why others, especially people affected by your actions, should accept your individual conclusion on how to act. Of course, if you have used the ethical cycle, you will be able to argue your choice, but given the nature of moral reflection and the diversity of ethical frameworks that might give conflicting advices, it seems doubtful that every person using the ethical cycle would come to the same conclusion as you did. The natural inclination of many ethicists would be to look for a better, overarching moral framework. Even if one would believe that such an endeavor is worthwhile, it certainly does not solve the problem if you have to act here and now. We therefore propose a more practical solution – engaging in a moral deliberation with other people involved and possibly affected.

Engaging in deliberation is also useful for other reasons. If you are confronted with moral problems, you often have to act in a situation in which you depend on others to achieve certain options for action. A certain support from others is therefore required to be able to act in a morally effective way. This is certainly true if you are working in a company. Therefore, deliberation and discussion with others are important additions to the ethical cycle.

The final step in the ethical cycle is reflection, leading to a well-argued choice for an option of action. This choice, however, needs not be your final choice; it can also be seen as a provisional choice that you can revise in discussion with others. The objective of such deliberations is to make public your reasons for a certain choice and to expose them to criticism by others. Such discussion and criticism can result in a revision of your choice, for example because your arguments turn out not to be adequate after all, or because certain arguments have been overlooked. So conceived, deliberation is mainly a tool to improve one's moral judgment.

However, as already suggested above, one could also argue that **moral deliberation** is essential for more fundamental reasons. Philosopher Jürgen Habermas has argued that moral judgments are legitimized by them being the result of a moral deliberation that meets certain standards (Habermas, 1981).

> **Moral deliberation** An extensive and careful consideration or discussion of moral arguments and reasons for and against certain actions.

This includes the standard that the discussion should not be decided on the basis of authority or power, but on the basis of arguments. Other requirements for rational discussion or deliberation are that people should be honest and sincere, and should argue their point of view. The idea is that if deliberation meets such requirements, we have good reason to believe that the outcomes are sound.

A somewhat comparable idea has been formulated by the political philosopher John Rawls. Rawls (1971) embraces the wide reflective equilibrium approach that we earlier described in this chapter (Subsections 4.3.5 and 4.4.5). However, he realizes that it is very well possible that people come to different reflective equilibriums (especially in his later work, see Rawls, 1993, 2001). He nevertheless believes that people might often agree on moral issues even if they disagree on how their moral judgments are exactly to be justified. He calls this situation an **overlapping consensus**. Here we shall understand an overlapping consensus as agreement on the level of moral judgments, while there may be disagreement on the level of moral principles and background theories. In the case of safety improvements described in Section 5.4, two people might for example agree that a choice should be made for site A but one person might justify this choice

> **Overlapping consensus** An agreement on the level of moral judgments, while there may be disagreement on the level of moral principles and background theories. Each of the participants should be able to justify the overlapping consensus in terms of their own wide reflective equilibrium.

in terms of utilitarianism while the other adopts a notion of fairness that requires a minimal level of safety that is in their view already met at both intersections so that a choice can be made for the one that is best in absolute numbers.

An overlapping consensus is different from a compromise because it requires that each of the discussants can justify the overlapping consensus in terms of their own reflective equilibrium. In case of a compromise, you sometimes accept an outcome because you think it is the best you can get given the preferences of the others involved. But how can we achieve an overlapping consensus? Rawls himself believes that the achievement of an overlapping consensus is easier if all parties involved accept a reasonable degree of pluralism among moral opinions. When we do, Rawls believes, we are also able to distinguish between private reasons and public reasons in moral discussions. Private reasons are reasons that are important for how I want to live my life and make private moral decisions that do not directly affect others. I may, however, recognize that others do not, and should not, share my private reasons. Public reasons are reasons that we think apply to everyone. Rawls believes that when our focus in moral discussions is, as much as possible, on public reasons we are more likely to achieve an overlapping consensus.

Both the perspectives of Habermas and Rawls stress the importance of procedural criteria for arriving at a moral judgment, and both require social arrangements that meet certain norms. In this respect, both fit well with approaches such as Constructive Technology Assessment (CTA, see Section 1.6), and the social ethics approach to engineering (Subsection 3.10.3) that were briefly discussed before. This emphasis is different from the ethical cycle in which the various substantive ethical frameworks play a much more important part. We think that this need not be seen as incompatible, however. To engage in a moral deliberation, it is desirable that you have a well-argued moral opinion. Of course, you should be willing to revise your opinion, but in order to have a debate at all, you should first have your own well-argued opinion. For this purpose, the ethical cycle, including the use of substantial ethical frameworks to arrive at a moral opinion, is very useful.

4.6 Chapter Summary

Moral problem-solving is a difficult and complex process because moral problems are usually ill-structured. They do not have a clear-cut problem formulation, need to satisfy different, often conflicting, moral constraints, and do not have one best solution. In these respects, moral problems are like design problems. Solving moral problems, therefore, does not only require analysis but also synthetic reasoning (devising new options) and creativity.

The complex nature of moral problem-solving does not preclude a systematic approach. The approach that we have introduced in this chapter is called the ethical cycle. It consists of five basic steps:

1 Formulating the moral problem;
2 Analyzing the problem in terms of stakeholders and their interests, values, and facts;
3 Identifying and devising options for action with the help of strategies such as the black-and-white strategy and the cooperation strategy;
4 Ethical evaluation of the various options for action with the help of various ethical frameworks;
5 Reflection on the outcomes of the evaluation phase, resulting eventually in a well-argued choice for one of the options for action.

With respect to the reflection step, we have proposed the wide reflective equilibrium approach that aims at coherence between moral beliefs at three levels: (1) considered moral judgments; (2) moral principles; and (3) background theories. The important thing is that by trying to achieve a wide reflective equilibrium you are forced to engage in a broader and more systematic theoretical consideration of the case, including a range of arguments and reasons. It is precisely because this reflection involves *theories*, that such reflection becomes broader and more encompassing. This suggests that theories have an important role to play in moral judgment. However, this role is more complex than simply applying the theory to the case at hand.

In addition to coming to an individual moral judgment by using the ethical cycle, discussion with others is important. The goal of such discussion is to make public the reasons you have for a certain judgment and possibly to revise those reasons and your conclusion in debate with others. Such discussion is also important because other stakeholders in technological development might not agree with your conclusions, and including their point of view may improve the decision made. Deliberating with them makes the final result more legitimate, because you are then also respecting the moral autonomy of other stakeholders, and are expressing due care to them. Moreover, it is likely to make you as an engineer more effective because you will often need the cooperation of others to live by your moral judgments.

Study Questions

1 Why are moral problems ill-structured problems?
2 Why are most moral problems not real dilemmas?
3 XYZ orders 5,000 custom made parts from ABC for one of its products. When the order is originally made ABC indicates it will charge $75 per part. This cost is based in part on the cost of materials. After the agreement is completed, but before production of the part begins, ABC engineer Christine Carsten determines that a much less expensive metal alloy can be used while only slightly compromising the integrity of the part. Using the less expensive alloy would cut ABC's costs by $18 a part. Christine brings this to the attention of ABC's Vernon Waller, who authorized the sales agreement with XYZ. Vernon asks, "How would anyone know the difference?" Christine replies, "Probably no one would unless they were looking for a difference and did a fair amount of testing. In most cases the performance will be virtually the same – although some parts might not last quite as long." Vernon says, "Great, Christine, you've just made a bundle for ABC." Puzzled, Christine replies, "But shouldn't you tell XYZ about the change?" "Why?" Vernon asks, "The basic idea is to satisfy the customer with good quality parts, and you've just said we will. So what's the problem?"

  Source: https://onlineethics.org/cases/cases-teaching-engineering-ethics/information-due-customer.

  a. What exactly is Christine's problem here? Explain why it is a moral problem.
  b. Which moral values or principles are at stake here?
  c. Mention three things Christine can do to deal with her problem.
  d. What would an analysis of this problem according to (Bentham's) classical utilitarianism look like? What would the utilitarian advice to Christine be?
  e. Would John Stuart Mill's modified form of utilitarianism lead to a different advice? Why or why not?
  f. What would a Kantian ethicist recommend to Christine? Motivate your answer.
  g. What do you think is the right thing to do for Christine? Motivate your answer and also explain why you do not accept (some of) the advice from d, e, and f.
4 Apply the ethical cycle to the Challenger case from Subsection 1.5.1. Explain how each step would be applied in this case.
5 Apply the ethical cycle to the BART case from Section 2.1. Explain how each step would be applied in this case.

Discussion Questions

1 What do you consider appropriate grounds for overriding someone's personal decisions? Would you, for instance, prevent the sale of home body piercing kits or child pornography, and if so, on what grounds?
2 Motivate what you would do if you were a VW engineer who knows about the company's decision to install secret software to cheat emissions tests directly affected public health (see Subsection 4.3.4).
3 According to the wide reflective equilibrium approach, people should aim at coherence between the different levels of their (moral) beliefs. Is coherence indeed as important as this approach presupposes? Are coherent beliefs never wrong? Can a belief be right but nevertheless be incoherent? Can you think of an approach to ethical judgment in which coherence is not important at all?
4 Do discussions with others (moral deliberation) lead to better moral judgments? Why or why not?

Notes

1 Whitbeck provides us with a compelling sketch of what a designer-perspective on moral problems could offer, but the analogy was not fully developed. In Dorst and Royakkers (2006) this analogy is constructed more carefully and completely.

2 This is case 98–5 of the Board of Ethical Review (BER) of the National Society of Professional Engineers (NSPE). Available at https://www.nspe.org/resources/ethics/ethics-resources/board-ethical-review-cases/public-health-and-safety-code (accessed February 20, 2023).

3 As an example we will figure out the gross benefit of site A according to the pricing scheme of NSC:

$$((50\% \text{ of} 2) \times \$52\,000) + ((50\% \text{ of} 6) \times \$3\,000) + ((25\% \text{ of} 40) \times \$440) = \$65\,400.$$

4 As an example we will figure out the current risk of fatality of site A:

$$2/((20\,000 + 4000) \times 365) = 2.3 \text{ E-07}.$$

5

Design for Values

Having read this chapter and completed its associated questions, readers should be able to:

- Distinguish between the intended, embedded, and realized values of a design;
- Identify the main stakeholders in a Design for Values project and describe their perspectives;
- Identify, conceptualize, and specify the main values to be designed for in a Design for Values project;
- Explain and apply the distinction between value conceptualization and value specification and the quality criteria for both;
- Identify value conflicts in a design process;
- Describe the various methods for dealing with value conflicts in design and their pros and cons;
- Apply these methods to engineering design problems;
- Describe the role of prototyping, simulation, testing, and monitoring in Design for Values.

Contents

5.1   **Introduction** 146

5.2   **Embedding Values in Technology** 147

5.3   **Designing for Values** 149

5.4   **Stakeholder Analysis and Value Identification** 150

    5.4.1   Stakeholder analysis 150
    5.4.2   Sources of value in design 153

5.5   **Conceptualization and Specification of Values** 153

    5.5.1   Conceptualization 154
    5.5.2   Specification 156

5.6   **Value Conflicts** 159

    5.6.1   Cost-benefit analysis 161
    5.6.2   Multiple criteria analysis 163
    5.6.3   Thresholds 166
    5.6.4   Respecification: reasoning about values 167
    5.6.5   Innovation 170
    5.6.6   A comparison of the different methods 171

5.7   **Prototyping and Monitoring** 172

5.8   **Chapter Summary** 178

Study Questions 179

Discussion Questions 180

Ethics, Technology, and Engineering: An Introduction, Second Edition. Ibo van de Poel and Lambèr Royakkers.
© 2023 John Wiley & Sons Ltd. Published 2023 by John Wiley & Sons Ltd.

5.1 Introduction

Case Robert Moses' Racist Overpasses

Figure 5.1 Low overpass. Oversized overturned truck on the BQE under the Brooklyn Bridge. Photo: B. Yanev.

Robert Moses (1888–1981) was a very influential and also contested urban planner of mid-twentieth century New York. In his 1974 book, *The Power Broker*, biographer Robert Caro argued that Moses also demonstrated racist tendencies. He designed several overpasses over the parkways on Long Island, which were too low to accommodate buses (see, for example, Figure 5.1). Only cars could pass below them and for that reason the overpasses complicated access to Jones Beach Island. Only people who could afford a car – and in Moses' days these were generally not Afro-American people – could easily access the beaches now.

This case has become especially famous due to the philosopher of technology Langdon Winner who mentioned it in his article "Do artifacts have politics?" (1980). As Winner notes although the overpasses are extraordinarily low, one would normally not be inclined to attach any special meaning to that fact. It turns out, however, that they were deliberately designed to achieve a specific social effect. In the words of Langdon Winner:

> Robert Moses, the master builder of roads, parks, bridges, and other public works from the 1920s to the 1970s in New York, built his overpasses according to specifications that would discourage the presence of buses on his parkways. According to ... Moses' biographer, Robert A. Caro, the reasons reflect Moses' social class bias and racial prejudice. Automobile-owning whites of "upper" and "comfortable middle" classes ... would be free to use the parkways for recreation and commuting. Poor people and blacks, who normally used public transit, were kept off the roads because the twelve-foot tall buses could not handle the overpasses.

Winner's analysis of these low-hanging overpasses has become a paradigmatic example, even though some objections have been raised against it because timetables show that it was actually possible to reach the beach by bus, bypassing the overpasses. The example shows how values may get embedded in the design of an artifact. In this case, it is a negative value or disvalue, racism, that is embedded in the artifact's design, but one may of course embed positive values in a technology. A speed bump, for example, may be said to embed the value of traffic safety as it encourages car drivers to drive slower in residential areas.

In this chapter, we will discuss how engineers can embed positive values in the technologies they design. Several approaches have been developed to achieve this like, for example, Value Sensitive Design (Friedman and Hendry, 2019), value-based design (Spiekermann, 2015), and ethically aligned design (IEEE Global Initiative on Ethics of Autonomous and Intelligent Systems, 2019). In this chapter, we will refer to these approaches as **Design for Values** (Van den Hoven et al., 2015).

> **Design for Values** Approach that aims at integrating values of ethical importance in a systematic way in (engineering) design.

In this chapter, we explain the Design for Values approach and how it can be used in the development and design of new technologies. We start with a discussion of what it means to embed values in a technological design (Section 5.2). We then introduce the Design for Values approach (Section 5.3), and explain its main components, namely the identification of stakeholders and values (Section 5.4), how to conceptualize and specify values in design (Section 5.5), how to deal with conflicting values (Section 5.6), and – finally – how to build prototypes and test them (Section 5.7).

5.2 Embedding Values in Technology

As the example of racist overpasses suggests, values may get embedded in technological artifacts through their design. However, the values that are eventually realized when a technological artifact is used might be different from those embedded. In the case of racist overpasses, we saw that there might well have been other possibilities for poor people and blacks to travel to the beach, for example, using public transport, so overcoming the discriminatory and racist effects of the overpasses. This does not mean that racist values were not embedded in the design of the overpasses, but rather that we should distinguish between embedded and realized values.

But first, it should be explained what exactly is meant by "embedded value." To do so, we introduce the notion of **use plan** (Houkes et al., 2002). We will be assuming that when designers design a technological artifact, they do not just design a material artifact but also, at least implicitly, a so-called use plan. A use plan is a bit like a manual as it describes how a technology is to be used and for what purposes. However, unlike a manual, a use plan need not be put on paper; it may be tacit and implicit.

> **Use plan** A use plan describes how a technical artifact is to be used and for what purposes. It may be proposed by the designers but also depends on what users accept or take to be the normal way of using the artifact. The use plan may be tacit and implicit.

Embedded value is the value that is realized if a design is used according to its use plan under normal circumstances.

Realized value is the value that is realized through the actual use of a technological artifact in actual circumstances.

Intended value is the value that the designers aim to realize through the technical artifact designed.

We can now define **embedded value** as the value that is realized when a technology is used according to its use plan under normal circumstances. Such embedded value is to be distinguished from **realized value**, that is the value that is realized in actual use. The realized value may be different from an embedded value because actual use may be different from the use plan or because the circumstances are unusual. For example, light bulbs that consume less energy may embed the value of sustainability, but if – in practice – they are used to have the light always on, they may not contribute to the value of sustainability.

Both embedded and realized value needs to be distinguished from **intended value.** Intended value is the value that designers intend to realize when they design their products. Such values may not always be explicit, but in this chapter, we will be concerned with the situation in which you as a designer aim to realize certain values with the product you are designing. The subsequent sections will discuss how to do that.

The intended value may be different from the embedded value because designers failed to properly design their products for the intended values. For example, pacemakers (devices that help to control the heartbeat to prevent the heart from beating too slow) are intended to contribute to values like human health and well-being, but when they are badly designed, they might not do so, even when properly used.

If you want to know whether the intended value is embedded in your designed artifact, you may want to build a prototype of it and use it according to its use plan and observe whether the intended values are indeed realized under these circumstances. Is the product indeed safe or sustainable as you intended it to be?

Figure 5.2 shows the relation between intended, embedded and realized value. If the embedded value is different from the intended value you can try to redesign your product,

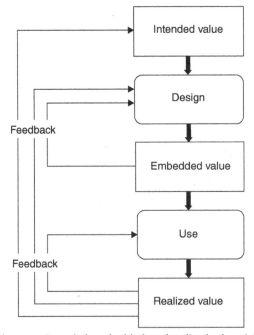

Figure 5.2 The relation between Intended, embedded, and realized value. Adapted from Van de Poel, (2020)/ Springer Nature/CC BY 4.0.

either the material artifact or its use plan. If the realized value is different from the embedded value you can either try to change how the product is used (e.g., by enforcing the use plan) or you can change the design of the product.

Sometimes realized values make you aware that you have forgotten an important value in the design of your product. This is the case because technologies can have unintended or unforeseen consequences. For example, the use of social media has led to hate speech and fake news. This was, of course, never the intention of the designers of the (algorithms for) social media. However, they also did not try to prevent such effects as they had not expected or anticipated them. Once such effects – like hate speech and fake news – occur, it would seem a good reason to redesign (algorithms for) social media, or at least to change their use. This means that designers should adopt new intended values like respect (to avoid hate speech) and veracity (to avoid fake news) for their adapted designs. This thus requires monitoring the effects of technologies in society and redesigning them if needed. We discuss this in more detail in Section 5.7.

5.3   Designing for Values

The design process is a central area where ethical considerations concerning technology arise. The reason for this is that crucial decisions regarding technology are made in the design process. To an important extent, the design of a technology determines how it will be produced and used, what maintenance will be required, and how the product is to be scrapped. Obviously, later choices by, for example, users are also of importance, but choices in the design process greatly influence the social consequences of a product. Therefore, nearly all ethical questions related to technology development that engineers are confronted with are reflected in the design process in some way or another.

Designing can be described as an activity in which engineers translate certain functions or aims into a working product or system. A ferry can be conceived of as the translation of the function "transporting people from one side of the river to the other." In most cases, a function or social goal can be translated into a technical solution in several ways. If you want to achieve transport between two riverbanks, you can choose among a series of possible technical solutions, such as a bridge, a tunnel, a ferry, or a cable-lift. The solution chosen does not only depend on the function to be realized, but on a series of additional design requirements, such as speed of transport, costs, building time, sustainability, and safety.

Engineering design is thus the process in which certain functions are translated into a blueprint for an artifact, system, or service that can fulfill these functions. Engineering design is usually a systematic process in which use is made of technical and scientific knowledge.

> **Engineering design**   The activity in which certain functions are translated into a blueprint for an artifact, system, or service that can fulfill these functions.

As we have seen in the previous section, values can get embedded in technological artifacts through their designs. Values always play a role in engineering design. For example, technical and economic values like costs, reliability, effectiveness, and efficiency are important in most design processes. But in addition, there are moral and social values that should be taken into account in the design of new technologies, like, for example, safety, sustainability, human well-being, and justice. In this chapter, we are interested in how we can systemically design for such moral values.

Value Sensitive Design (VSD) is an approach that aims at integrating values of ethical importance in a systematic way within engineering design (Friedman et al., 2006). The approach aims at combining three kinds of investigations: conceptual, empirical, and technical:

- *Empirical investigations* aim at understanding the contexts and experiences of the people affected by technological designs. This is relevant to appreciating precisely what values are at stake and how these values are affected by different designs.
- *Conceptual investigations* aim at clarifying the values at stake, and at making tradeoffs between the various values. Conceptual investigations in Value Sensitive Design are similar to the kind of investigations described in Subsection 5.5.1.
- *Technical investigations* analyze designs and their operational principles to assess how well they support particular values, and, conversely, to develop new innovative designs that meet particular morally relevant values particularly well. The second is especially interesting and relevant because it provides the opportunity to develop new technical options that more adequately meet the values of ethical importance than do current options.

VSD is one of the approaches that have been developed in the past decades for integrating values of moral importance in (engineering) design. There are several other general approaches like ethics-by-design, value-based design, ethically aligned design, and approaches for more specific values like privacy-by-design, safety-by-design, design for sustainability or green design, and inclusive design. In this chapter, we will refer to all these approaches as Design for Values.

5.4   Stakeholder Analysis and Value Identification

5.4.1   Stakeholder analysis

Direct stakeholders are the users and those who directly interact with a technology, like, for example, the operators of a technology.

Indirect stakeholders are those affected by (the use of) technology but not using it or directly interacting with it.

Design for Values usually starts with making an inventory of the relevant stakeholders and an investigation of their needs and values. One can distinguish between direct and indirect stakeholders (Friedman and Hendry, 2019). **Direct stakeholders** are those who directly use or interact with the technology; **indirect stakeholders** are indirectly affected by the technology. It is important to take into account both direct and indirect stakeholders.

Stakeholders may have different perspectives on the product being designed and it is useful to collect information on these perspectives. What are the needs and expectations of different stakeholders? How do they expect the designed product to affect their life or professional activities? What values do they consider important in the design? How would they understand these values?

There are different ways in which you can collect information from or about stakeholders. You can talk to them and interview them; sometimes there may be documentation you can read from or about them. There might be earlier studies that you can consult and sometimes it might be needed to set up a more extensive empirical study yourself to collect views and values from stakeholders.

If it is not possible to talk to stakeholders yourself or collect information from them, you might want to create a so-called **persona**. A persona is a fictional character that represents a stakeholder or group of stakeholders. You can, for example, imagine a day in the life of such a persona with the technology you are designing. How

Persona: A fictional character that represents a stakeholder or stakeholder group and that can be used to think about how a technology will be used, but also for investigating relevant values to its design.

would that persona be confronted with the technology? What could go well and what could go wrong? In this way, you might be able to find important values for the design of a technology.

The perspectives of the different stakeholders should be accounted for when you are designing for values. However, this does not mean that you should focus on the most important or influential stakeholders. Oftentimes, it may be instructive to investigate the perspective of marginalized groups, or groups that could use the technology but are currently excluded, or are not fully accounted for, in the design (see box). There may also be groups whose perspective you should *not* take into account. For example, when designing tools to increase cybersecurity, cyber criminals are strictly speaking (indirect) stakeholders, and while it may be instructive to know their viewpoint and practices in order to design better tools, you would not necessarily want to accommodate their (criminal) values in your design.

Sexist and racial bias in design

Figure 5.3 The humanoid robot Sophia, ITU Pictures / Wikimedia Commons / CC BY 2.0.

Design often reflects the tacit assumptions and biases of designers. This may unintentionally lead to, for example, racist or sexist biases. Oftentimes, such biases result from a failure to properly consider all relevant stakeholders in design.

Some examples:

- Crash tests for cars have been using dummies since the 1950s, and for decades these were modelled based on an average male.[1] The most commonly used dummy is 1.77 cm tall and weighs 76 kg, which is significantly taller and heavier than an average woman. Since 2011, female crash dummies have been introduced in the US. However, oftentimes female dummies are just scaled-down versions of male dummies, not representing other physiological differences between women and men. Moreover, female dummies are sometimes only used for the passenger seat rather than the driver seat. Consequently, when a woman is involved in a car crash, she is 71% more likely to be moderated injured, 47% more likely to be seriously injured, and 17% more likely to die.

- Several studies suggest there are serious demographic differences in the accuracy of face recognition algorithms. In most cases, such algorithms are less reliable for women of color (Buolamwini and Gebru, 2018; Grother et al., 2019). Underlying causes of such biases may be the composition of the training set (in which often white males are overrepresented) and default camera settings, which are usually not optimized for darker skins.[2]

- Some studies suggest that when computer assistants employ a female voice, the users start to associate female voices with being an assistant (Nass and Brave, 2005). Consequently, voice assistants that often use female voices promote an image of a docile woman, always available and never saying no (West et al., 2019).

- Intelligent machines are often portrayed as white (Cave and Dihal, 2020). An example is the humanoid robot Sophia developed by Hanson Robotic (Figure 5.3). Sophia is able to imitate human gestures and facial expressions and can have simple conversations with humans. She was also the first robot to receive citizenship in any country when she was granted Saudi Arabian citizenship in 2017.[3] But also other intelligent robots, or portrayals of them in fiction, are often white.

Relevant perspectives from stakeholders can be accounted for in different ways in the design process. Minimally, you should consider the values of stakeholders when deciding what values to design for. This does not mean that you should simply design for any value that stakeholders find important. As we have seen in Chapter 1, an engineer or designer is more than a hired gun. Nevertheless, you should at least consider the values of relevant stakeholders as relevant input (see also Subsection 5.4.2). It is also important to be aware that even when stakeholders agree about values, they may still have different conceptions and specifications of them (see Section 5.5).

There might also be ways to involve stakeholders more closely in the design process. For example, you may ask them to try out a prototype of the design and collect their experiences as input for revising your design (see Section 5.7), or you may involve stakeholders in decisions about conflicting values (see Section 5.6). You can also follow a **participatory design** approach, in which you involve stakeholders in all steps of the design process, including the development of alternative technological options or designs.

Participatory design is an approach to design in which you involve (direct) stakeholders actively in all phases of the design process.

5.4.2   Sources of value in design

In general, one can distinguish between four sources of value in a specific design project. One is the design project itself and more specifically its design brief, in which the goal of the project is further described and constraints and requirements may be given. Projects may be explicitly formulated to serve certain values, but even if this is not the case the design brief will often implicitly contain certain values.

A second source is the values of direct and indirect stakeholders. Empirical investigations may be required to identify what users and stakeholders consider relevant values for the design of a technology. Users and stakeholders may have different values from those articulated in the design brief and different values from each other.

A third source is the values of the designers and the engineering profession. When designers design a product they try to put, at least implicitly, specific value into a product which they hope will be realized in practice. Not only individual designers may be driven by certain values, but also engineering as a profession is committed to certain values such as human well-being, health, safety, and sustainability as we have seen in Chapter 2.

A fourth source of values is the general social values expressed in, for example, relevant laws. Next to general legislation, there often are **technical codes and standards** that are relevant to the design process. Technical codes are legal requirements that are enforced by a governmental body to protect safety, health, and other relevant values. An example is building codes that may contain requirements with respect to sustainability (e.g., a minimal degree of isolation) and accessibility (e.g., accessible by wheelchair for public building).

> **Technical codes and standards**
> Technical codes are legal requirements that are enforced by a governmental body to protect safety, health, and other relevant values. Technical standards are usually recommendations rather than legal requirements that are written by engineering experts in standardization committees.

Technical standards are usually recommendations rather than legal requirements that are written by engineering experts in standardization committees. Codes and standards have two main functions (Hunter, 1997). The first is standardization and the promotion of compatibility. This results in, for example, the design drawings being understandable and clear for others and spare parts being compatible. The second aim of codes and standards is to guarantee a certain quality or protect public values. Though ethical considerations usually are not explicitly stated in codes and standards, ethical considerations concerning matters like safety, health, and the environment often are the foundation for the content of codes and standards.

These four sources help to distinguish relevant values in a concrete design project. They may also be instrumental in avoiding overlooking certain relevant values. However, they do not provide an answer to the question: What values *should* be included in the design of this technology? This is a *normative* question that cannot be answered by simply citing possible sources of value in a design project. Answering this question requires a normative theory or point of view. Here the theories on normative ethics from Chapter 3 may be of help.

5.5   Conceptualization and Specification of Values

Identifying relevant values is a first step in any systematic attempt to take values into account in the design of new technology. However, for these values to impact the actual design process, they need to be made more specific and operational so that they can guide design choices.

Conceptualization of values is the providing of a definition, analysis or description of a value that clarifies its meaning and often its applicability.

Conceptualization of values is the providing of a definition, analysis, or description of a value that clarifies its meaning and often its applicability (Van de Poel, 2013, p. 261). Conceptions typically provide an explanation or reasons why a value is good or positive, as well as an interpretation of how to understand the value. Conceptualization of values is largely a philosophical activity that is independent of the specific context where the value is applied, although some conceptions may be more appropriate or adequate for certain contexts than for others. For example, privacy may be conceptualized in terms of the "right to be left alone" (Warren and Brandeis, 1890), but also in terms of control over what information about oneself is shared with others (cf. Koops et al., 2017). Although both conceptions are general, in the design of a specific information system, the second one may be more relevant than the first.

Specification of values is the translation of values into more specific norms and design requirements that can guide the design process of new technology.

Specification of values refers to the translation of values into more specific norms and design requirements that can guide the design process of new technology. In specification, contextual information is added that makes it more specific what it means to strive for (or respect or meet) a certain value in a certain context. For example, in the context of designing a chemical plant, safety may be specified in terms of minimizing explosion risks, as well as in terms of providing containment so that in case of an accident it is less likely that hazardous materials leak into the environment of the plant.

We will now discuss the conceptualization and specification of values in more detail.

5.5.1 Conceptualization

Concerning conceptualization, it is often the case that different philosophical conceptions of a value are available in the literature. Take, for example, well-being. In the philosophical literature, three main theories of well-being have been articulated (Crisp, 2021):

- Hedonism, which understands human well-being in terms of pleasurable experiences.
- Desire satisfaction accounts which conceptualize well-being in terms of the fulfillment of people's desires.
- Objective list accounts which understand well-being in terms of a list of general prudential values (like, for example, accomplishment, autonomy, liberty, friendship, understanding, and enjoyment) (cf. Griffin, 1986).

Each of these conceptions suggests another approach to how to design for well-being. On a hedonistic account of well-being, design should be aimed at creating products that give users and other stakeholders pleasurable experiences. On a desire-satisfaction account, people's (actual) desires should be the starting point of design. For objective list accounts, one can, for example, aim at a design that increases human capabilities as distinguished by the philosopher Martha Nussbaum (see Chapter 3).

Some relevant values for design and their conceptualization

Safety: Safety may be conceptualized as "the absence of risk," but since technological risks can usually not be completely avoided (see Chapter 6), it might also be understood as "the reduction of risks in as far as reasonably feasible."

Health: Health may be understood as the "absence of disease," but the World Health Organization (WHO) has suggested that it might be better understood as the "state of complete physical, mental and social well-being and not merely the absence of disease or infirmity" (World Health Organization, 2006).

Security: Whereas the value of safety is usually understood as the absence of, or protection against, unintentional harm, security is usually understood in terms of absence of, or protection against, *intentional* harm (Hansson, 2009).

Privacy: Privacy may be understood as "the right to be left alone" (Warren and Brandeis, 1890), but in the information age it is often understood in informational terms. Such informational privacy may be conceptualized in terms of confidentiality and secrecy (i.e., lack of access to personal information), but also in terms of control over your personal information. In the latter case, privacy means that personal information can only be shared if you freely and knowingly consent to it (informed consent). A third understanding of informational privacy is in terms of contextual integrity (Nissenbaum, 2010). According to this conception, it depends on the context what information can be shared with whom. For example, most people consider it appropriate to share their medical information with their doctor but not with their employer.

Justice: On a very general level, justice may be understood as "giving people what they deserve." So understood, justice may (sometimes) be different from equality, as not in every situation equal treatment may be the most just. Think, for example, of positive discrimination which may sometimes be just to correct historical injustices. A distinction might be made between *distributive justice*, which concerns the just distribution of benefits and costs (risks, disadvantages); *procedural justice*, which concerns just procedures (e.g., to make a decision), and *recognition justice*, which concerns the recognition of particularly vulnerable groups, or groups that has suffered from historical injustices. For each of these, further conceptions can be found in the philosophical literature.

Environmental sustainability may be understood in terms of protection of the natural environment or the sustenance of (nonrenewable) environmental resources. An influential definition of sustainable development has been provided by the Brundlandt commission: "Sustainable development is development that meets the needs of the present without compromising the ability of future generations to meet their own needs" (WCED, 1987). This definition relates to a broader range of values than just protection of the environment. Consequently, sustainability may be conceptualized as an overarching value that is composed of values like "care for nature," intergenerational justice (i.e., justice between generations) and intragenerational justice (i.e., justice within the current generation) (Van de Poel, 2017a).

Inclusiveness refers to the value of being inclusive to people of all ages and backgrounds. In terms of the design *process*, it might be understood in terms of the need to include all relevant stakeholders and their perspectives (see also Chapter 9); in terms of designed *products* (or services, systems, etc.), it might be understood ensuring that these work for, and are equally accessible to, all people independent of age, abilities, race, gender, etc.

The fact that there are various, often competing, philosophical conceptions of many values (like well-being, privacy, sustainability, and justice) raises the question of which conception practitioners who want to design for a certain value should adopt. There is no straightforward answer to this question, but it is important to be aware that two types of considerations are relevant in answering this question in a specific case.

One type of considerations is more general philosophical considerations about the adequacy of certain conceptions. The mentioned philosophical theories of well-being have each been criticized which in response has resulted in more sophisticated accounts that try to meet some of the raised criticism, and one might have good philosophical reasons to prefer one account over the others.

In addition to such philosophical considerations, there are also more practical considerations. Not every technology will (potentially) affect the same dimensions of well-being, and depending on how a technology may or may not impact humans, there may be good reasons to focus on a specific notion of well-being in the design of a specific technology. For example, if a technology mainly affects human experiences and to a lesser extent other aspects of well-being (according to the other two accounts), it may be justified to focus on a more hedonistic notion of well-being, while in other cases such a notion may be too narrow, for example, because a technology also affects friendship or personal relations.

5.5.2   Specification

Specification involves translating values into more concrete norms and design requirements that can guide the design of new technology. A useful tool here might be the so-called values hierarchy (Van de Poel, 2013). Figure 5.4 shows the general lay-out of a values hierarchy.

> **Values hierarchy**   A tool for specifying values in the design process. It consists (at least) of three layers: value, norms and design requirements. It can be constructed bottom-up as well as top-down.

A **values hierarchy** consists of three main layers, that is, values, norms, and design requirements; each layer may, in turn, have several sublayers. A values hierarchy is held together by two relations. Top-down, where the lower-level elements in the values hierarchy are specifications of higher-level elements. Bottom-up, where the relation between the elements can be characterized as "for the sake of," that is, the lower level elements are strived for the sake of higher-level elements. For example, in Figure 5.6, the design requirement that a chicken need at least 450 cm^2 floor area is strived for the sake of enough living space for chickens, which in turn is strived for the sake of animal welfare.

A values hierarchy can be construed top-down as well as bottom-up. In the top-down construction, we start with values that are then specified in terms of norms and design requirements. In the bottom-up construction, we start, for example, with already formulated design requirements for a design task, and ask the question for the sake of what these design requirements are strived for, in order to reconstruct underlying values.

In practice, the construction of a values hierarchy will usually be an iterative process, consisting of moving bottom-up as well as top-down. For example, if values are first reconstructed bottom-up (on basis of a set of given design requirements), we can then ask the question of whether the set of design

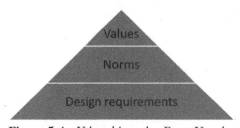

Figure 5.4   Values hierarchy. From Van de Poel (2013)/Springer Nature.

requirements is indeed the most appropriate specification of these values or should perhaps be adapted.

If you make a values hierarchy for a specific value it has to meet the following quality criterion: you need to ensure that meeting the design requirements would count as an instance of properly meeting the value you are specifying for the specific context you are considering. If this is not the case, you need to add additional norms and design requirements, or you need to revise some of the existing ones in your hierarchy. An example is given in the case of Design of housing systems for laying hens (see box).

Case Design of housing systems for laying hens

Figure 5.5 Battery cage (1990). Credit: Otwarte Klatki/Flickr.

Since the early twentieth, laying chickens have been held in so-called battery cages, a system that allows producing eggs in a very efficient way (Figure 5.5). However, the emphasis on the value of efficiency in designing battery cages came at the cost of other values, particularly *animal welfare*. In 1964, Ruth Harrison published her book *Animal Machines*, which is widely regarded as the point of departure for resistance to the battery cage. In her book Harrison attacked the drive for efficiency in animal husbandry on the modern farm. She showed that people have a quite unrealistically romantic picture of animal life on the farm. In fact, as she argued, animals on the modern farm are reduced to production machines. This publication and public protests have led to more attention to the value of animal welfare in the design of chicken husbandry systems.

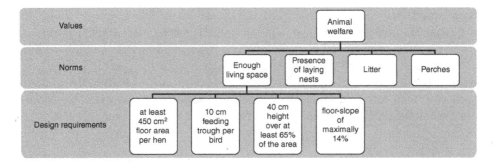

Figure 5.6 Specification of animal welfare following Directive 88/116/EEC (From Van de Poel (2013)/Springer Nature).

In order to design chicken husbandry systems for animal welfare, this value needed to be conceptualized. The relevant conceptualization of animal welfare that has become dominant over time, at least for chicken husbandry systems, is in terms of so-called ethological needs. Ethology is a branch of biology that studies the behavior of animals in their natural environment. This "natural" behavior can be used as a kind of reference point with respect to which one can discern "abnormality" in the behavior of chickens in battery cages. Deviant or absent behavior was then interpreted as possible failure of the animal to adapt itself to the new environment. So, ethology as a science provided a normative standard by which to judge the suffering of animals.

The UK's Brambell committee, installed six weeks after the appearance of Ruth Harrison's book, was probably the first to advance several criteria for animal welfare in housing systems based on the concept of ethological needs: "An animal should at least have sufficient freedom of movement to be able, without difficulty, to turn around, groom itself, get up, lie down and stretch its limbs."[4] These and other criteria based on ethology have in the course of time also resulted in specification of the value of animal welfare in terms of design requirements, some of which have now become legally binding.

We can reconstruct how the value of animal welfare has been translated into design requirements by making a values hierarchy. An example is given in Figure 5.6, which is based on EU Council Directive 88/116/EEC that at the time posed legally binding minimal requirements for new battery cages coming in use after January 1, 1988. This figure also illustrates how a values hierarchy can be used to judge the adequacy of a specification of a value. The question to be asked is whether chicken husbandry systems that meet the low-level requirements indeed count as meeting the value of animal welfare. Many people would probably answer this question with a "no"; also because, as can be seen, some of the norms proposed by ethologists were not translated into design requirements at all (for economic reasons). Indeed in the course of time, stricter legal requirements have been introduced, aiming at a de-facto phase-out of the traditional battery cage in the EU.

Oftentimes, different adequate specifications of a value for a given context will be possible. These specifications may all be acceptable as long as they meet the mentioned quality criterion: each of them should count as an instance of properly addressing the relevant value in the given context. As we will see in Subsection 5.6.5, the fact that there may be several acceptable specifications may sometimes help to deal with value conflicts in design.

The criterion that the design requirements should count as an instance of properly addressing the value in the given context may sometimes be hard to apply and people may disagree whether it is fulfilled or not in a specific case. The values hierarchy as a tool does not solve that disagreement, but it makes it possible to have focused discussions on how to specify values. So instead of as a tool for the designers to specify values, the values hierarchy can also be used together with the relevant direct and indirect stakeholders to have a discussion about how to specify values. It can then be used to map how different stakeholders would specify values differently, and at a later stage it might be tried to come to an agreement about these differences.

5.6 Value Conflicts

In Design for Values, the aim is to respect all relevant values. However, sometimes it might be impossible to meet all values simultaneously. In such cases, we will speak of a value conflict.

Value conflicts can be morally problematic because they may well result in the situation where the designers cannot do justice to all relevant moral values simultaneously. In such cases, a value conflict amounts to a moral problem (see Subsection 4.3.1). An example of

> **Value conflict** A value conflict arises if in a situation it is not possible to (fully) respect all relevant values simultaneously.

a value conflict in design is given in the case study about alternative coolants for CFC 12 (see box). A crucial question in the latter is: how should environmental concerns regarding the design of new coolants for refrigerators be weighed against safety concerns? Below, we shall discuss five ways in which this evaluation can take place: cost-benefit analysis; multiple criteria analysis; the determination of thresholds for value; respecification; and the search for new technical solutions (innovation). For each method, we shall present the main advantages and disadvantages.

Case   Household Refrigerators – An Alternative for CFC 12

In the 1930s chemists at General Motors developed the so-called chlorofluorocarbons (CFCs) – hydrocarbons in which some of the hydrogen (H) atoms are replaced by chlorine (Cl) or fluorine (Fl) atoms. Due to their thermodynamic properties, CFC turned out to be excellent coolants. Moreover, they are non-toxic and non-flammable. For household refrigerators, the most commonly used CFC coolant became CFC 12. In the 1980s it was discovered that CFCs are the main contributors to the hole in the ozone layer. In 1987, the Montreal Protocol called for a worldwide reduction in the production and use of CFCs. Subsequently, in the 1990s, CFCs were forbidden in many countries.

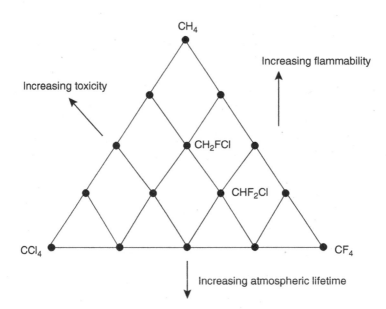

Figure 5.7 Properties of refrigerants. From McLinden and Didion (1987). Copyright ASHRAE: ASHRAE, 1791 Tullie Circle, N.E., Atlanta, GA 30329, ©1987, ASHRAE (www.ashrae.org). Used with permission from ASHRAE.

As a consequence of the ban on CFCs in the 1990s, an alternative had to be found to replace CFC 12 as a refrigerant in household refrigerators. Three moral values played an explicit role in the formulation of design requirements for alternative coolants: safety, health, and environmental sustainability. In the design process, safety was mainly interpreted as non-flammability, and health as non-toxicity. Environmental sustainability was equated with low ODP (Ozone Depletion Potential) and a low GWP (Global Warming Potential). Both ODP and GWP mainly depend on the atmospheric lifetime of refrigerants. In the design process, a conflict arose between those three values. This value conflict can be illustrated with the help of Figure 5.7.

Figure 5.7 is a graphic representation of CFCs based on a particular hydrocarbon. At the top, there is methane or ethane, or another hydrocarbon. If one moves to the bottom, the hydrogen atoms are replaced either by chlorine atoms (if one goes to the left) or fluorine atoms (if one goes to the right). In this way, all the CFCs based on a particular hydrocarbon are represented. The figure shows how the properties of flammability (safety), toxicity (health), and environmental effects depend on the exact composition of a CFC. As can be seen, minimizing the atmospheric lifetime of refrigerants means maximizing the number of hydrogen atoms, all of which increases flammability. This means that there is a fundamental trade-off between flammability and environmental effects, or between the values of safety and sustainability.

The main alternatives to CFC 12 (CCl_2F_2) that were considered are HFC 134a ($C_2H_2F_4$) and hydrocarbons like isobutane (HC 600a or C_4H_{10}). Table 5.1 shows the ODP and GWP of these substances. Initially, the industry preferred the alternative HFC 134a. Although this substance has a larger GWP, and thus contributes more to the greenhouse effect, than hydrocarbons like HC 600a, it is inflammable while hydrocarbons are flammable. HFC 134a was also attractive for the chemical

Table 5.1 ODP and GWP of some coolants

Coolant	ODP (compared to CFC 12)	GWP (compared to CO_2)
CFC 12	1	8 500
HFC 134a	0	1 300
HC 600a (isobutane)	0	3

industry because it could be patented, in contrast to the existing hydrocarbons, and therefore would be much more profitable to produce.

In parts of Europe, the tide has turned against HFC 134a since Greenpeace in the early 1990s found a refrigerator firm from former East Germany, Foron, willing to develop a refrigerator with hydrocarbons as coolant. When Greenpeace and Foron in August 1992 succeeded in collecting more than 50 000 orders for Foron's so-called *Greenfreeze*, within months the main German refrigerator firms switched to the hydrocarbon isobutane as coolant. In December 1992, the *Greenfreeze* acquired safety approval from the German certification authorities. Although there has been some discussion about the energy consumption of refrigerators with hydrocarbon as coolants, current studies seem to suggest that refrigerators with isobutane as coolant are at least as energy efficient as those using HFC 134a.

Source: Adapted from Van de Poel (2001).

5.6.1 Cost-benefit analysis

Cost-benefit analysis is a general method that is often used in engineering. What is typical of cost-benefit analysis is that all values that are relevant for the choice between different options are eventually expressed in one common unit, usually a monetary unit, like dollars or euros. There are various types and variants of cost-benefit analysis (see, e.g., Mishan, 1975). If we consider the costs and benefits for society as a whole, this is usually referred to as a social cost-benefit analysis. Cost-benefit analysis can also be limited to the costs and benefits of a company that is developing a product and looking to market it.

> **Cost-benefit analysis** A method for comparing alternatives in which all the relevant advantages (benefits) and disadvantages (costs) of the options are expressed in monetary units and the overall monetary cost or benefit of each alternative is calculated.
>
> **Discount rate** The rate that is used in cost-benefit analysis to discount future benefits (or costs). This is done because 1 dollar now is worth more than 1 dollar in 10 years time.

Cost-benefit analysis may be an appropriate tool if one wants to optimize the expected economic value of a design. Still, even in such cases, some additional value-laden assumptions and choices need to be made. One issue is how to discount future benefits from current costs (or vice versa). One dollar now is worth more than one dollar in 20 years' time, not only because of inflation but also because a dollar now could be invested and would then yield a certain interest rate. To correct this, a **discount rate** is chosen in cost-benefit analysis. The choice of discount rate may have a major impact on the outcome of the analysis. Another issue is that one might employ different choice criteria once the cost-benefit analysis has been carried out. Sometimes all of the options in which the benefits are larger than the costs are considered to be acceptable. However, one can also choose the option in which the net benefits are highest, or the option in which the net benefits are highest as a percentage of the total costs.

Contingent validation An approach to express values like safety or sustainability in monetary units by asking people how much they are willing to pay for a certain level of safety or sustainability (for example, the preservation of a piece of beautiful nature).

Cost-benefit analysis is more controversial if non-economic values are also relevant. Still, the use of monetary units does not mean that only economic values can be taken into account in cost-benefit analysis. In fact, approaches like **contingent validation** have been developed to express values like safety or sustainability in monetary units. Contingent validation proceeds by asking people how much they are willing to pay for a certain level of safety or, for example, the preservation of a piece of beautiful nature. In this way,

a monetary price for certain safety levels or a piece of nature is determined. Approaches like contingent validation have serious limitations and are often criticized, because they are based on the assumption that all values can be expressed in monetary terms. However, it would be premature to conclude that cost-benefit analysis necessarily neglects non-monetary or non-economic values. When employing cost-benefit analysis, different ethical criteria might be used to choose between the options (Kneese et al., 1983; Shrader-Frechette, 1985). One might, for example, choose an option with which nobody is worse off. By selecting a specific choice criterion, ethical considerations beyond considering which options bring the largest net benefits might be taken into account.

In terms of values, cost-benefit analysis might be understood to be the maximization of one overarching or super value. Such a value could be an economic value like company profits, or the value of the product to users but it could also be a moral value like human happiness. If the latter is chosen, cost-benefit analysis is related to the ethical theory of utilitarianism. With Bentham's classical variant of utilitarianism (see Subsection 3.7.1), for example, the assumption is that all relevant moral values can eventually be expressed in terms of the moral value of human happiness. One might question this assumption, however. One issue is that it is often difficult to indicate to what extent values like safety, health, sustainability, and aesthetics contribute to the value of human happiness, and to furthermore express this in monetary terms. A second, more fundamental issue is that such an approach treats all these values as instrumental values, whose worth should ultimately be measured based on their contribution to the intrinsic value of human welfare. One might wonder whether values like human health, sustainability, and aesthetics do indeed have only instrumental value or are intrinsically valuable.

It might, however, be possible to employ cost-benefit analysis more instrumentally, that is, as a mere technical way to compare alternatives in the light of heterogeneous considerations or values. Although expressing everything in terms of money presupposes a common value, it could be maintained that this value is only a means of comparison, rather than a substantial value like human happiness. Nevertheless, this still presupposes that various criteria can be measured or expressed on a common scale (Hansson, 2007b). According to some ethicists, the existence of such a common scale is problematic because some values are incommensurable (see box). Moreover, different people think differently about the relative importance of values like safety, welfare, and sustainability.

Value Incommensurability and Trade-Offs

Two or more values are **incommensurable** if they cannot be expressed or measured on a common scale or in terms of a common value measure. Incommensurable values cannot be traded off directly. It has been suggested that incommensurability and a resistance to certain trade-offs are constitutive of certain values or goods (Raz, 1986). Consider, for

> **Incommensurability** Two (or more) values are incommensurable if they cannot be expressed or measured on a common scale or in terms of a common value measure.

example, the following trade-off: for how much money are you willing to betray your friend? It may well be argued that accepting a trade-off between friendship and financial gain undermines the value of friendship. On this basis, it is constitutive of the value of friendship to reject the trade-off between friendship and financial gain. It has also been suggested that values may resist trade-offs because they are "protected" or "sacred" (Baron and Spranca, 1997; Tetlock, 2003). This seems especially true of moral values and values that regulate the relations between, and the identities of, people. Trade-offs between protected values create an irreducible loss because a gain in one value may not always compensate or cancel out a loss in the other. The loss of a good friend cannot be compensated by having a better career or more money.

Some philosophers have denied the existence of value incommensurability. They believe that all values can ultimately be expressed in terms of one overarching or super value. Utilitarianism often attributes such a role to the value of human happiness, but a similar role may be played by the value of "good will" in Kantianism. The notion that there is ultimately only one value that is the source of all other values is known as value monism. Value monists do not necessarily deny the existence of more than one value but they believe that value conflicts can essentially be solved by having recourse to a super value.

5.6.2 Multiple criteria analysis

A second method to weigh different values is **multiple criteria analysis**. Similar to cost-benefit analysis there are various types and variants. We shall restrict ourselves to the main outlines. Multiple criteria analysis is based on a comparison of different options with each other with respect to several criteria or values. Usually, the relative importance of the criteria or values is determined first, because usually not all values are equally important. Next, each option is weighed for all the values and a numeric value is

> **Multiple criteria analysis** A method for comparing alternatives in which various decision criteria are distinguished on basis of which the alternatives are scored. On basis of the score of each of the alternatives on the individual criteria, usually a total score is calculated for each alternative.

awarded, on a scale from 1 to 5 for example. Finally, the overall value for each option is calculated according to the following formula: $w_j = \Sigma g_i \times v_{ij}$ over the set I of criteria, where w_j is the value of the jth option, g_i is the relative weight of the ith criterion, and v_{ij} is the score of the jth option on the ith criterion. The option with the highest value is then selected.

Multiple criteria analysis does not demand that all criteria are translated into one over-arching criterion or value, such as human happiness or welfare. However, multiple criteria analysis does demand that we determine the relative importance of the different design criteria in some way or another. Like cost-benefit analysis, multiple criteria analysis thus presupposes the commensurability of values. Compared to cost-benefit analysis, the comparison between options in multiple criteria analysis is vaguer because no explicit attempt is made to translate all criteria to a common unit (like money), which may result in flawed decision-making because the result depends on the scale chosen, as can be shown with the help of an example.

Let us have a look at the example of the coolants in refrigerators. Say that we assess the options HFC 134a and isobutane for the criteria of safety (flammability) and environmental impact (ODP and GWP). Moreover, we feel both criteria are equally important. One possible result from this multiple criteria analysis could be as in Table 5.2, in which higher numbers mean that the option scores better on that value, that is, it is safer or more environmentally friendly. Here, the score for the criteria options ran from 1 to 3. Usually, this scale is understood as an ordinal scale, in which only the order of the items is relevant (see also the box for explanation). So the only relevant information that Table 5.2 contains is that HFC134a is safer than isobutane (because 3 is larger than 2) and that isobutane is more environmentally friendly than HFC134a (because 3 is larger than 1). If we convert this to an assessment on an interval scale from 1 to 3, we get Table 5.3, which suggests that isobutane is the best option. However, we can also convert this to an interval scale from 1 to 5, with Table 5.4 as a result. Note that Table 5.3 and Table 5.4 contain the same ordinal

Table 5.2 Ordinal scale

	HFC134a	Isobutane
Safety	3	2
Environment	1	3

Table 5.3 Interval scale ranging from 1 to 3

	HFC134a	Isobutane
Safety	3	2
Environment	1	3
Total score	4	5

Table 5.4 Interval scale ranging from 1 to 5

	HFC134a	Isobutane
Safety	5	2
Environment	2	4
Total score	7	6

information as Table 5.2: HFC 134a is safer than isobutane and isobutane is more environmentally friendly than HFC134a. In Table 5.4, however, the total score of HFC 134a is larger than that of isobutane.

This example shows that by changing solely the choice of scale, we can change the chosen option. So our choice depends on the chosen scale instead of on the inherent properties of the different options, which is undesirable. Avoiding this would require measuring both criteria (safety and the environment) on a ratio scale with the same unit (see box). It seems, however, unlikely that we can measure safety and environmental effects on the same scale due to value incommensurability. The box explains the background of this flaw in more detail.

Multiple Criteria Analysis and Measurement Scales

To understand why the results of multiple criteria analysis sometimes depend on the measurement scale chosen, we first need to distinguish different measurements scales:

- An **ordinal scale** is a scale in which only the order of the items of the scale has meaning. An example is an ordering of the tastefulness of meals.
- An **interval scale** is a scale in which in addition to the order of items also the distance between the items has meaning. An example is the temperature scale Celsius (or Fahrenheit). It is meaningful to say that the difference between 10°C and 20°C is the same as between 20°C and 30°C.
- A **ratio scale** is a scale in which also the ratio between items on a scale has meaning. An example is distance measured in meters (or feet). It makes sense to say that 2 m is twice as long as 1

> **Ordinal scale** A measurement scale in which only the order of the items of the scale has meaning.
>
> **Interval scale** A measurement scale in which in addition to the order of items also the distance between the items has meaning.
>
> **Ratio scale** A measurement scale in which the ratio between items on a scale has meaning.

m, whereas it does not make sense to say that 20°C is twice as hot as 10°C. The reason for this is that the Celsius scale lacks an absolute point of zero. It would be different if we measure temperature in Kelvin because 0 Kelvin is defined as the lowest possible temperature. It is theoretically impossible to have a temperature below 0 Kelvin (as it is impossible to have a distance below 0 meter).

Each of these scales allows for different sets of mathematical operations:

- On an ordinal scale, arithmetical operations like addition, subtraction, multiplication, and division are not allowed. Options can only be compared in terms of better and worse.
- On an interval scale, the arithmetical difference between two options has meaning, so that addition and subtraction are allowed, while multiplication and division are not.
- On a ratio scale, all arithmetical operations (addition, subtraction, multiplication, and division) are allowed.

In the variety of multiple criteria analysis, we consider here the overall worth of an option is calculated with the formula $w_j = \Sigma g_i \times v_{ij}$ over I, where w_j is the value of the jth option, g_i is the relative weight of the ith criterion, and v_{ij} is the score of the jth option on the ith criterion. This formula is only meaningful if v_{ij} is measured on a ratio scale. Multiple criteria analysis, therefore, places great demands on how precisely we can measure the value of options on individual criteria. In many cases, it is not that hard to determine which option scores better for which criterion, that is, to order the options on an ordinal scale. However, the method does require that we can express the value of the options for each criterion on a ratio scale. It should also be noted that this ratio scale should have the same unit for all criteria; otherwise, we cannot meaningfully add up the scores on individual criteria. (One cannot add up meters and degrees Celsius for example.)

Take, for example, the design of an elevator. One relevant design criterion is the travelling time with which we can move from one floor to the next. It is not hard to rank various potential elevator designs according to this criterion. However, can we unequivocally translate this ranking into a relative valuation on a ratio scale? We can obviously measure traveling time in seconds which is a ratio scale. This is, however, not enough. Suppose that one of the other criteria is maintenance costs measured in Euros. Obviously, we cannot add up seconds and Euros; so we either need to convert seconds to Euros or the other way around or to convert both to a third scale like "goodness." But can we measure the travelling time of the elevator in terms of "goodness" on a ratio scale? For example, if traveling time is reduced from 30 seconds to 20 seconds, is that as good as saving 10 seconds by reducing traveling time from 20 seconds to 10 seconds, and is 10 seconds twice as good as 20 seconds? In most cases, it is not.

5.6.3   Thresholds

Threshold   The minimal level of a value that an alternative has to meet in order to be acceptable with respect to that value.

A third way to cope with conflicting values is to set a **threshold** for each value. For each separate value (e.g., safety, health, costs, and sustainability) a threshold is determined for what is acceptable. Setting thresholds not only occurs in the design process, but also in legislation (standardization) and in technical codes and standards. A minimal level of safety is often defined this way for example.

An example of setting thresholds is to be found in the case of the design of new refrigerants. In this case, the engineers McLinden and Didion (1987) from the *National Bureau of Standards* in the USA drew Figure 5.8 with respect to the properties of CFCs. According to McLinden and Didion, the blank area in the triangle contains refrigerants that are acceptable in terms of health (toxicity), safety (flammability), and environmental effects (atmospheric lifetime). Note that by drawing the blank area in the figure, McLinden and Didion – implicitly – establish threshold values for health (toxicity), safety (flammability), and the environment.

An advantage of setting thresholds is that the acceptable threshold is considered for each criterion without making direct trade-offs between different design requirements. This may, for example, be helpful to guarantee a minimal level of, for example, safety in the design process. However, the question is whether it is possible or desirable to determine thresholds in complete isolation from other concerns. If, for example, the government draws up safety standards, this usually takes place in the light of the costs involved to

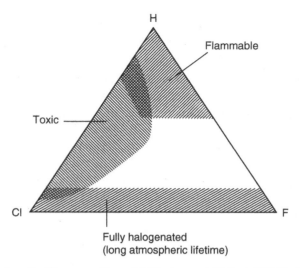

Figure 5.8 Properties of refrigerants. From McLinden and Didion (1987). Copyright ASHRAE: ASHRAE, 1791 Tullie Circle, N.E., Atlanta, GA 30329, ©1987, ASHRAE (www.ashrae.org). Used with permission from ASHRAE.

achieve that level of safety and what else we have to sacrifice in terms of other moral values such as welfare, a good life, or sustainability. Setting a threshold per value in the design process also occurs with reference to other values and what is technically feasible. Taking into account other values in setting thresholds may make sense, but there is a danger that thresholds are selected that make a design possible under all circumstances. The question is whether this is ethically acceptable. It would for example not always be desirable to adapt your assessment of the desired degree of sustainability to what is achievable.

Another possible disadvantage of setting thresholds is that you limit yourself as an engineer to realizing these values, while more can be achieved with a given design in terms of environmental impact or sustainability for example. This was also the case with the design of coolants for refrigerators. The alternative HFC 134a did meet the thresholds set by the engineers McLinden and Didion, but according to others the chemical's environmental impact was too large.

5.6.4 Respecification: reasoning about values

The approaches to dealing with value conflicts that have already been discussed are all calculative approaches. They strive to operationalize and measure the value of a design in one way or another. Of these approaches, the setting of thresholds does not aim at calculating the overall value of an option, but it does presuppose that the value of an option can be measured for each of the individual design criteria. We will now look at an

> **Respecification** is the revision of the design requirements in the light of the values of which they are a specification. Respecification may sometimes be used to overcome conflict between design requirements.

approach that does not share this calculative assumption, but that emphasizes judgment and reasoning about values. This approach, which is called **respecification**, aims at respecifying

---

Case   Automatic Seatbelts

A car with automatic seatbelts will not start if the automatic seatbelts are not put on. This forces the user to wear the automatic seatbelt. One could say that the value of driver safety is built into the technology of automatic seatbelts. This comes at a cost, however: the user has less freedom. Interestingly, there are various seatbelt designs which exist that would imply that there are different trade-offs in terms of safety and user freedom. The traditional seatbelt, for example, does not enforce its use, but there are various systems that give a warning signal if the seatbelt is not being worn. This does not enforce seatbelt use, but it does encourage drivers to wear their seatbelt.

---

the values that underlie the conflicting design criteria. The advantage of this approach is that it might solve a value conflict while still doing justice to the conflicting values and without the need to make the values commensurable or to define thresholds for them.

The first thing to do when one wants to exercise judgment in cases of value conflicts is to identify what values are at stake and to gain a better understanding of these. What do these values imply and why are these values important? Take the value of freedom in the case of safety belts (see box). Freedom can be construed as the absence of any constraints on the driver; it then basically means that people should be able to do whatever they want. Freedom can, however, also be valued as a necessary precondition for making one's own considered choices; so conceived freedom carries with it a certain responsibility. In this respect, it may be argued that a safety belt that reminds drivers that they have forgotten to use it does not actually impede the freedom of the driver but rather helps them to make responsible choices. It might perhaps even be argued that automatic safety belts can be consistent with this notion of freedom, provided that the driver has freely chosen to use such a system or endorses the legal obligation for such a system, which is not unlikely if freedom is not just the liberty to do what one wants but rather a precondition for autonomous responsible behavior. One may thus think of different conceptualizations of the values at stake and these different conceptualizations may lead to different possible solutions to the value conflict.

Another approach might be to respecify the values in order to solve the value conflict. This may be possible because usually more than one specification of a value will be tenable. Value conflicts in design are usually conflicts between specifications of the values at stake because abstract values as such are too general and abstract to guide design or to choose between options. So if there is room for different possible specifications of the values at stake, it might be possible to choose that set of the specifications of the various values at stake that is not conflicting. Sometimes, it will only become apparent during the design process, when the different options have been developed and are compared that certain specifications of the values at stake are conflicting. In such cases, it may sometimes be possible to respecify the values at play so as to avoid the value conflict.

An interesting example of respecification took place in the refrigerant example.[5] In the first instance, the industry preferred HFC134a as an alternative to CFC 12, basically following the satisficing reasoning as explained in the previous section. This led to the specification shown in Figure 5.9. However, environmental groups were against this specification, and the consequent choice for HFC 134a, as they viewed the threshold for environmental sustainability (at least one H atom) too lenient, especially because it resulted in much higher GWPs (Global Warming Potentials) than if a flammable coolant was chosen. As we saw in

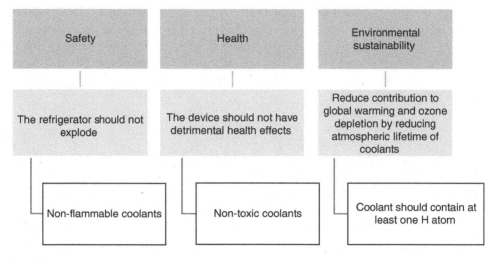

Figure 5.9 The initial specification of safety, health, and environmental sustainability for alternative coolants. This specification is similar to Figure 5.8.

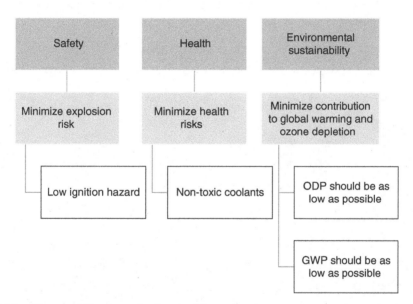

Figure 5.10 The respecification of safety, health, and environment suggesting isobutane as the best coolant rather than HFC 134a.

the case description, at some point Greenpeace succeeded in convincing a former East German refrigerator producer of using a flammable coolant in its new design. The refrigerator was also able to acquire the safety approval of the German certification institute TUV. Following the success of this refrigerator, German, and later other European, refrigerator producers also switched to flammable coolants like propane and isobutane. Such coolants were seen as acceptability despite their flammability because a new specification of safety was developed. Where safety was first specified as non-flammability of coolants, it now came to be specified as a low explosion risk of the whole refrigerator. It turned out to be possible to achieve a low explosion risk even with flammable coolants. This respecification is shown in Figure 5.10. It led to the choice for isobutane as coolant instead of HFC 134a.

Although it might be possible to solve a value conflict in design through respecification, this will not always be possible. Even in cases in which it is possible, it may not always be acceptable. It may especially be unacceptable if respecification leads to a serious weakening of one of the values compared to the original specification (Hansson, 1998). Still, solving a value conflict through respecification does not necessarily or always imply a weakening of one of the values (Van de Poel, 2017b).

5.6.5   Innovation

The previous approach treats the occurrence of value conflict merely as a philosophical problem to be solved by philosophical analysis and argument. However, in engineering design value conflicts may also be solved by technical means. That is to say, in engineering it might be possible to develop new, not yet existing, options that solve or at least ease the value conflict. In a sense, solving value conflicts by means of new technologies is what lies at the heart of engineering design and technological innovation. Engineering design is able to play this part because most values do not conflict as such, but only in the light of certain technical possibilities and engineering design may be able to change these possibilities. An interesting example is the design of a storm surge barrier in the Eastern Scheldt estuary in the Netherlands.

Case   The Design of the Storm Surge Barrier in the Eastern Scheldt

Figure 5.11   Eastern Scheldt storm surge barrier. Photo: Daniel Täger/Adobe Stock.

After a huge flood disaster in 1953, in which a large number of dikes in the province of Zeeland, the Netherlands, gave way and more than 1800 people were killed, the Delta Plan was drawn up. Part of this Delta Plan was to close off the Eastern Scheldt. From the end of the 1960s, however, there was growing societal opposition to closing off the Eastern Scheldt. Environmentalists, who feared the loss of an ecologically valuable area because of the desalination of the Eastern Scheldt and the lack of tides, started to resist its closure. Fishermen also were opposed to its closure because of the negative consequences for the fishing industry. As an alternative, they suggested raising the dikes around the Eastern Scheldt to sufficiently guarantee the safety of the area.

In June 1972, a group of students launched an alternative plan for the closure of the Eastern Scheldt. It was a plan that had been worked out as a study assignment by students of the School of Civil Engineering and the School of Architecture of the Technical University of Delft and the School of Landscape Architecture of the Agricultural University of Wageningen. The aspects the students focused on were safety and ecological care. On the basis of these considerations, they proposed a storm surge barrier, that is, a barrier that would normally be open and allow water to pass through, but that could be closed if a flood threatened the hinterland. The flood barrier was a creative compromise to balance the two moral values, safety and ecological care, that were at stake.

At first, the Rijkswaterstaat, the governmental body responsible for waterways in the Netherlands, discarded the idea because it was not considered feasible technically. However, pressure from political developments – parliament too started to resist closing off the Eastern Scheldt – made the Rijkswaterstaat take the option more seriously and after some time it was decided to build a storm surge barrier (see Figure 5.11). Though the storm surge barrier turned out to be much more expensive than the original solution – and also exceeded the original budget – many still consider the design to be a creative and acceptable compromise between safety and ecological values.

Source: Adapted from Van de Poel (1998).

5.6.6 A comparison of the different methods

We discussed five methods for dealing with value conflicts. We saw that each method has its pros and cons: these are summarized in Table 5.5. What is striking is that none of the methods reaches a definite solution for value conflicts. Both cost-benefit analysis and multiple criteria analysis suppose the commensurability of values, which might be problematic. Respecification through reasoning might help to solve some value conflicts, but probably not in all cases. Similarly, technical innovation often is useful, but in practice it usually does not lead to a definite solution to the problem. Thresholds have the disadvantage that sometimes less is achieved in a given situation than could be the case. Moreover, you still have to weigh the various criteria and values while drawing up the thresholds.

We should not conclude from the above that the choice between alternatives that score differently for various values is random. The methods are useful. However, which method is best will depend on the situation. The discussion of pros and cons can help you to make a choice based on proper reasons. Moreover, it is good to be aware of the shortcomings of the various methods, so that you can try to limit these shortcomings in a concrete situation.

Table 5.5 Overview of methods for making trade-offs in design

Method	How are the values weighted?	Main advantages	Main disadvantages
Cost-benefit analysis	All values are expressed in monetary terms	Options are made comparable	• Values are treated as commensurable • May be difficult to adequately express all relevant (moral) values in monetary terms
Multiple criteria analysis	Trade-offs between the different values	Options are made comparable	• Values are treated as commensurable • Result depends on measurement scale
Thresholds	A threshold is set for each value	The selected alternatives meet the thresholds No direct trade-off between the values	• Can thresholds be determined independently from each other? • Less achieved than possible
Respecification	Not directly	Might solve value conflict by reason and judgment	• Not all value conflicts can be solved in this way
Innovation	Not applicable	Can lead to alternatives that are clearly better than all of the present alternatives	• Does not solve the choice problem in many cases

5.7   Prototyping and Monitoring

The design process often does not end with choosing a design alternative and detailing it, but may also involve the building of prototypes and testing them out with users and stakeholders, for example, in a pilot project or through field tests. Moreover, it is often advisable to keep monitoring the effects of the technology used, once it has come into production, and is used on a large(r) scale.

Activities like prototyping, testing, and monitoring are important for answering three types of questions:

1 Does the designed artifact indeed embed the intended values?
2 Does the designed artifact indeed realize the intended values in actual use?
3 Are there any additional values that should be taken into account apart from the (explicitly) intended values? Or are there any other adjustments to the design needed on basis of the experience of using the product?

The first question might be answered by checking whether the design requirements, that specify the relevant values, are indeed achieved by your design. This may be done through calculations, **simulations,** or by **testing**.

In some engineering areas, calculations, simulations, or tests are required for the **certification** of products. Certification is the process in which it is judged whether a certain technology meets the applicable technical norms and standards and is, for example, safe enough. Certification is for some products legally required before they may enter the market.

However, even if your design meets applicable technical norms and standards, and the design requirements, you might still wonder whether you have adequately conceptualized and specified the relevant values. You might want to gather additional information about that by testing a **prototype** of your design with relevant users and stakeholders and collecting their experiences. This may lead to new information that may sometimes be a reason to revise your design.

Simulation refers to trying out your design by building a (computer) model of it in order to simulate its behavior under different circumstances.

Testing refers to the execution of a technology in circumstances set and controlled by the experimenter, and in which data are gathered systematically about how the technology functions in practice.

Certification The process in which it is judged whether a certain technology meets the applicable technical codes and standards.

Prototype A sample or model of the designed product that is used for testing it out and so gathering new information and insights

Another issue is that designed products may be used in other ways that are intended or foreseen by their designers. As we have seen before, this may mean that the realized values are different from the values embedded in the product (and the value intended by the designers). To deal with this, designers may try to anticipate how their products may change the perception and behavior of users, for example by taking into account the mediating role of technological artifacts in human perception and action. They may even deliberately try to change the behaviors of users through their design to achieve certain values; see the box on technological mediation and the Cubicle Warrior case.

Technological mediation and the moralization of technology

When technologies are used, they also shape the actions and experiences of their users. This phenomenon is called **technological mediation** (Ihde, 1990; Latour, 1992; Verbeek, 2005). Technologies are not neutral "intermediaries," that simply connect users with their environment; they are impactful mediators, that help to shape *how* people use technologies, how they experience the world and what they do.

Technological mediation The phenomenon that technologies shape the actions and perceptions of their users.

Technologies may mediate both human perception, that is, how human perceive the world around them, as well as human action, that is, how humans acts in, and on the world.

There are two ways to take mediation analyses into ethics and design. First of all, such analyses can be used to anticipate, and morally assess, the mediating roles of technologies in human practices and experiences. Second, they can be used to design technologies that *deliberately* aim to shape the (moral) actions and decisions of their users. The latter has been called the **"moralization of technology"** (Achterhuis, 1995).

Moralization of technology The deliberate development of technologies in order to shape moral action and decision-making.

Moralization of technology can be done in a variety of ways. Sometimes, designers may try to force users to act in a certain way. An example is the lock-out switch on a chain saw, which tries to force users to use two hands in using it, so that they cannot inadvertently saw off their own hand, so contributing to the value of safety (see Figure 5.12).

In other cases, designers may try to nudge users into certain behavior. Nudges do not ban certain options or possibilities for behavior, but make use of the fact that how choices are presented will typically affect what people choose (Thaler and Sunstein, 2009). An example is putting healthy food in front in the canteen, but without banning junk food. In still other cases, technologies may give users feedback on their behavior or show consequences of their behavior. Think of apps that show you the daily distance you walked or give insights in the carbon footprint of different transportation options.

Moralization of technology may, at least sometimes, diminish the freedom of users. As we have seen in the earlier example of safety belts, there may be debate about how to understand and conceptualize freedom. One particular worry is that moralizing technologies may undermine the moral autonomy of users, which is their ability to make their own reasoned moral choices. Moralization of technology may also raise questions about paternalism (see Chapter 1), as it seems to suggest that designers know better how users should behave than these users themselves. On the other hand, it might be argued that technologies mediate the behavior of the users anyway and that it may be better to make deliberate choices about how they should mediate user perception and behavior rather than leaving it to chance.

Figure 5.12   Chain saw: Unless the lock-out switch above the rear handle is also pressed, the throttle cannot move from the idle position, and the chain saw cannot be used. Credit: Unknown Author/Wikimedia Commons/CC SA 1.0.

Case   Cubicle Warrior

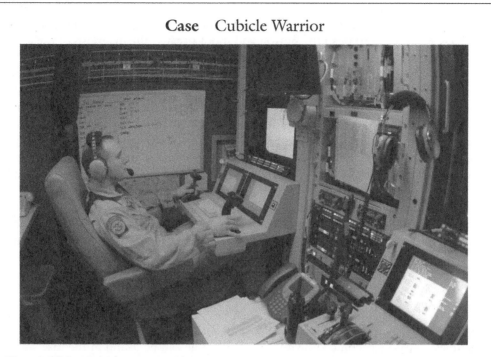

Figure 5.13 Cubicle warrior. Photo: Senior Master Sgt. David H. Lipp/North Dakota National Guard/U.S. Air Force.

The deployment of military robots is growing rapidly. Presently, more than 17,000 military robots are active in the US military. Most of these robots are unarmed, and are mainly used for clearing improvised explosive devices and reconnaissance; however, over the last years the deployment of armed military robots is in the increase. One of the most widely used military robots is the unmanned combat aerial vehicle *Predator*. This unmanned airplane which can remain airborne for 24 hours, and can fire Hellfire missiles. The *Predators* are flown by pilots located at the military base in the Nevada desert, thousands of miles away from the battlefield. They connect the cubicle warriors (human operators) – who remotely control these military robots behind visual interfaces – with the war zone; they are the eyes of the tele-soldier. These robots can precisely determine a certain target and send the GPS-coordinates and camera images back to the operator. Based on the information projected on their computer screen the cubicle warrior has to decide, for example whether or not to launch a missile. Their decision is *mediated* by a computer-aided diagnosis of the war situation. Future military robots will have built into their design ethical constraints, the so-called ethical governor which will suppress unethical lethal behavior. Although these ethical governors are not very sophisticated yet, current research shows some major progress in this development. For example, research has been done – sponsored by the US Army – to create a mathematical decision mechanism consisting of constraints represented as prohibitions and obligations derived directly from the laws of war (Arkin, 2007). Moreover, a future goal is that military robots can refuse orders of a cubicle warrior which according to the ethical governor are illegal or unethical. For example, a military robot might advise a cubicle warrior not to push the button and shoot because the diagnosis of the camera images tells the operator they are about to attack non-combatants, that is, the software of the military robot that diagnoses the war situation provides the cubicle warrior with ethical advice. An ethical governor helps to shape moral decision-making. In other words, the task to see to it that no Rules of Engagement[6] are violated could be delegated to a military robot. A consequence is that humans then simply show a type of behavior that was desired by the designers of the technology instead of explicitly choosing to act this way. This will also be the case with ethical governors, since an ethical governor may form a "moral buffer" between cubicle warriors and their actions, allowing them to tell themselves that the military robot has taken the decision (Cumming, 2006). The consequence of the moralization of military robots is that the decision of a cubicle warrior is not the result of moral reflection, but is mainly determined or even enforced by a military robot (Royakkers and Van Est, 2010).

However, you may not be able to anticipate all social consequences of your design, and sometimes they only become known later. An example can be found in the case study on ride-sharing platforms. While freedom from bias was a value accounted for in the design of such platforms, some effects were insufficiently recognized initially, like, for example, violence against women. Moreover, it was also not recognized that a lack of transparency and accountability in the rating system could make these rating systems a "vehicle of bias" (Hanrahan et al., 2017). Several design adaptions have been suggested to address these new insights like a more transparent rating system and tracing whether certain users systematically give biased ratings and lower the weight of their ratings in the overall rating score (Hanrahan et al., 2017). It has also been suggested that Uber could do better in stricter selecting of drivers and could require tamper-free video cameras in all Uber vehicles.[7] In 2018, Uber indeed announced better screening of drivers, the ability to share rides with trusted contacts, and an emergency button.[8]

Case   Ride-sharing platforms[9]

Ride-sharing platforms are offered by so-called transportation network companies (TNCs) like Uber, Lyft, and Didi. A range of values is relevant for the design of such platforms including transparency, accountability, safety, environmental sustainability, privacy and freedom from bias (see, e.g., Royakkers et al., 2018). Our focus here will be on the latter value, and particularly on issues related to gender and race. We largely focus on Uber.

In some neighborhoods, it is much harder, or takes much longer, to get a taxi than in others, and there is evidence of several forms of discrimination by taxi drivers (e.g., Brown, 2018). It might be argued that platforms like Uber have the potential to reduce such discrimination and bias. One particular technical feature of Uber is particularly relevant here: drivers and passengers do not get identifying information about each other (like race or gender) at the moment the system proposes certain rides.[10] It is only after both parties have accepted an offer that this information is exchanged.[11] This feature is intended to limit bias on the part of consumers and drivers concerning gender and/or race.

A study by Brown (2018) suggests that TNCs like Uber and Lyft have improved taxi services in less well-off neighborhoods in Los Angeles. She finds that "ridehailing extends reliable car access to travelers and neighborhoods previously marginalized by the taxi industry" (Brown, 2018, p. iii), and she suggests that "ridehailing provides auto-mobility in neighborhoods where many lack reliable access to cars" (Brown, 2018, p. iii). She also finds that racial-ethnic differences in service quality (e.g., waiting times) in Los Angeles are much lower for TNCs than for traditional taxis, although some differences remain.

This does not mean that the system has no inbuilt bias at all; one relevant feature is that the system requires users to have a credit card,[12] which obviously creates a bias against certain groups, although there are ways to use the system without a credit card.[13] A study by Ge et al. (2016) carried out two randomized controlled trials in the Boston area and Seattle to investigate whether TNCs treat passengers of all race and gender equally. They found "a pattern of discrimination, which we observed in Seattle through longer waiting times for African American passengers – as much as a 35 percent increase. In Boston, we observed discrimination by Uber drivers via more frequent cancellations against passengers when they used African American-sounding names. Across all trips, the cancellation rate for African American sounding names was more than twice as frequent compared to white sounding names. … We also find evidence that drivers took female passengers for longer, more expensive, rides in Boston" (Ge et al., 2016).

Interestingly, the study, which looked at UberX, Lyft, and Flywheel, also revealed a difference between Uber and Lyft: Lyft drivers see both the name and photo before accepting, or denying, a ride; while Uber drivers see these only after acceptance of a ride. As a consequence, in the study, Uber drivers much more often cancelled rides for passengers with African American-sounding names. The authors of the study suggest that TNCs like Uber might choose to completely omit personal information about potential passengers. They recognize, however, that this would leave open (and might even exaggerate) other forms of discrimination like not taking rides from certain neighborhoods.

A study by Hanrahan et al. (2017) relates bias to the rating system of Uber. After a ride, the driver and passenger rate each other. This rating is, however, anonymous and ratings may be given without further explanation or justification. If drivers receive biased ratings from passengers (for example, because of their skin color), such a rating will propagate through the system as Uber assigns work, and allow passengers to ask for services, based on past ratings of drivers. Consequently, drivers are worried about low ratings and may suspect these to be based on bias. In response, some develop strategies to avoid biased ratings, like avoiding rides from certain areas or demographic groups (Hanrahan et al., 2017, pp. 11–12). In this way, the suspicion of bias by drivers may lead to an increased bias toward passengers, fueled by the rating system.

Another issue that has raised concern is violence against certain groups of passengers, in particular women. In 2017, the US law firm Wigdor LLP filed a lawsuit against Uber on behalf of women sexually assaulted by Uber drivers. They claim that "[s]ince Uber launched in 2010, thousands of female passengers have endured unlawful conduct by their Uber drivers including rape, sexual assault, physical violence and gender-motivated harassment"[14]. They do not mention a source for these numbers, although there is indeed quite some anecdotal evidence (e.g., Guo et al., 2019).[15] Female drivers have faced similar issues.[16]

Evidence from other countries also suggests that violence against women is a serious issue. In China two women were raped and killed after using the platform Didi.[17] As a consequence the platform suspended its car-pooling service Hitch in August 2018 and started to implement additional safety features.[18]

While some of these issues might perhaps have been anticipated by ride-sharing companies like Uber, it seems likely that at least some of them were hard to exactly foresee beforehand. It may therefore be required to **monitor** the effects of your product over time. This may lead to the discovery of new values that should be taken into account in your design. It is also conceivable that relevant values in the rest of society change, which may require a revision of your design.

5.8   Chapter Summary

Design is at the heart of engineering. It is at the design stage that values get embedded in new technologies. Such embedded values should be distinguished from the values intended by the designers (intended values) and those realized in actual use (realized values).

In this chapter, we have discussed Design for Values as a systematic approach for integrating values of moral importance in (engineering) design. The main activities that need to be carried out if one wants to systematically design for values are:

- *Stakeholder analysis.* Design for values typically starts with identifying the main stakeholders, distinguishing between direct stakeholders (users) and indirect stakeholders (those somehow affected by the use of the product). It is important to describe the values, needs, and perspectives of different stakeholders. You may, for example, use *persona* to do so.
- *Value identification.* Relevant values may be found by considering four sources of value: (1) the design brief; (2) stakeholder values; (3) professional values; and (4) ethical theories, laws, ethical codes, and technical codes and standards. In addition, you need to answer the *normative* question: what values *should* be taken into account in this design project? Here the ethical theories from Chapter 3 may be helpful.

- *Conceptualization of the relevant values.* Relevant conceptions of values may be found in the philosophical literature. Value conceptualization is largely independent from the specific context; still for some technologies some conceptions may be more relevant or appropriate than others.
- *Specification of value.* Values may be specified by translating them into norms and design requirements using a values hierarchy. A specification of a value is adequate if meeting the design requirements would count as an instance of meeting the specified value.
- *Value conflict.* It is important to identify potential value conflicts in the design and to decide how to deal with them.
- *Prototyping and monitoring* are important to validate whether the designed product indeed embeds (and realizes) the values intended by the designers. It may also help to find new relevant values and to revise your design.

Without doubt, some of the most important ethical decisions during design are related to value conflicts. A value conflict is a situation in which the various relevant (moral) values select different options as the best one and there is no clear hierarchy between the values. We have discussed five methods for making decisions in cases of value conflict in design:

- Cost-benefit analysis. The main disadvantage of this method is that, by expressing everything in monetary units, it treats all relevant values as commensurable. Nevertheless, the method is systematic and various ethical concerns can be included by adding certain ethical decision criteria.
- Multiple criteria analysis. Like cost-benefit analysis, values are treated here as commensurable, albeit not by expressing everything in terms of money. The method can also be plagued by methodological problems related to the choice of measurement scales.
- Setting thresholds. This method avoids direct trade-offs between the relevant values and may be helpful to set, for example, a minimal level of safety.
- Reasoning about values, which is useful to get a better grip on the values at play and how they are to be understood in the context of your design project. Sometimes, it may be possible to respecify the relevant values in a way that resolves a value conflict while still doing justice to all the values at stake.
- Innovation can help to solve value conflicts by technical means, that is, by developing new (technical) opportunities that overcome value conflicts.

Study Questions

1 Why can embedded values not be directly perceived? How can we nevertheless empirically investigate what values are embedded in a technological design?
2 What is the difference between direct and indirect stakeholders?
3 What are the four sources of values in a design project? Explain why discerning values using these four sources will not be enough for answering the normative questions what values *should* be taken into account in design.
4 What is the difference between the conceptualization and specification of values?
5 What is the question to be answered if you want to check whether a set of design requirements is an adequate specification of a certain value? Why might it sometimes be hard to agree on an answer to that question?

6 Why does multiple criteria analysis presuppose the commensurability of design criteria?

7 Why does the solving of value conflicts by means of new technologies lie at the heart of engineering design and technological innovation? Provide an example illustrating this.

8 How can respecification be used so solve a value conflict in design? What are the pitfalls of using respecification as a method for solving value conflicts in design?

9 How might the values for which a technology should be designed change after that technology has been introduced into society?

Discussion Questions

1 Some people argue that ethical issues in technology arise due to how technologies are used and can, therefore, not be addressed in design. Do you agree? Is design necessarily irrelevant when most ethical issues arise due to how technologies are used?

2 Some philosophers believe that there cannot be incommensurable values because even if values are seemingly incommensurable we actually choose an option and this choice reveals the relative importance of the values. Do you agree with this argument? To assess the argument, you can, for example, consider the dilemma in *Sophie's Choice* (see Subsection 4.3.1): If Sophie chooses one child rather than the other does this show that she loves that child more than the other? Is the only way to show that she loves both children equally to refuse to make a choice (so that both die)?

3 In cost-benefit analysis, human lives are often expressed in money. Do you consider this an acceptable practice? If it is not acceptable, how should we then determine how much money to spend on increasing, for example, human safety?

4 Choose a product that is designed in your own discipline.
   a. What are relevant direct and indirect stakeholders for this product? What are relevant needs and values of these stakeholders?
   b. What are the values this product should be designed for in your view?
   c. Provide a conceptualization and a specification for at least some of these values.
   d. Give an example of a value conflict a designer may face when designing this product. How should the designer deal with this value conflict in your view?

5 In the case described in the text of alternative coolants, the engineers McLinden and Didion eventually chose for HFC 134a.
   a. Do you agree with their choice?
   b. What do you think of the decision procedure (setting thresholds) by which they came to this choice?
   c. Argue your answers.

6 Read the case on the design of a suicide barrier for the Golden Gate Bridge (see Section 6.3).
   a. What values are at stake in the design of the barrier?
   b. Are these values conflicting? If so, how and why?
   c. What would in your view be the best way to deal with this value conflict?

7 In many mobile phones and laptop computers, the material tantalum is used.[19] Most of the worldwide tantalum supply comes from legitimate mining operations in Australia, Canada, and Brazil, but tantalum is also extracted from the metallic ore coltan that is mined in Congo. Rebel groups exploit coltan mining to raise funds for the civil war that is going on in Congo and in which thousands of people have been killed. Some organizations have alleged that there is a direct link between the mining of coltan in Congo and the human right abuses. Coltan mining by rebels in Congo also causes environmental degradation. Some manufacturers have shown concerns about the use of coltan from Congo or have declared that they will no longer use coltan from Congo, but such policies are hard to implement because in many cases it is hard to find out where coltan that is traded on the world market is actually coming from. It might well be

smuggled to another African country from Congo before it is sold. Banning coltan from such other African countries would, however, affect countries that are legitimately mining coltan and for which the product may be an important source of income.

   a. What are the main stakeholders and values at stake in this case?

   b. Do designers have a responsibility to avoid the negative consequences associated with the use of tantalum and coltan? Or is this more a management decision?

   c. Should phone manufacturers try to avoid the use of tantalum? At the cost of what (other) values might this choice come?

8 In the Ford Pinto case (see Section 3.1), an issue was how to trade off safety and economic considerations.

   a. Explain how this trade-off would be made with each of the five methods discussed in Section 5.6.

   b. What are the advantages and disadvantages for each of these methods in this specific case?

   c. Which method (or methods) do you consider most appropriate to trade off safety and economic considerations in this case and why?

Notes

1 This description is based on Criado-Perez (2019).

2 https://sitn.hms.harvard.edu/flash/2020/racial-discrimination-in-face-recognition-technology, accessed June 22, 2022.

3 https://en.wikipedia.org/wiki/Sophia_(robot), accessed June 22, 2022.

4 Cited in Harrison (1993, p. 120).

5 Based on Van de Poel (2001), and Kroes and Van de Poel (2015).

6 Rules of Engagements compromise directives issued by competent military authorities that delineate both the circumstances and the restraints under which combat with opposing forces is joined.

7 https://www.wigdorlaw.com/uber-class-action-sexual-assault, accessed June 7, 2022.

8 https://money.cnn.com/2018/06/04/technology/uber-passenger-safety/index.html, accessed June 7, 2022.

9 Drawn from De Reuver et al. (2020).

10 https://help.uber.com/riders/article/requesting-a-specific-driver?nodeId=1aaf0913-484f-4695–9042-e61fc7613f24, accessed June 1, 2022; https://help.uber.com/partners/article/getting-a-trip-request?nodeId=e7228ac8-7c7f-4ad6-b120-086d39f2c94c, accessed June 1, 2022.

11 https://help.uber.com/riders/article/how-to-identify-a-driver-and-vehicle?nodeId=02746faf-1bc6-4d3f-8ba2-ab35f36d7191, accessed June 1, 2022.

12 https://www.dallasobserver.com/news/southern-dallas-leaders-say-uber-is-profiling-customers-and-they-want-city-hall-to-act-7134986, accessed June 1, 2022.

13 https://www.wikihow.com/Use-Uber-Without-a-Credit-Card, accessed June 1, 2022.

14 https://www.wigdorlaw.com/uber-class-action-sexual-assault, accessed January 13, 2023.

15 See also https://www.independent.co.uk/news/uk/uber-drivers-accused-of-32-rapes-and-sex-attacks-on-london-passengers-a7037926.html, accessed June 1, 2022.

16 https://www.bbc.com/news/technology-46990533, accessed June 1, 2022.

17 https://qz.com/1370345/another-didi-hitch-ride-hailing-driver-murdered-a-female-passenger-in-china, accessed June 1, 2022.

18 https://en.wikipedia.org/wiki/DiDi, accessed June 1, 2022.

19 Description is based on https://www.humanityinaction.org/knowledge_detail/your-phone-coltan-and-the-business-case-for-innovative-sustainable-alternatives/?lang=nl (accessed June 12, 2022).

6

Ethical Aspects of Technical Risks

Having read this chapter and completed its associated questions, readers should be able to:

- Discuss why engineers are responsible for safety and how they can apply this responsibility in engineering practice;
- Describe the main approaches to risk assessment;
- Describe the main ethical considerations for judging the moral acceptability of risks and apply these to concrete cases;
- Argue why risks that are of similar magnitude are not necessarily equally acceptable;
- Identify ethical issues in risk communication and to judge different ways of dealing with them;
- Explain what is meant with engineering as a societal experiment and to reflect on the conditions under which such experiments are morally acceptable.

Contents

6.1 Introduction 183

6.2 Definitions of Central Terms 186

6.3 The Engineer's Responsibility for Safety 187

6.4 Risk Assessment 191

    6.4.1 The reliability of risk assessments 193

6.5 When Are Risks Acceptable? 194

    6.5.1 Informed consent 197
    6.5.2 Do the advantages outweigh the risks? 198
    6.5.3 The availability of alternatives 199

    6.5.4 Are risks and benefits justly distributed? 200

6.6 Risk Communication 201

6.7 Dealing with Uncertainty and Ignorance 203

    6.7.1 The precautionary principle 203
    6.7.2 Engineering as a societal experiment 206

6.8 Chapter Summary 209

Study Questions 210

Discussion Questions 212

Ethics, Technology, and Engineering: An Introduction, Second Edition. Ibo van de Poel and Lambèr Royakkers.
© 2023 John Wiley & Sons Ltd. Published 2023 by John Wiley & Sons Ltd.

6.1　Introduction

Case　Boeing MAX 737 crashes (Photo Boeing 737 MAX)

Figure 6.1　A Boeing 737 MAX. Photo: Stephen Brashear/Getty Images.

In August 2011, Boeing Commercial Airplanes, a subsidiary of Boeing, announced the launch of its new 737 MAX aircraft as the fourth generation of the 737 line. Initial deliveries of the aircraft took place in May of 2017, and the plane entered commercial service shortly thereafter. Among the first passenger carriers to run the 737 MAX commercially were Lion Air, of Indonesia, and Norwegian Air, of Norway. Within a year of its launch, 130 737 MAX aircraft were delivered to 28 Boeing customers, and in total 387 aircraft were eventually delivered.

On October 29, 2018, a Boeing 737 MAX aircraft operated by Lion Air crashed thirteen minutes after takeoff, killing all 189 individuals on board. Responding to the incident, Boeing issued guidance on its operational manual to advise airline pilots regarding procedures for handling so-called erroneous cockpit readings. On March 10, 2019, only five months after the Lion Air disaster, a Boeing 737 MAX aircraft operated by Ethiopian Airlines crashed six minutes after takeoff, killing all 157 individuals on board. After this second tragic incident, all Boeing 737 MAX planes were grounded worldwide. Both accidents were characterized by a sudden nose dive at high speed with which the aircraft bores mercilessly into the sea and the ground.

Initially, Boeing was reluctant to admit to a design flaw in the Boeing 737 MAX, instead of blaming the pilots' alleged inability to control the MAX. Following the second crash, Boeing acknowledged that the extra software designed for the MAX was the primary cause of the crashes. The House Transportation and Infrastructure Committee (2020) conducted an independent 18-month investigation – based on interviews with two dozen Boeing and agency employees and an estimated 600,000 pages of records – in which they condemned both Boeing and the Federal Aviation Administration (FAA, the aviation regulator in the US) for safety failures, and in which they ruled out wrongdoing by the pilots. The report argues that Boeing emphasized profits over safety and that the FAA granted the company too much sway over its own oversight.

The decision to reduce costs and speed up 737 MAX production led directly to the crashes. Boeing initially planned to build a new plane to replace the aging 737. However, when American Airlines announced their plans to purchase 460 more fuel-efficient airplanes from Airbus, Boeing scrapped the ideas for a new airplane and made a plan to re-engineer the Boeing 737 instead, which became the Boeing 737 MAX. In this way, Boeing did not have to go through the more stringent FAA certification requirements for a brand new airplane. Had Boeing released a new aircraft, pilots, for example, would have had to train for it by spending time in flight simulators, which would make the purchase of the MAX considerably more expensive, up to USD 1 million per MAX aircraft.

To compete with the Airbus A320-NEO, the MAX had larger engines than the previous 737 models. They were designed for greater range and fuel efficiency but came with a tradeoff. The new engines were too big to fit in their traditional spot under the wings. To combat the problem, Boeing had to change the position of the engines on the wing to give the plane ground clearance and account for the extended length of the fuselage. Moving the engine position forward shifted the plane's center of gravity, which altered the aerodynamics of the aircraft. The position of the new engines pulled the 737 tail down, pushed its nose up, and put it at risk of stalling. When an aircraft stalls, it begins to fall because the wings stop creating lift. Slow airspeed and high nose position are the most common causes of stalls.

Boeing designed extra software for the MAX to prevent stalls by compensating for the position of the engine on the wing, and forcing the aircraft's nose down automatically when the sensors determined the airplane was flying at a dangerous angle. The crashes were precipitated by a failure of an Angle of Attack (AOA) sensor and the subsequent activation of this new flight control software, the Maneuvering Characteristics Augmentation System (MCAS). In both incidents, the AOA sensor falsely indicated that the plane was close to stalling, triggering MCAS, pushing down the nose of the plane, and going into a dive from which they never recovered. The pilots were unable to regain control of the planes. The existence of the software was not disclosed until after the first crash. A November 2018 FAA analysis, after the first crash, reportedly found that without design changes, a 737 Max plane could be expected to crash every two to three years. But in explaining its decision not to ground the plane, the FAA said in its statement that the actual risk at the time, considering the number of planes in the air, was as close to zero as their calculations allowed. The agency had given Boeing 150 days to fix MCAS and issued official directives to pilots, however, pilots were not required to undergo simulation training on the MAX which also was discouraged by Boeing because of time and expenses.

Boeing had managed to influence the FAA to be lenient about the design approval of the Boeing 737 MAX. For example, the aviation regulator was persuaded not to classify the automatic correction system MCAS as "critical to safety." Pilots were therefore not aware of the existence of this system, let alone its operation and impact. Boeing also hid an important test result of the updating system from the FAA. Namely, internal test data showed that a pilot must intervene within 10 seconds if the MCAS incorrectly pulls down the nose of the aircraft, otherwise the consequences can be "catastrophic." However, in a flight simulator, it took a Boeing test pilot more than 10 seconds to diagnose and respond to uncommanded MCAS activation. Furthermore, an alarm for the MCAS did not function in most of the global fleet of 737 Max aircraft. This alarm should warn pilots that there may be a problem with the sensor (AOA) that measures whether the aircraft is stalling.

In November 2020, the FAA lifted its 20-month ban on the Boeing 737 MAX, after Boeing presented key software and other changes, such as a new pilot training, and unparalleled scrutiny by the FAA to the Boeing 737 MAX of these safety measures.

Sources: Herkert et al. (2020), House Committee on Transportation and infrastructure (2020), Englehardt et al. (2021), and the Netflix documentary *Downfall: The Case Against Boeing* (2022).

This case illustrates that new technologies may come with new hazards. In this case, the new hazards were known beforehand. Boeing, as well as, the FAA, had warnings about the inadequacy of the MCAS's design, and about the lack of communication to the pilots about its existence and functioning. The risks were – in retrospect – regarded as unacceptable by the engineers involved as can be seen from the report of the House Transportation and Infrastructure Committee (2020). However, the relevant risks were not removed or diminished.

Negative effects of technology do not only emanate from known risks as in the case of the Boeing 737 MAX but also from unknown hazards. Sometimes such hazards have to do with the fact that certain technology can fail in a way that had not been foreseen beforehand, such as in the Tacoma Narrows Bridge case (see box). Sometimes they are due to unforeseen side effects of technology, as with asbestos. Asbestos is a product that started to come into large-scale use at the beginning of the twentieth century. Due to several positive characteristics, such as heat resistance, durability, and good insulation properties, it was applied in a large number of products. However, over time asbestos proved to have some extremely harmful side effects. Inhaling asbestos fibers can lead to asbestos-related diseases such as asbestosis and mesothelioma (cancer of the lung and stomach lining), which can be lethal.

Case   Tacoma Narrows Bridge Collapse

In 1940 in the United States, the Tacoma Narrows Bridge collapsed (Petroski, 1982, Chapter 13). It was an innovative bridge design, but the bridge started to vibrate when it was hit by side winds. The bridge collapsed when it was closed for safety reasons. Although the bridge designers were aware that bridges could start to vibrate with certain winds, they had not considered the possibility of flutter, which is the phenomenon that under certain wind conditions, the energy fed in the bridge deck gets an oscillating structure so that the vibration amplitude increases and the bridge deck collapses. From experiences with previous bridges, they had wrongly concluded that narrower suspension bridges could be built.

The existence of certain hazards can sometimes be controversial. Even if most scientists now agree that the intensified greenhouse effect exists and that the earth is warming up, for a long time this was controversial. There is still no agreement on what exactly the consequences of an intensified greenhouse effect will be. The greenhouse effect is also different from the other mentioned hazards in the sense that it is not precisely clear how this hazard can be attributed to a specific technology. More so than with other hazards it is a hazard that arises in the use phase of technology rather than in the design or production phase.[1] Furthermore, less so than in several other cases, it is hard to ascribe the possible hazard to one specific technology. It is connected more with the large-scale use of a wide range of technologies.[2]

In this chapter, we discuss the moral issues that are raised by the risks and hazards of technologies, and how engineers can deal with those issues. We will discuss the responsibility of engineers for safety (Section 6.3) and the current methods for assessing risks (Section 6.4). Section 6.5 discusses the moral acceptability of technological risks and the next section focuses on risk communication. Then, we will pay attention to situations of uncertainty and ignorance (Section 6.7). Finally, in Section 6.8 some conclusions are drawn regarding the responsibility of engineers. However, to begin with, we shall define some of the key terms.

6.2   Definitions of Central Terms

Hazard   Possible damage or otherwise undesirable effect.

We speak of a **hazard** if a technology, or its use, can cause damage or otherwise undesirable effects. The term *risk* is a specification of the term hazard. It is an attempt to name or specify the phenomenon of a hazard, which is often done in quantitative terms. In this chapter, we shall mainly concentrate on safety risks (risks of events in which there can be fatalities or injured) and health risks (risks in which the health of people is endangered). We shall not consider environmental risks or social risks.

Risk   A specification of a hazard. The most often used definition of risk is the product of the probability of an undesirable event and the effect of that event.

The same hazard can be expressed as a risk in various ways. In this chapter, the term **risk** will be defined as the product of the probability of an undesirable event and the effect of that event, unless stated otherwise. This probability is often taken to be the relative frequency of an undesirable event, such as "once every ten years." The effect is often expressed as the number of fatalities. With this definition, the term "risk" is a measure for the expected number of fatalities per time unit. Other definitions of risk are used too, like:

- The probability of an undesirable event taking place.
- The maximum negative effect of an undesirable event.

Safety   The condition refers to a situation in which the risks have been reduced as far as reasonably feasible and desirable.

Safety is sometimes defined as the absence of risk and hazards. Usually, a technological product cannot be made absolutely safe in this sense. Safety therefore also often refers to the situation in which the risks have been reduced in as far that is reasonably feasible and desirable. So conceived, safety is related to the notion of **acceptable risk**. We will discuss the acceptability of risks in more detail in Section 6.5.

For several reasons, it is not always possible to predict the hazards of a technology beforehand and to express them reliably as risks. One reason is the *complexity* of causal relations between potential harmful agents or events and specific undesirable effects. Complexity may be due to such factors as interactions between different substances or between substances and specific environments, long delay times, and intervening variables. The impossibility of expressing hazards in risks may also be due to **uncertainty**, that is, a lack of knowledge. Uncertainty may, in turn, be caused by a number of underlying factors, like modeling errors, indeterminacy, and the drawing of system boundaries (Renn, 2005, p. 30). In a more circumscribed sense, the notion of uncertainty is often used to refer to situations where we know the type of consequences, but cannot meaningfully attribute probabilities to the occurrence of such consequences (Felt et al., 2007, p. 36). In cases of uncertainty, we can therefore not calculate the risks. Sometimes, we do not even know that something can go wrong, that there is a hazard. In such cases, the term **ignorance** is often used (Felt et al., 2007, p. 36). What is typical of ignorance is that we do not know what type of things we do not know. Therefore, it is extremely hard, if not impossible, to anticipate the consequences of ignorance because often we do not know what we have to be prepared for.

The impossibility of expressing hazards in risks may also be due to **ambiguity**. Ambiguity refers to the fact that different interpretations or meanings may be given to the measurement, characterization, aggregation, and evaluation of hazards. The International Risk Governance Council (IRGC) distinguishes between interpretive ambiguity – referring to different interpretations of scientific data (for example, how to extrapolate dose-response relations to low doses for which no data are available) – and normative ambiguity – referring to disagreement about the relevant (moral) values and their relative importance (Renn, 2005).

> **Acceptable risk** A risk that is morally acceptable. The following considerations are relevant for deciding whether a risk is morally acceptable: (1) the degree of informed consent to the risk; (2) the degree to which the benefits of a risky activity weigh against the disadvantages and risks; (3) the availability of alternatives with a lower risk; and (4) the degree to which risks and advantages are justly distributed.

> **Uncertainty** A lack of knowledge. Refers to situations in which we know the type of consequences, but cannot meaningfully attribute probabilities to the occurrence of such consequences.

> **Ignorance** Lack of knowledge. Refers to the situation in which we do not know what types of things we do not know.

> **Ambiguity** The property that different interpretations or meanings can be given to a term.

6.3   The Engineer's Responsibility for Safety

From where does the engineer's responsibility for safety come? Many codes of conduct for engineers attribute responsibility for safety to engineers (see Chapter 2). For example, according to the first "fundamental canon" of the National Society of Professional Engineers (NSPE) Code of Ethics, engineers "shall hold paramount the safety, health, and welfare of the public." In the case of the Boeing 737 MAX, we can see that the Boeing engineers by ignoring the warnings about the inadequacy of the MCAS's design, and about the lack of communication to the pilots about its existence and functioning, did certainly not "hold paramount the safety, health, and welfare of the public." Although some engineers had already concerns about the safety of the airplane, they did only express these concerns in public testimonies after the two crashes, which conflicts with the NSPE Code, which clearly states that "[i]f engineers' judgment is overruled under circumstances that endanger life or property, they shall notify their employer or client and such other authority as may be appropriate." In this case, they had to inform the FAA, and one could argue also the pilots. Besides the codes of conduct, there are legal obligations concerning the safety of products or technical codes and standards, in which safety often plays an important role. In the case of Boeing, the airplane manufacturer had to apply for the FAA Airworthiness Certification for their Boeing 737 MAX, which they received in March 2017. In addition to the law and codes of conduct, the ethical frameworks that were dealt with in Chapter 3, provide arguments why engineers should strive for safe products; the exact arguments differ for each ethical framework. Utilitarianism, for example, states that engineers must strive for good consequences: safe products definitely fall into that category. The desirability to design safe products is sometimes described as "do no harm." This can be defended in terms of utilitarianism with the freedom principle of Mill (see Subsection 3.7.2). It is a kind of minimum standard that applies to striving for good consequences. In duty ethics the notion "you should not harm anyone" can be seen as a general norm. In care ethics, care for the users or your customers who suffer the consequences of your design is an important virtue. Striving for safe products, therefore, is a moral obligation.

However, it is not always clear whether safety of a design is the responsibility of engineers. In the case of the Golden Gate Bridge, there was a lot of discussions about who was responsible to prevent suicides (see box).

Case   The Golden Gate Bridge Suicide Barrier

Figure 6.2   Golden Gate Bridge in San Francisco. Photo: Heyengel/Adobe Stock.

When the Golden Gate Bridge in San Francisco was finished in 1937 it was – with its span of 1280 m (4200 ft) – the world's longest suspension bridge. It is considered to be one of the best and most beautiful examples of bridge design. The American Society of Civil Engineers has included the Golden Gate Bridge in its enumeration of the seven wonders of the modern world. However, the bridge is also the US's most popular place to commit suicide. Since 1937, at least 1300 people killed themselves by jumping off the bridge; an average of 20 to 25 per year. Suicide prevention has been a concern ever since the bridge was designed. Joseph Strauss, the chief engineer is quoted as saying in 1936 (a year before the opening of the bridge): "The guard rails are five feet and six inches high [i.e. about 1.7 m] and are so constructed that any persons on the pedestrian walk could not get a handhold to climb over them. The intricate telephone and patrol systems will operate so efficiently that anyone acting suspiciously would be immediately surrounded. Suicide from the bridge is neither possible nor probable." However, for some reason, the height of the railing was reduced in the detail design to 1.2 meter (4 feet), so making it not too difficult to climb the railing and jump off the bridge.

Already in 1940, the Board of Directors discussed an "anti-suicide screen" but decided against it based on aesthetic and financial considerations. Another concern was that the screen might create dangerous wind resistance and make the bridge structurally unstable. Since then, proposals for a suicide barrier have been made every decade but were unsuccessful until 2008.

There have been debates about whether engineers have a responsibility to try to prevent suicides through bridge design. Some have argued that it is not the responsibility of society or the designers of the bridge to try to prevent suicides. Moreover, they suggest that it is wrong to use architecture to solve a social problem. In the words of architect Jeffrey Heller: "You can't correct all of the sadness and evil in the world." In addition, the Golden Gate Bridge has been praised for its transparency and openness. Any barrier design would destroy the view. Jeffrey Heller, for example, stated: "When you look straight out, you'll see through all this mesh, which will be sad enough, but looking straight down the roadway, it will become a cage. ... That is far too high a price for our society to pay." This kind of utilitarian argument could be strengthened by adding that the barrier is not worth the costs: even if the barrier would save 20 to 25 lives a year, it is not worth 50 million building costs plus the annual operation and maintenance costs.

Proponents of adjusting the design have argued that it is the responsibility of society and designers of the bridge to act. As Jerome Motto, a past president of the American Association of Suicidology expressed it: "If an instrument that's being used to bring about tragic deaths is under your control, you are morally compelled to prevent its misuse. A suicide barrier is a moral imperative. It's not about whether the suicide statistics would change, or the cost, or whether [it] ... would be as beautiful ... A barrier would say, 'Society is speaking, and we care about your life.'"

On October 10, 2008, the Board of Directors voted 14 to 1 to install a stainless steel net which would be placed 6 m (20 ft) below the deck, extending 6 m out from the Bridge. The estimated costs are 50 million dollars. Net construction began in 2018 and is expected to be completed in 2023.

Source: https://www.newyorker.com/magazine/2003/10/13/jumpers; https://www.sfgate.com/bayarea/article/LETHAL-BEAUTY-The-Critic-2562374.php (accessed November 17, 2022).

During the design process, engineers can follow different strategies for ensuring safe products, such as:

Inherently safe design An approach to safe design that avoids hazards instead of coping with them, for example by replacing substances, mechanisms, and reactions that are hazardous with less hazardous ones.

Safety factor A factor or ratio by which an installation is made safer than is needed to withstand either the expected or the maximum (expected) load.

Negative feedback mechanism A mechanism that if a device fails or an operator loses control assures that the (dangerous) device shuts down.

Multiple independent safety barriers A chain of safety barriers that operate independently of each other so that if one fails the others do not necessarily also fail.

1 **Inherently safe design**: avoid hazards instead of coping with them for example by replacing substances, mechanisms, and reactions that are hazardous by less hazardous ones.
2 **Safety factors**: constructions are usually made stronger than the load they probably have to bear. Adding a safety factor to the expected load or maximum load is an explicit way of doing this.
3 Negative feedback: For cases where a device fails or an operator loses control, a **negative feedback mechanism** can be built in that causes the device to shut down. An example is the dead man's handle that stops the train when the driver falls asleep or loses consciousness.
4 **Multiple independent safety barriers**: A chain of safety barriers can be designed that operates independently so that if the first fails the others still help to prevent or minimize the effects. This can, for example, be achieved through redundancy in design (see box). Also, emergency escapes can be quite useful (Hansson, 2007a).

Redundant Design

The failure of a component or sub-system can often be compensated by producing systems with redundant designs. Nuclear reactors have redundant systems to sustain electricity (to operate pumps etc.) and to cool the reactor core (see, e.g., Mostert, 1982). If one system drops out, one can in principle fall back on the redundant system so that the reactor can continue to operate safely. The O-rings of the *Challenger* were designed redundantly too. There were two O-rings for each connection, so that if the first were to fail the second would compensate. The redundancy of the O-rings was one of the arguments given to allow the flight to continue on the evening before the fatal flight (Vaughan, 1996). Airplanes often have redundant systemsi too, for example, for the control of the plane.

For the system as a whole, it can also be useful to have some fallback options. Take for example a back-up system for an electronic databank or spare capacity reserves if an electricity plant shuts down. The latter can be very important in preventing large areas of the grid from dropping out.

What is important is that these strategies do not only address known risks but also to some extent uncertainties.[3] Negative feedback mechanisms may also, for example, be effective if the causes of a certain accident are not foreseen or are unknown.

In the case of the Boeing 7373 MAX, some of the mentioned strategies for safe design were not employed, while these strategies might either have decreased the probability of the accident or the consequences of it, even if the exact risks were unknown. Boeing installed the automatic correction system MCAS to make the updated 737 MAX fly like traditional ones. We could, however, argue that this system was an *inherently unsafe design*: it was even called a "death trap" by pilot Sullenberg (2019), who famously crash-landed an Airbus A320 in the Hudson River in 2009. The internal test data showed that a pilot must intervene within 10 seconds if the MCAS incorrectly pulls down the nose of the aircraft, otherwise the consequences can be "catastrophic." However, in a flight simulator, it took a Boeing test pilot more than 10 seconds to diagnose and respond to uncommanded MCAS activation. Nothing was done to address this problem, let alone to inform the pilots about the system. With respect to the strategy "multiple independent safety barriers," Boeing typically uses two or even three separate components as fail-safes for crucial tasks to reduce the possibility of a disastrous failure. However, with the new design of the MCAS they abandoned the principles of component redundancy, ultimately entrusting the MCAS to just one sensor. This angle-of-attack (AOA) sensor sends data to the MCAS that pushes the nose of the airplane down if it senses an imminent stall. The crashes were precipitated by a failure of this sensor and the subsequent activation of the MCAS.

6.4 Risk Assessment

To judge whether certain hazards are acceptable, an attempt is usually first made to map them and express them as risks. This takes place by carrying out so-called **risk assessments**. In engineering, there are many types and methods for risk assessment. The exact methods

> **Risk assessment** A systematic investigation in which the risks of a technology of an activity are mapped and expressed quantitatively in a certain risk measure.

differ from one engineering domain to the other. We shall not attempt to give an overview of all the methods for risk assessment used in engineering, but limit ourselves to a general overview.

A risk assessment usually consists of four steps:

1 Release assessment
2 Exposure assessment
3 Consequence assessment
4 Risk estimation. (Covello and Merkhofer, 1993)

Release assessment
Releases are any physical effects that can lead to harm and that originate in a technical installation. Examples are shock waves, radiation, and the spread of hazardous substances. In general, we can distinguish between two kinds of releases: incidental and continuous. Incidental releases are usually unintended and are due to, for example, an explosion in a

chemical plant or an accident with a nuclear power plant. Such releases can often cause immediate and major harm. Continuous releases are often anticipated and may be accepted as side-effects of, for example, production processes. Continuous releases do not necessarily or always lead to exposure or harm.

Failure mode   Series of events that may lead to the failure of an installation.

Event tree   Tree of events in which one starts with a certain event and considers what events will follow.

Fault tree   Tree of events in which we move backwards from an unwanted event (a fault) to the events that could lead to the undesirable event.

In the case of incidental releases, an important step is the detection of so-called **failure modes** and accident scenarios. These are series of events that lead to the failure of the installation or to an accident. The probability of the occurrence of such scenarios is calculated too. This takes place in two ways.[4] In the first method, the probability of certain accidents occurring is calculated using statistical data about accidents in the past. If such statistical data are absent event trees and fault trees are often used to calculate the probability of an accident. For **event trees,** we start with a certain event and consider what events will follow. For **fault trees,** we move backwards from an unwanted event (a fault) to the events that preceded and could have led to the undesirable event. To each event in the event or fault tree a probability value is attached based on failure data concerning components. Next, the probability of a specific accident scenario is calculated.

Exposure assessment

In this step, the aim is to predict the exposure of vulnerable subjects like human beings to certain releases. Exposure assessment usually describes what vulnerable subjects (human beings, animals, the environment) are exposed to a certain release, through what mechanisms (for example, inhalation of toxic substances by humans), and the intensity, frequency, and duration of the exposure.

Consequence assessment

In the third step, the focus is on determining the relationship between exposure and harmful consequences. In some risk assessments, the analysis is limited to acute harm or to the number of direct fatalities. In other cases, long-term effects on health or the environment are also considered. An important part of this step is usually the determining of dose-response relationships. Such relationships can be established through tests on animals, epidemiology, and models (Covello and Merkhofer, 1993, pp. 127–178).

In the case of **animal tests,** the harmful effects are tested by exposing animals to dosages. Different animals are given different dosages. Next to that, there is a control group. By comparing the groups, a dose-response relationship can be determined for the type of animal involved. The idea behind this is that it tells us something about the dose-response relationship in humans.

Animal tests   Tests for determining dose-response relationships by exposing animals to various dosages and assessing their response.

In **epidemiological research,** we use population data to find out what the relationship is between the occurrence of certain diseases or mental deviations and certain factors that may cause these deviations. The advantage of epidemiological research is that we do not need to translate the effects on animals to the effects on humans. Epidemiological research has its disadvantages too, however. First, it can only occur

> **Epidemiological research** Research in which population data is used to find out what the relationship is between the occurrence of certain diseases or certain mental deviations and certain factors that may cause these deviations.

after the fact, when certain health effects have already occurred. Second, it requires reliable statistical data. This often requires extensive empirical research. Often the time and money for this are lacking. Third, only statistical correlations are usually established. Demonstrating a statistical correlation however is not enough to prove the existence of a causal relationship. A nice example is that an empirical study found that married people eat statistically significantly less candy than unmarried people.[5] An analysis of the data showed, however, that both being married and eating less candy were strongly correlated to the underlying variable age. Older people are more likely to be married and are more likely to eat less candy. So it would have been wrong to conclude in this case that being married is a cause of eating less candy. To establish a cause, we thus need to exclude all other possible causal factors, which is often difficult in practice.

Many **models for dose-response relationships** have a hypothetical and descriptive nature. They presuppose a certain relationship between dose and effect, but they hardly or do not explain how a dose of a harmful substance leads to consequences.

> **Models for dose-response relationships** Models that presuppose or predict a certain relationship between dose and response.

Risk estimation

In the fourth step, the risk is determined and presented using the results obtained earlier. In this step, we determine in what measure the risk is expressed. This can be done using the number of expected fatalities per time unit, for example, or the reduced lifespan of people that work or live in the neighborhood of an installation.

6.4.1 The reliability of risk assessments

In many cases, risk assessments only have limited reliability. This is because the results often depend on the original assumptions made, as the box on the estimated risks of dioxin shows. In connection with this, it is striking that many risk assessments do not give an estimate of the accuracy and reliability of the final result. One may well wonder whether it would not be more responsible to list uncertainty intervals for results, or to state explicitly under which conditions results apply.

Case The Risks of Dioxin

In 1978, a test was carried out with rats to determine the health effects of dioxin. Rats that were given 100,000 picograms of dioxin per kilogram of body weight per day developed cancer significantly more often than the control group. At 10,000 picogram/kg per day there was only a small increase in cancer and at 1000 picogram/kg per day no effects could be measured.

Based on these data, regulatory bodies in the United States and Canada made very different assessments concerning which concentrations are acceptable. In the United States, the assumption was made that animal tests are not sensitive enough to measure the health effects at low doses. Moreover, a more or less linear relationship was assumed between dose and effect. As a result, they concluded that a dose of 0.006 picogram/kg per day in humans would lead to an individual risk of less than 1 in 1 million. That means one fatality per million people exposed. In Canada, it was assumed that dioxin was not an initiator of cancer but a promoter. In contrast with cancer initiators, cancer promoters are assumed to have a no-effect level, so that no harmful effect can be found below a given level. Applying this model led to the conclusion that around 1 to 10 picogram/kg per day would be safe for people. Based on these conclusions, the acceptable exposure to dioxin would be about one thousand times higher in Canada than in the United States.

Source: Adapted from Covello et al. (2020).

Type I error The mistake of assuming that a scientific statement is true while it actually is false. Applied to risk assessment: The mistake that one assumes a risk when there is actually no risk.

Type II error The mistake of assuming that a scientific statement is false while it actually is true. Applied to risk assessment: The mistake that one assumes that there is no risk while there actually is a risk.

One relevant issue is also the degree of evidence that is needed to establish a risk during a risk assessment based on, for example, epidemiological data. In establishing a risk based on a body of empirical data one might make two kinds of mistakes. One can establish a risk when there is actually no risk (a so-called **type I error**) or one can mistakenly conclude that there is no risk while there actually is a risk (a so-called **type II error**). Science traditionally aims at avoiding type I errors because one usually does not want to assume a hypothesis as knowledge unless there is strong evidence for it. Several authors have argued that in the specific context of risk assessment it is often more important to avoid type II errors (Cranor, 1990; Shrader-Frechette, 1991). The reason for this is that risk assessment not just aims at establishing scientific truth but also has a practical aim, that is, to provide the knowledge based on which decisions can be made about whether it is necessary to protect the public against certain risks. It might be worse not to protect the public against a risk than to take unnecessary precautions against a risk that turns out not to exist.

6.5 When Are Risks Acceptable?

Some engineers and scientists believe that if the risks of two different activities are the same according to risk assessments, the activities are equally acceptable. In other words, if one activity is acceptable the other (which has the same risk) must be acceptable too. This argument is flawed for a number of reasons. First, the question of whether the *risks* of technology A are acceptable is not the same question as the question of whether technology A is acceptable. This will become clear when we consider the ethical objections to human cloning (see box).

Ethical Objections to Cloning

Philosopher of technology Tsjalling Swierstra has reconstructed which arguments played a role in the social discussion on human cloning. He made the following list of arguments:

Cloning:

1 undermines the uniqueness of humans;
2 is contrary to human dignity;
3 leads to psychosocial problems in the cloned child;
4 suffers from numerous scientific and technical risks, such as health problems of poor copies, reduced human biodiversity, or an increased risk of contagious diseases;
5 will lead to misuse and has unforeseen and undesirable consequences;
6 is unnatural;
7 is based on an overreaching desire for manipulation and leads to the abject instrumentalization of people;
8 will lead to undesirable changes in our self-image.

Source: Based on Swierstra (2000, p. 42).

Objection 4 is already formulated in terms of risks. The third and fifth objections can easily be reformulated in terms of risks. However, objections like cloning undermine the uniqueness of humans (the first) and can lead to an undesirable change in our self-image (the eighth), cannot so easily be understood in terms of risks. This is because these kinds of objections cannot be justified on the basis of consequentialism; they stem from a duty ethics or virtue ethics approach. Especially the argument that cloning leads to a kind of human or kind of society that is not virtuous and therefore is undesirable, to which the second and seventh objections refer, is clearly based on virtue ethics (Swierstra, 2000). The question of whether cloning is ethically acceptable is thus more encompassing than the question of whether the risks of the technology are acceptable.

Even if we restrict our analysis to the acceptability of risks, it is a fallacy to conclude that if the magnitude of the risks of two technologies is the same these risks are equally acceptable. A number of reasons why the conclusion that equally large risks are equally acceptable is flawed are given in the box.

Why the Magnitude of the Calculated Risk Does Not Tell Us Everything about the Acceptability of the Risk

There are a number of arguments why equally large risks are not necessarily equally ethically acceptable:

As not all risk assessments are equally *reliable*, the results of risk assessments are not easily comparable. Take for example the comparison between the risks of nuclear power plants and traffic risks. Traffic risks are usually calculated using a large number of statistical data based on years of experience. In the case of nuclear power plants, the risks cannot be calculated using statistical data and all sorts of assumptions have to be made to estimate the risk in question. That is why the estimations of traffic risks are usually more reliable than the estimates of the risks for nuclear power plants.

Risks are often *multi-dimensional*, while only one dimension is used in the comparison of risks in many cases. This dimension often is the number of expected fatalities per time unit. Fatalities can occur both through accidents in traffic and because of accidents in nuclear power plants. In many other aspects, however, the risks in question are not that easy to compare. Take, for example, the lasting risk of nuclear waste.

It is not obvious that *a small probability of a major accident is as acceptable as a large probability of a small accident*, even if the product of probability and effect is the same. In this respect, the risks of traffic and of nuclear power plants differ. The probability of having an accident in traffic is far higher than the probability of a nuclear accident at a power plant. However, there are far more fatalities with nuclear accidents compared to traffic accidents. Accidents in which multiple deaths occur – even if the total risk is the same as that of another accident – are often considered less acceptable, because the degree of social disruption is much higher. Whether or not this is a good argument is a matter of debate.

The acceptability of a risk partly depends on the degree to which people *voluntarily* take a risk or consent to a risk.[6] Still, the distinction between voluntary and imposed risk is not always clear-cut. To what extent are traffic risks voluntary if someone has to travel a lot for work? Nevertheless, traffic risks are more voluntarily taken than the risks of a nuclear power plant being built near you without any consultation. The risks of skiing are more voluntary than traffic risks.

Risks as such are not acceptable, but they can be acceptable because risky activities bring certain *benefits*. Instead of assessing isolated risks, it is, therefore, a better idea to assess risky activities. This way it is possible to weigh the risks and the benefits. If an activity does not result in any advantage in someone's eyes then it is reasonable that they will reject the activity if it involves a risk, however small.

In extension of this, it makes sense to weigh *different options* – to achieve the same goals – when the acceptability of risks is being assessed. This means that the acceptability of nuclear energy as a technology for generating energy not only depends on the question of whether the advantages of nuclear energy weigh against the risks (and other negative effects), but also depends on the question of whether other technologies – like wind energy – are more attractive.

The acceptability of certain risks also depends on how justly the risks and advantages of a specific risky activity are *distributed*. Take, for example, a chemical plant that is being built in a ghetto in a Third World country. If the advantages like employment, profit, and useful products do not go to the people living in the ghetto, the risk may be unacceptable.

Source: Based on Otway and Von Winterfeldt (1982), Slovic et al. (1990), Shrader-Frechette (1991), and Stern and Feinberg (1996).

The above arguments refute the proposition that immediate conclusions can be drawn from risk assessments about the acceptability of risks. Risk assessments are nevertheless an important source of information for judging the acceptability of risks. Besides that, ethical considerations play a role too. Though there is no full agreement and an exhaustive enumeration is impossible (see, e.g., Lave, 1984; Shrader-Frechette, 1991; Hansson, 2003; Harris et al., 2005), we can at least mention the following four ethical considerations. We will elaborate on these points in the following sections:

1 the degree of informed consent to the risk;
2 the degree to which the benefits of a risky activity weigh against the disadvantages and risks;
3 the availability of alternatives with a lower risk; and
4 the degree to which risks and advantages are justly distributed.

6.5.1   Informed consent

Risks are more acceptable if those who run the risk consent to the risk in question. Some posit that risks are only acceptable if those running the risk have agreed to the risk after having received complete information concerning the risk. This principle is known as **informed consent**. The principle of informed consent originally stems from medical practice; it is closely related to ideas from normative ethics (see box).

> **Informed consent**   Principle that states that activities (experiments, risks) are acceptable if people have freely consented to them after being fully informed about the (potential) risks and benefits of these activities (experiments, risks).

Informed Consent and Normative Ethics

The principle of informed consent can be justified on the basis of ideas from normative ethics, as discussed in Chapter 3. It is a good match for Mill's freedom principle, which states everyone is free to lead their own life as long as it does not harm others. Informed consent is aimed at creating conditions through which people can act according to the freedom principle. This principle posits that risks are only acceptable if people have chosen them freely.

Besides these points,, informed consent closely ties in with Kant's second formulation of the categorical imperative: "Act as to treat humanity, whether in your own person or in that of any other, in every case as an end, never as means only." As we saw in Section 3.8, this means that we must respect the moral autonomy of others to reach their own choice. In other words, informed consent is aimed at creating the conditions under which people can make an autonomous choice. People must decide for themselves whether a risk is acceptable or not.

There are different ideas about how the principle of informed consent should be applied in technology. One idea is to allow this to occur through the economic market. In such a scenario, it is assumed people will decide for themselves which risks they wish to take concerning the purchase of certain risky or dangerous products. The choice of many individuals results, through a kind of invisible hand, in an optimal risk level. However, it is doubtful whether market transactions lead to informed consent in practice. First, consumers are often insufficiently or incompletely aware of the risks of technical products. So we cannot speak of real consent to the risks involved. People, for example, buy cell phones without being aware of any possible health risks. The principle of informed consent is not as easy to apply in engineering contexts as in, for example, medical contexts: the doctor–patient relationship is more direct and predictable than the engineer–user relationship (Peterson, 2017). Engineers rarely interact directly with the user and technological devices are sometimes (mis)used in ways that engineers could not reasonably have foreseen. In the case of Boeing 737 MAX, however, we could argue that the engineers had a moral obligation to inform the pilots about the automatic correction system MCAS. By the violation of this principle of informed consent, the pilots were not aware of the existence of the system MCAS, let alone its operation and impact. Pilots responsible for the safe operation of airplanes need to be properly informed about all critical systems, and consent to the use of systems that take away control from the pilots who are ultimately responsible for passenger safety.

Second, technical products often introduce risks that affect people other than the buyers or sellers of the products in question without their consent. One example is paint with organic solvents that contribute to environmental problems such as smog. On top of this, the economic market does not behave as ideally as many economists would have us believe. The choice of consumers is often limited because of the existence of monopolies, for example. In a number of cases, safer technologies simply do not reach the market even if they are technically feasible.

The Ford Pinto and Informed Consent

In Chapter 3 we discussed the Ford Pinto case. We saw that Ford decided not to alter the design of the Pinto, because the benefits to society from such an alteration supposedly did not weigh against the costs to society.

What is striking about the way, in which Ford dealt with this problem, is that the company believed that it had to make the choice of whether reducing the risk weighed up against the extra costs. There is, however, another way to deal with this problem: leave the choice to the consumer. Ford could have given consumers the choice to have an improved tank installed at limited costs, or to settle for the original design. In this way, the consumer could make their own choice and the extra risk would have been voluntary.

Informed consent can also be applied to technology by asking everybody who is potentially suffering from a risk for their consent. Unless everybody agrees that the risk be taken is considered unacceptable. A major disadvantage of this approach is that it gives almost unlimited veto power to individuals and will in many cases lead to a situation in which no risk is accepted at all, eventually making everybody worse off. One could try to avoid this situation from occurring by introducing the possibility of offering compensation for certain risks or by introducing the possibility of trading risks against each other.

6.5.2   Do the advantages outweigh the risks?

An important reason why risks can be morally acceptable is that risky activities can have advantages. More generally we could argue that risky activities and thus the risks that are linked to these activities are acceptable if the benefits of the activities outweigh the costs. These ideas are in agreement with consequentialism of which utilitarianism is a specific type (see Chapter 3). Methods have been developed based on such utilitarian arguments to determine the most desirable level of risk for a technology. One example of this is **risk-cost-benefit analysis** (Fischoff et al., 1981; Lave, 1984; Shrader-Frechette, 1985, 1991). This is a variant of

> **Risk-cost-benefit analysis** This is a variant of regular cost-benefit analysis. The social costs for risk reduction are weighed against the social benefits offered by risk reduction, so achieving an optimal level of risk in which the social benefits are highest.

the regular cost-benefit analysis (see Subsection 5.6.1). In risk-cost-benefit analysis, the social costs for risk reduction are weighed against the social benefits offered by risk reduction. The optimal risk level of a certain technology or product is where the net social benefit – the social benefits minus the social costs – is as high as possible. The basic ethical ideal behind this is that we should strive for the greatest happiness for the greatest number (Bentham). Such risk-cost-benefit analyses are often carried out by engineers.

There are two important objections to risk-cost-benefit analysis. First, such an analysis may commit the fallacy of pricing: it is not always possible to express all the relevant costs, benefits, and risks in money in a comparable fashion. People, especially, tend to have reservations about expressing the value of human life in monetary terms (as happened in the Ford Pinto case). Second, little attention is paid to informed consent and to the just distribution of costs and benefits, although these are important ethical considerations too. The second objection can in part be met by taking certain moral principles into account in risk-cost-benefit analysis (see, e.g., Kneese et al., 1983; Shrader-Frechette, 1985). A risk is only accepted, for example, if no one suffers from it – after compensation for those who might – or if it proves that those who were worst off are now better off.[7]

6.5.3   The availability of alternatives

The acceptability of the risks of a technology also depends on the availability of alternatives with lower risks. Suppose that two technologies, say A and B, introduce the same risks. Now also suppose that for technology A, an alternative is available with lower risk and no major other disadvantages, while for technology B such an alternative is absent. In this case, we may well conclude that the risks of technology A are unacceptable (because an alternative with lower risks is available) while the risks of technology B are acceptable (since no alternative is available and the technology is in other respects acceptable).

The importance of the availability of alternatives is also reflected in some environmental laws, like the Integrated Pollution Prevention and Control Directive of the European Union and the Clean Air Act and Clean Water Act in the US, which all use the notion of **best available technology** or a comparable notion. The idea is that environmental emissions should be reduced to the degree that is possible with the best available technology. This approach does not prescribe a specific technology but uses the best available technological alternative as yardstick.

> **Best available technology** As an approach to acceptable risk (or acceptable environmental emissions), best available technology refers to an approach that does not prescribe a specific technology but uses the best available technological alternative as yardstick for what is acceptable.

Best Available Technology

The European Union defines best available technology as follows:[8]

"the most effective and advanced stage in the development of activities and their methods of operation which indicates the practical suitability of particular techniques for
providing the basis for emission limit values and other permit conditions designed to
prevent and, where that is not practicable, to reduce emissions and the impact on the
environment as a whole:

- 'techniques' include both the technologies used and the way in which they are
  designed, built, maintained, operated and decommissioned;
- 'available means developed on a scale which allows implementation in the relevant
  industrial sector, under economically and technically viable conditions, taking into
  consideration the costs and advantages, whether or not the techniques are used or
  produced inside the Member State in question, as long as they are reasonably accessible to the operator;
- 'best' means most effective in achieving a high general level of protection of the environment as a whole."

6.5.4   Are risks and benefits justly distributed?

An important ethical consideration in accepting risks is the degree to which risks and benefits of risky activities are justly distributed. It is, for example, not just if certain groups of
people always have to carry the load of certain activities, while other groups reap the benefits. Some people believe that everyone should be treated equally with regard to risks.
This argument can be supported by Kant's first categorical imperative, which states that
you should act only according to the maxim whereby you can at the same time will that it
should become a universal law (see Section 3.8). This implies an equality principle.

Equal treatment concerning risks can be achieved by setting standards. Standards treat
everybody equally, which does not mean that everyone runs the same risk. It means that
the maximum permissible risk is equal for everybody in principle. Though standardization
can be defended by appealing to the ethical principle of equity, there are ethical objections
to be raised as well. A first possible objection is that little account is given to the pros and
cons of risky activities. This utilitarian objection is clarified by means of Figure 6.3 (Derby
and Keeney, 1990). Say that technology 1 is characterized by a curve through points A and
B in Figure 6.3. To allow technology 1 to meet the standards, considerable costs are
involved (point A). These high costs only reduce the risk slightly compared to point B.
Some utilitarians will therefore find point B more desirable than meeting the standard
against high costs in point A.

What is also possible is that standardization leads to higher risks than the situation in
which cost considerations are taken into account. Take technology 2, for example, which
is characterized by a curve through points X and Y in Figure 6.3. Technology 2 meets the
standard (point Y), but a much safer product can be designed against little additional cost
(point X). Some utilitarians will find point X a much more desirable result than point Y.

A second possible ethical objection against standardization is that it is paternalistic (see
Subsection 1.5.2 on paternalism). People do not get to choose which risks they find

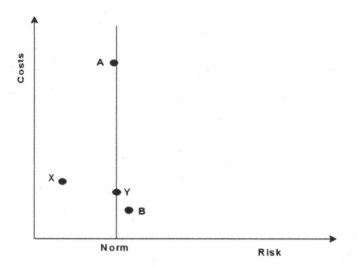

Figure 6.3 Costs for risk reduction for different technologies. The points A and B represent technology 1. Points X and Y represent technology 2. Adapted from Derby and Keeney (1990).

acceptable – the regulator, often being the government, does that for them instead. This objection is especially applicable to **personal risks**, that is, risks that only affect the buyer, user, salesperson, or producer. Market regulation for such products can be an option provided certain conditions are met, such as the availability of full information about risks and the freedom of choice between different products, because this leads to informed consent. For **collective risks**, that is, risks that affect larger groups, for example, floods, market regulation does not work. There can be no informed consent through market regulation because the risks affect other people besides the user and producer. These kinds of situations demand a collective decision about what acceptable risks are. Such a collective decision can – but need not – result in standardization.[9]

Personal risks Risks that only affect an individual and not a collective. For example, the risk of smoking. The relevant distinction with collective risk is whether individuals can stop or avert the risks for them individually. We can individually decide not to smoke but cannot individually prevent flooding for ourselves.

Collective risks Risks that affect a collective of people and not just individuals, like the risks of flooding.

One can also wonder whether an equal treatment of people concerning risk standards leads to a just division of risks. This is doubtful because the degree to which people benefit from a risky activity differs. An equal distribution of risks is, for example, not justified if only a limited group has the benefits of a specific risky activity. Justice and equality are related, but are not the same.

6.6   Risk Communication

According to some professional codes of conduct, engineers must inform the public about risks and hazards (see Chapter 2). In some cases, specialists are used, who are called **risk communicators**. Risk

Risk communicators Specialists that inform, or advise how to inform, the public about risks and hazards.

communication raises a number of ethical questions (Morgan and Lave, 1990; Valenti and Wilkins, 1995; Jungermann, 1996; Johnson, 1999). A first question is whether risk communication should only inform or also (try to) persuade. Can the government discourage smoking, or should it only inform about the risks of smoking? Another question is whether people should always be informed about risks even if it is not always in their best interests or if it is likely that they will interpret the information the wrong way. As a risk communicator, should you inform people of the risk of burglary if people have to leave their homes as quickly as possible because of the safety risk as a result of a coming hurricane?

In both examples, the contrast between duty ethics and consequentialism plays a role in the background. From the perspective of duty ethics, the consequences of risk communication are not relevant for the question "What is responsible risk communication?" Risk communication must first be honest (do not lie). Next to that, it must respect the freedom of choice and autonomy of people and hence not be paternalistic (see also Subsection 1.5.2 on paternalism). Here, the principle of informed consent is of importance too. It implies that you must not try to convince people but only inform them. From the perspective of consequentialism, the considerations and conclusions would be quite different in some cases. Risk communication is judged by means of the goodness of the consequences. Attempts to convince people by means of risk communication or withholding certain information can be morally right if it results in good consequences.

In risk communication ethical questions about the amount of information you give and how you present and structure your information also arise. How detailed should the information be for example? If a risk assessment was carried out, should you just give the result or should you talk about the uncertainty margin too? Should you explain how the risk assessment was carried out, so that people can check how reliable it is? From psychological research, we know that the way in which risks are presented has a great influence on the way an audience interprets these risks (see, for example, Martin and Schinzinger, 1996, pp. 134–136). It is even possible for people to take opposite decisions on the basis of the same information framed differently. The risk measure used can also influence how people interpret risks. Ethical judgments are often hidden in the use of a certain risk measure. In the risk measure "number of deaths per time unit," the assumption made is that each death has the same value. If the risk is expressed as a decrease in life expectancy, the implicit assumption is that the death of a young person is worse than the death of an old person. This is the case because the death of a young person weighs more heavily in this risk measure than the death of an old person.

Risk Perception

From psychological research on risk perception among the public, it appears a large number of factors plays a role in how people perceive risks:[10]

- The (perceived) voluntariness of the risk;
- The expected benefits of the activity or technology that causes the risk and the distribution of these effects;
- The maximal occurring negative effects and the possible controllability of these effects (degree of social disruption);
- The situation the risk is related to. Many people estimate risks related to work as smaller compared to other situations;

- The proximity and visibility of risks. Close risks are usually experienced as being greater than risks further away, or those that are less visible or imaginable;
- The way risks are presented and the risk measure that is used.

Source: Based on Slovic et al. (1990), and Martin and Schinzinger (1996, pp. 134–137).

Given the importance of the way in which risks are presented upon how they are perceived, the question to consider is what types of presentation are acceptable. It is important to realize that it usually is impossible to present risks neutrally. It makes a difference whether you express the maximum dosage of dioxin per day in picograms, milligrams, or kilograms. The latter presentation – maybe unintentionally – gives the impression that the risk is far smaller than in the first case. Nearly always, a certain method of presenting will intentionally or unintentionally make a certain interpretation more probable than another. That is not to say that you should consciously strive for a particular interpretation, but you should consider how you can present data in the most honest and best way.

6.7 Dealing with Uncertainty and Ignorance

Up to now, we have presupposed that it is possible to predict and express risks related to the hazards of technologies to a certain extent. But what happens if that is not the case? Consider the supposed health issues surrounding cell phones and the possible negative health and environment effects of growing and consuming genetically manipulated crops. In some of these cases, risks are calculated, but the question is whether all possible hazards have been assessed in a reliable way. A crucial question thus arises: Is it acceptable to introduce a new technology with potential hazards into society when there is scientific controversy or uncertainty about these hazards? This is a moral and political question because, on the one hand, it is desirable to protect society from hazards, while, on the other hand, outright forbidding a technology may also be undesirable. Much of the current debate on this question somehow focuses on the precautionary principle.

6.7.1   The precautionary principle

The **precautionary principle** was initially proposed to deal with environmental problems (see box). The principle can, however, also be applied to unknown risks and we will discuss that application here. The principle is mainly suitable for situations in which we cannot fully express hazards as risks because we have insufficient scientific knowledge. In general, the precautionary principle states that precautionary measures must be taken *if there are indications of certain hazards*, despite the fact that the hazards cannot be completely scientifically proven. So, it is not a general principle of cautiousness or "better safe than sorry" (Hansson, 2020).

Boeing clearly violated the precautionary principle. For example, after the two crashes, Boeing CEO Dennis

Precautionary principle   Principle that prescribes how to deal with threats that are uncertain and/or cannot be scientifically established. In its most general form the precautionary principle has the following general format: If there is (1) a threat, which is (2) uncertain, then (3) some kind of action (4) is mandatory. This definition has four dimensions: (1) the threat dimension; (2) the uncertainty dimension; (3) the action dimension; and (4) the prescription dimension.

Muilenberg reportedly put in a call with President Donald Trump to try to convince him not to ground his company's embattled 737 Max.[11] According to Muilenburg, it was still unclear what had caused the crashes, and there was no need for grounding the airplanes. There was, however, enough evidence, although uncertain, that there was something wrong with the automatic correction system MCAS. The same could be said for the FAA, which resisted grounding the airplane until March 13, 2019, when it received evidence of similarities between the two crashes. By then, 51 other regulators had already grounded the plane as a precautionary measure.

The Precautionary Principle

The precautionary principle originates from the Rio Declaration, the closing statement of the first conference of the United Nations on sustainable development, which was held in Rio de Janeiro in 1992:

> In order to protect the environment, the precautionary approach shall be widely applied by States according to their capabilities. Where there are threats of serious or irreversible damage, lack of full scientific certainty shall not be used as a reason for postponing cost-effective measures to prevent environmental degradation.[12]

Irreversible damage is usually understood in terms of environmental resources than cannot be replaced or restored. In the above formulation, the precautionary is primarily an *argumentative* principle: it indicates that certain reasons are invalid for arguing against environmental measures. The principle has also been formulated as a prescriptive principle that prescribes certain actions. A well-known formulation here is the so-called Wingspread Statement:

> When an activity raises threats to the environment or human health, precautionary measures should be taken, even if some cause-and-effect relationships are not fully established scientifically. (Raffensperger and Tickner, 1999, pp. 354–355)

The precautionary principle is hotly debated. Some philosophers and legal scholars have argued that the principle is basically a form of practical rationality or what Aristotle called practical wisdom (Adorno, 2004; Hansson, 2009). Others have argued that the principle is incoherent because "it forbids the very measures it requires" (Sunstein, 2005, p. 366). Partly the controversy seems to be based on different understandings of the "precautionary principle." Those who argue that the principle is basically a form of practical wisdom see the principle as an open-ended principle that can be specified in various ways (see the main text on the four dimensions of the precautionary principle that can be further specified). The main thrust of the principle is, according to them, that decisions are not only based on known risks but also on what we have called uncertainty above. As we have seen, engineers in fact already do that in safety engineering. Those who argue that the principle is incoherent have in mind a strong version of the principle that forbids any activities that potentially raise risks that have not yet been established scientifically. Since such potential risks are often inherent both to doing something and refraining from that something, they consider the principle incoherent (see also Miller and Engemann, 2019).

The precautionary principle has also led to debates between the United States and Europe in the World Trade Organization (WTO), for example on genetically modified organisms (GMOs). With reference to the precautionary principle, the

European Union introduced a de facto moratorium on new GMOs in 1998. In 2003, this was replaced by labeling requirement for genetically modified food. The US has opposed these measures because there was no conclusive scientific evidence of the risks of GMOs and the measures were seen as trade barriers. In 2009 the WTO ruled that the de facto moratorium was illegal under the WTO rules.[13]

Per Sandin has argued that the precautionary principle as a prescriptive principle contains four dimensions (Sandin, 1999): If there is (1) a threat, which is (2) uncertain, then (3) some kind of action (4) is mandatory.

The four dimensions in this formulation are:

1 the threat dimension;
2 the uncertainty dimension;
3 the action dimension; and
4 the prescription dimension, expressed here in the phrase "is mandatory."

One technology to which the precautionary principle has been applied is nanotechnology, in particular nanoparticles (see box). Application of the precautionary principle to this technology seems to imply that the potential hazards of nanoparticles should first be properly assessed before they are introduced on a large scale in society. Although this is a sensible strategy, one could raise the question of whether the strict separation that this strategy proposes between toxicity testing and introduction into society is tenable. The point is that in many cases reducing complexity and uncertainty is not possible without the introduction of the new technology into society. It is often only *after* introduction into society that the hazards of certain technologies can be properly assessed as we will see next.

Case   Nanoparticles

Figure 6.4   Nanoparticles. Photo: Annie Cavanagh/Wellcome Images.

Nanoparticles are very small particles with a size in the range of 1 to 100 nanometer (10^{-9} meter). Such small-scale particles often have quite different mechanical, optical, magnetic and electronic properties than the bulk material they are made of. Nanotechnology, therefore, makes it possible to create products with new characteristics and functions. Currently, nanoparticles are already used for sporting goods, tires, stain-resistant clothing, shoe polish, sunscreens, cosmetics, electronics, and self-cleaning windows.

Although some of the new properties of nanoparticles may prove very useful and economically important, they also imply a potential hazard because the toxicological properties of nanoparticles may be different from that of the bulk material. Currently, the toxicity of nanoparticles can be determined with tests on cultured cells or isolated organs (in vitro), with animal tests (in vivo), or with theoretical models based on knowledge about the effects of other small particles (e.g., ultrafine particles) on the human body (The Royal Society & The Royal Academy of Engineering, 2004, and Oberdörster et al., 2005). However, the current knowledge of potential hazards of synthetic non-biodegradable nanoparticles is limited. The philosophers John Weckert and James Moor have therefore proposed to apply the precautionary principle to the risks of nanoparticles (Weckert and Moor, 2007). They propose the following more precise formulation of the principle:

> If an action A poses a credible threat P of causing some serious harm E, then apply an appropriate remedy R to reduce the possibility of E. (Weckert and Moor, 2007, p. 144)

P causing harm E is here the threat component (1), "credible" refers to the uncertainty component (2), remedy R is the action component (3) and "apply" refers to the prescription component (4). Weckert and Moor argue that their version of the precautionary principle applied to nanoparticles "will require at least a concerted attempt at establishing the risks to health and the environment and perhaps slowing the development of products until the threats have been properly assessed" (Weckert and Moor, 2007, p. 144). The Health Council of the Netherlands comes to a similar conclusion in applying the precautionary principle to nonbiodegradable nanoparticles: before such particles are taken into production and are brought onto the market on a large scale, their toxicological properties should be properly investigated (Health Council of the Netherlands, 2006, p. 15).

6.7.2   Engineering as a societal experiment

Like science, engineering has an experimental nature. In science, hypotheses are deduced from theories, which are then tested in experiments. A hypothesis can be confirmed or falsified through the experiment. Analogous to this, newly designed products can be seen as hypotheses for properly functioning products. The hypothesis that they are properly functioning is usually tested through simulations and experiments in the laboratory, in small-scale field tests, or in clinical trials (in the case of medicine). Although such simulations, experiments, tests, and trials are very important in engineering, they do not always provide complete and reliable knowledge of the functioning of technological products and the hazards and risks involved. For a variety of reasons, it is often not possible to completely predict the possible hazards of new technologies before they enter society (Krohn and Weyer, 1994):

1 For carrying out a risk assessment, one first needs to identify possible hazards and failure mechanisms. It is, however, possible that certain hazards are overlooked. In laboratory experiments as well, certain hazards may not surface, so they become only apparent after the technology has been introduced into society. Examples of partly unknown failure mechanisms are the Tacomo Narrows Bridge and the Boeing 737 MAX.

2 Laboratory and field tests are not always representative of the circumstances in which the product eventually has to function. You need to know what circumstances are relevant in actual practice and which are irrelevant for performing a good test. This knowledge may only become available after a product has been introduced into society. An example is the case of 2,4,5-T (see Appendix VI).

3 Risks may be due to long-term cumulative effects of substances, sometimes in interaction with other substances. Examples are the possible effects of DDT, dioxin, and radioactive radiation. Cumulative or interactive long-term effects can hardly be studied in the laboratory; so the testing actually takes place during the use of the product in society.

4 Natural and socio-technical systems may be characterized by recursive and nonlinear dynamics. Even if a recursive natural system is deterministic, due to such dynamics future developments and possible hazards may be impossible to predict. The only way to find out the effects may be to actually introduce a new technology or substance into society.

The above implies that at least in some cases society has become the laboratory for engineering experiments. This means that it is ultimately during actual implementation of a technology in society that its functioning and possible hazards and risks are tested.

However, unlike traditional scientific experiments, **societal experiments** are usually difficult to terminate if something goes wrong. Moreover, in societal experiments with new technologies, the consequences can be much larger and can have an impact on third parties. If an elementary particle in a particle accelerator does not behave as expected, it is simply an interesting fact. If a nanoparticle turns out to be toxic, after it has been introduced into society, it is a potential disaster

> **Societal experiments** We speak of the introduction of new technology in society as a societal experiment if the (final) testing of possible hazards and risks of a technology and its functioning take place by the actual implementation of a technology in society.

as well as an interesting technical fact. Societal experiments in technology are nearly always large scale, can have irreversible negative consequences, and usually involve people as experimental subjects. In other words, they are experiments with major social consequences. Despite, or maybe because of, the mentioned characteristics of societal engineering experiments, it is still hardly recognized that society has become the laboratory for new technologies. As the European Expert Group on Science and Governance writes:

> [W]e are in an unavoidably experimental state. Yet this is usually deleted from public view and public negotiation. If citizens are routinely being enrolled without negotiation as experimental subjects, in experiments which are not called by name, then some serious ethical and social issues would have to be addressed. Even if no simple or accessible solutions exist to this problem, if our concern is public trust, surely a minimal requirement is that we acknowledge the public predicament. (Felt et al., 2007, p. 68)

A crucial question then is under what conditions it is acceptable to carry out societal experiments with new technologies. One important principle that has been proposed to judge this is informed consent (see also Subsection 6.5.1). Since World War II, informed consent has become the leading principle for experiments involving human subjects (Whitbeck, 1998). Before that date, the leading moral principle was whether experimenters would be willing to subject themselves to the experiment (Whitbeck, 1998). For medical experiments involving human subjects, informed consent is usually legally required. An example is medical experiments on human subjects that are regularly advertised in local media. For other experiments, it might not be legally required but can still be considered an important moral yardstick.

The engineering ethicists Mike Martin and Roland Schinzinger have proposed to apply the principle of informed consent also to societal experiments in engineering (Martin and Schinzinger, 1996). Such an application raises at least three issues. One issue is whether it makes sense to ask people to consent to *uncertain* hazards or even to hazards about which the experimenters are *ignorant*. What does consenting to an experiment which has an unknown risk amount to? It seems that it would imply having to accept all risks that emerge from the experiment because unknown risks potentially cover *any* risk. It is hard to see how people could rationally accept such experimental conditions. But if they do not give their informed consent (which seems the only rational thing to do), any societal experiment involving ignorance would be unacceptable.

This brings us to the second issue: is the principle not too restrictive? As soon as one individual objects to a certain societal experiment, it should be abandoned even if this experiment might bring large benefits to the rest of society. This seems unfair, at least in some cases, for example, if the actual hazard for the person objecting to the experiment is small and the social benefits are large. Of course, societal experiments in engineering can be carried out on a smaller scale than the whole of society, but still informed consent may be sometimes too restrictive a condition if large potential benefits are involved. Even for medical experiments, doubts have been raised whether the principle of informed consent does not unjustly exclude experiments from which large segments of society would profit (Hansson, 2004).

A third issue is how to deal with people who are indirectly involved in the experiment but are not able to give their informed consent. One specific example is future generations. For example, the introduction of nuclear energy in society amounts to a societal experiment that involves future generations because nuclear waste that remains radioactive (with current technology) for thousands of years is generated; even if this waste is stored or disposed of as well as possible, it will introduce uncertain and possibly unknown hazards to future generations. Some might conclude that the principle of informed consent shows that the introduction of nuclear energy is thus unacceptable, at least as long as it generates hazards for future generations. This judgment may, however, be a bit too quick as other sources of energy, especially fossil fuels, also generate hazards for future generations due to their contribution to the greenhouse effect or may not be able in the near future to meet the world energy demand (sustainable energy sources like solar and wind energy). One way to deal with this issue may be to introduce the notion of **hypothetical consent,** in contrast to actual consent: under which conditions would it be safe to assume that future generations consent to an experiment that involves hazards for them?

Hypothetical consent   Hypothetical consent refers to a form of informed consent in which people do not actually consent to something but are hypothetically supposed to consent if certain conditions are met, for example, that it would be rational for them to consent or in their own interest.

The mentioned issues raise serious doubts about whether it is desirable to apply the principle of informed consent to societal experiments with technology. It might, nevertheless, be possible to reformulate the principle to address these issues. Alternatively, if one rejects informed consent as a leading principle for societal experiments, one needs to propose an alternative approach that addresses at least the main underlying moral concern that people may not be subjected to societal experiments without knowing and without the ability to have a say, that is, without respect for their moral autonomy. More specifically, one could think of a set of principles, for example along the following lines:

- Experimental subjects are to be informed about the experiment, its set-up, risks and potential hazards, uncertainties and ignorance, and expected benefits.
- Societal experiments should be approved by democratically legitimized bodies. This can for example be parliament but also a governmental body that is controlled by parliament or the government.
- Experimental subjects should have a reasonable say in the set-up, carrying out, and (rules for) stopping of the experiment.
- Experimental subjects that are especially vulnerable to the hazards involved in the experiment should either not be subject to the experiment or be additionally protected.
- The experiment should entail a fair distribution of risks and benefits among different groups and among different generations.

6.8 Chapter Summary

As an engineer, you have a moral responsibility to ensure the safety of the technologies you design. Safety should here not be understood as the absence of risk but rather as the reduction of hazards and risks to an acceptable level. In this chapter, a number of methods have been reviewed that engineers can apply to live by this responsibility in professional practice. You can employ a number of different strategies for safe design including inherently safe design, adding safety factors to your design, adding negative feedback mechanisms, and providing multiple independent safety barriers. In addition, risks can be assessed through risk assessment, a systematic process consisting of four steps: (1) release assessment; (2) exposure assessment; (3) consequence assessment; and (4) risk estimation. On the basis of such risk estimations, one can reflect on the acceptability of certain risks. However, the question of whether a risk is acceptable depends on more than just the magnitude of the risk. More specifically, the following ethical considerations were identified that are important to judge the acceptability of risks:

1 the degree of informed consent to the risk;
2 the degree to which the benefits of a risky activity weigh against the disadvantages and risks;
3 the availability of alternatives with a lower risk; and
4 the degree to which risks and advantages are justly distributed.

Reflection on the acceptability of risks not only demands technical and scientific expertise, but ethical expertise too. Engineers do not have this expertise to any greater extent than other people. That is why the question of whether a risk is acceptable cannot usually be answered by engineers alone. This limits the responsibility of engineers on the one hand, but it gives them additional responsibility on the other, that is, they must properly inform others of the risks (risk communication) and involve them in decisions concerning the acceptability of risks.

Engineering often takes place under conditions of partial ignorance – in circumstances in which not all the risks of technology can be foreseen beforehand. To deal with potential hazards in such cases, one might employ the precautionary principle. In its most general formulation, this principle says that: If there is (1) a threat, which is (2) uncertain, then (3) some kind of action (4) is mandatory. This principle can be made more concrete in engineering in several ways. It can for example call for additional safety measures (along the lines discussed earlier) but also call for more risk assessment before a technology is introduced into society. Although the precautionary principle is useful, it cannot take away all the uncertainty with which the introduction of a new technology into society is accompanied. In that respect, the introduction of technology into society amounts to a societal experiment and the question is under what conditions such experiments are ethically acceptable. An often mentioned criterion here is informed consent, although it may be doubted whether this criterion can be usefully applied to societal experiments. Nevertheless, it is important that societal experiments somehow respect the moral autonomy of potential victims of such experiments.

At first sight, the role of uncertainty and ignorance in engineering seems to restrict the responsibility of engineers. As we saw in Section 1.3, knowledge of the consequences is a condition for responsibility. The fact that technology development always involves uncertainty and ignorance, therefore, diminishes the responsibility of engineers. On the other hand, science teaches us that there always is ignorance and unknown risks, and thus there is a special responsibility for engineers because they can indicate where there is uncertainty and ignorance and what the potential hazards may be. So engineers must not only communicate about what they know but also about what they do not know – a task that many engineers find very difficult. Nevertheless, this competence can be viewed as an important virtue for engineers.

Study Questions

1 Why do engineers have a responsibility for safety?
2 What is the difference between uncertainty and ignorance?
3 Give five arguments why the argument "if technology x with risk r is acceptable then a technology y with the same risk r is also acceptable" is not sound.
4 What is the difference between type I errors and type II errors? What type of error is worse in your view during a risk assessment? Argue your answer.
5 What is the difference between personal and collective risks? Is the risk of nuclear energy an individual or collective risk? Could this risk be dealt with by informed consent and, if so, how?
6 Why is an equal distribution of risks not always just? Give an example to illustrate your answer.
7 What is meant by engineering as a societal experiment? Can you give an example of the introduction of a technology in society that is clearly experimental in this sense? Argue your answer.
8 Mention a technology in which the risks are acceptable while the technology itself is unacceptable. Can you also think of an example of a technology that is acceptable while its risks are unacceptable?
9 Consider the following situation. A country is preparing for the outbreak of a rare disease.[14] If the disease arrives in the country and if it is not abated 600 people will die. To abate the disease, the following abatement programs are available:
   - Program A saves 200 people;
   - Program B in which there is a probability of 1/3 that 600 people are saved and 2/3 that nobody is saved;

- Program C in which 400 people die; and
- Program D in which there is a probability of 1/3 that nobody dies and 2/3 that 600 people die.
    a. Are programs A and C different in terms of expected fatalities and people saved? And programs B and D?
    b. Which program do you think will be preferred by most people?
    c. Is it possible to present the risks and advantages of the programs neutrally? If so, how? If not, what would be the best way to present the risks and advantages of the various programs?

10 In a risk assessment of genetically modified corn, it is argued that: "Since we have not been able to show scientifically that there are adverse health effects, genetically modified corn does not pose a health risk."
    a. Is this argument sound?
    b. Suppose that the precautionary principle is applied to the introduction of genetically modified corn as consumer food. What would that then imply?

11 As an engineer you are responsible for the safety of a new train tunnel. The tunnel consists of twin train tunnels. Every x meters a connection will be established between both train tunnels, so that in cases of an accident (for example the outbreak of fire), train passengers can more easily escape. This will reduce the number of expected fatalities and injuries as the result of accidents. You need to make a design decision about the desirable value of x. Possible values for x are 50, 100, or 250 meters. Relevant data are given in Tables 6.1 and 6.2.

A relevant design standard prescribes that the probability of an individual train passenger being fatality injured due to an accident should be lower than 10^{-7} per trip. This design standard is not legally binding. In a handbook, you have found the data for the societal costs of fatalities and injuries as result of a train accident (Table 6.3).
    a. What is the maximum distance x if the design standard is applied?
    b. How large should x be on basis of the ethical framework of classical utilitarianism?
    c. How should in your view a decision be made about the desirable distance x?

Table 6.1 Pertinent data for the safety of a train tunnel

Connection every x meter	*Average number of fatalities per accident*	*Average number of injuries per accident*	*Additional costs per year (construction and maintenance)*
x = 50	5	100	500,000 Euros
x = 100	10	200	300,000 Euros
x = 250	20	200	200,000 Euros
No connections	40	500	No additional costs

Table 6.2 Pertinent data of train accidents

Probability of an accident (per year)	0.01
Average number of passengers per train	400
Average number of trains per year	2500

Table 6.3 Estimated societal costs of fatalities and injuries as result of a train accident

Fatality	500,000 Euros per death
Injured	50,000 Euros per person injured

Discussion Questions

1   Who should in your view decide about the acceptability of risks? Engineers? Politicians? Company managers? The public? Argue your answer and discuss what your view would imply for the responsibility of engineers with respect to safety.

2   Do you consider informed consent a good principle for deciding about the acceptability of technological risks and hazards? Argue why or why not. If you do not consider informed consent a good principle, indicate how the moral autonomy of possible victims should then be protected. Or is this moral autonomy not important in your view?

3   Should risk communicators take into account the effect of their information on the public or should they solely try to ensure that people interpret the information in the right way and can make their own decision?

4   Do you agree that the precautionary principle is incoherent because it forbids the very measures it requires? Explain why you think that the principle is coherent or incoherent and what this (in) coherence implies for the acceptability of the principle.

Notes

1   Which does not mean to say that design and production adaptations are unimportant when it comes to striving to reduce CO_2 emissions.

2   Although it is obviously so that one technology contributes much more to CO_2 emissions than another.

3   The ostrich's fallacy states that if a product, activity or technology X does not give rise to any detectable risk or there is no scientific proof that X is dangerous, then X does not give rise to any unacceptable risk.

4   A third method that is sometimes used is to ask experts to estimate the risks.

5   The example is taken from Simon (1974).

6   It has been estimated that people are willing to accept voluntary risks that are up to 1000 times larger than involuntary risks. See, for example, Starr (1990).

7   The First principle is known as the Pareto Principle (cf. Zandvoort, 2000); the second one as the Difference Principle (Rawls, 1971).

8   Article 3(10) of the Industrial Emissions Directive (2010/75/EU).

9   It can also be based on informed consent of all involved but is then likely to sustain the status quo, which is often concerned morally problematic.

10   Some of these factors, especially the first three, are closely related to the acceptability of risks. This implies that people might well implicitly use a definition of risk that maintains a certain relationship between the magnitude of a risk and its acceptability.

11   https://www.nytimes.com/2019/03/12/business/boeing-737-grounding-faa.html, accessed April 21, 2022.

12   www.gdrc.org/u-gov/precaution-7.html, accessed June 14, 2022.

13   https://cordis.europa.eu/article/id/25179-wto-panel-rules-eu-moratorium-on-gmos-was-illegal, accessed June 14, 2022.

14   The example is based on Tversky and Kahneman (1981, p. 453). See also Martin and Schinzinger (1996, p. 134).

7

The Distribution of Responsibility in Engineering

Having read this chapter and completed its associated questions, readers should be able to:

- Describe the problem of many hands and explain how it applies to engineering;
- Judge responsibility distributions by the moral fairness and by the effectiveness requirement;
- Explain the difference between moral responsibility and legal liability;
- Distinguish different notions of legal liability and discuss their pros and cons;
- Describe the different models for allocating responsibility in organizations, to discuss their pros and cons, and to apply them;
- Describe how engineering designs may affect the distribution of responsibility;

Contents

7.1 **Introduction** 214

7.2 **The Problem of Many Hands** 217

    7.2.1 The CitiCorp building 218
    7.2.2 Causes of the problem of many hands 221
    7.2.3 Distributing responsibility 221

7.3 **Responsibility and the Law** 222

    7.3.1 Liability versus regulation 224
    7.3.2 Negligence versus strict liability 224

7.3.3 Corporate liability 227

7.4 **Responsibility in Organizations** 227

7.5 **Responsibility Distributions and Technological Designs** 231

7.6 **Chapter Summary** 236

Study Questions 237

Discussion Questions 238

Ethics, Technology, and Engineering: An Introduction, Second Edition. Ibo van de Poel and Lambèr Royakkers.
© 2023 John Wiley & Sons Ltd. Published 2023 by John Wiley & Sons Ltd.

7.1   Introduction

Case   Horizon Deepwater

Figure 7.1   Horizon Deepwater Photo. Photo: Chris Graythen/Getty Images.

On April 20, 2010 the crew of the Deepwater Horizon drilling rig, stationed in the Gulf of Mexico for exploratory drilling on the Macondo well, was testing the well that had been drilled and prepared for future production. During these tests, control of the well was lost, resulting in a large flow of well fluids up the riser. The fluids were diverted to a mud/gas separator that vents above the main deck. This resulted in a large release of hydrocarbons onto the main deck which quickly engulfed the rig in a hydrocarbon gas cloud. At approximately 9:49 pm, ignition of the gas cloud occurred, resulting in several explosions and a fire on the Deepwater Horizon drilling rig which led to the deaths of 11 workers and at least a dozen serious injuries. It is estimated that over 4 million barrels of hydrocarbons were released into the gulf over a period of nearly three months after the blowout. It is considered the largest accidental marine oil spill in the history of the petroleum industry.

Deepwater Horizon was a mobile, temporary rig to drill the well, to identify a viable reservoir of hydrocarbons, and then to make it safe and ready for more permanent production. This involves drilling a deep bore hole in stages and filling the casing with cement. The main companies involved in the Deepwater Horizon operations were BP (the well owner, and also responsible for the design of the well and for leasing the rig), Transocean (the owner and operator of the rig, and providing the rig crew), and Halliburton (responsible for the cement operations).

On the 9th of April, the final section of the well was drilled to the total depth of 18,360 feet to the hydrocarbon reservoir, and the day after the cement job was started to seal the well bore from the reservoir sands. The well had been sealed with casing and cement on the 17th of April. On the 20th of April, pressure tests were conducted to demonstrate well integrity. Among others, they carried out a negative pressure test, which places the well in a controlled underbalanced state to test the integrity of the mechanical barriers. An underbalanced state is one in which there is more pressure on the reservoir side. This ensures that the hydrocarbons will flow out of the well. The negative pressure test showed that hydrocarbons were leaking into the well, but BP's well site leaders misinterpreted the result. It appears that they did so in part because they accepted a facially implausible theory suggested by certain experienced members of the Transocean rig crew. Transocean rig personnel then missed a number of further signals that hydrocarbons had entered the well and were rising to the surface during the final hour before the blowout actually occurred. By the time they recognized a blowout was occurring, they tried to activate the rig's blowout preventer to prevent an explosion, but the blowout preventer emergency function failed to seal the well. Hydrocarbons flowed past the blowout preventer and were rushing upward through the riser pipe to the rig floor. Given the large quantity of hydrocarbon released, ignition was most likely; explosions followed.

The root technical cause of the blowout was that the cement that BP and Halliburton pumped to the bottom of the well on the 17th of April did not seal off hydrocarbons in the formation. The cement slurry was poorly designed, which was known since Halliburton's own internal tests showed that the design was unstable.

According to the Chief Counsels Report, "all of the technical failures at Macondo can be traced back to management errors by the companies involved in the incident" (National Commission on the BP Deepwater Horizon Oil Spill and Offshore Drilling, 2011a, p. x). The presented list contains many errors, such as

- BP did not adequately supervise the work of its contractors, who in turn did not deliver to BP all of the benefits of their expertise;
- BP personnel on the rig were not properly trained and supported, and all three companies [BP, Halliburton, and Transocean] failed to communicate key information to people who could have made a difference;
- BP did not adequately identify or address risks created by last-minute changes to well design and procedures;
- Halliburton appears to have done little to supervise the work of its key cementing personnel and does not appear to have meaningfully reviewed data that should have prompted it to redesign the Macondo cement slurry;
- Transocean did not adequately train its employees in emergency procedures and kick detection, and did not inform them of crucial lessons learned from a similar and recent near-miss drilling incident.

The National Commission on the BP Deepwater Horizon Oil Spill and Offshore Drilling established by President Obama released a final report on January 5, 2011. It made recommendations for preventing and mitigating the impact of any future spills that result from offshore drilling. In the report, the Commission stated that "[t]hough it is tempting to single out one crucial misstep or point the finger at one bad actor as the cause of the Deepwater Horizon explosion, any such explanation provides a dangerously incomplete picture of what happened – encouraging the very kind of complacency that led to the accident in the first place" (National Commission on the BP Deepwater Horizon Oil Spill and Offshore Drilling, 2011b, p. viii). Rather, the Commission concluded that "the accident of April 20 was avoidable. It resulted from clear mistakes made in the first instance by BP, Halliburton, and Transocean, and by government officials who, relying too much on industry's assertions of the safety of their operations, failed to create and apply a program of regulatory oversight that would have properly minimized the risk of deepwater drilling" (National Commission on the BP Deepwater Horizon Oil Spill and Offshore Drilling, 2011b, p. 127). In their accident investigation report, published on September 8, 2010, BP also concluded that no single action caused the incident, "[r]ather a complex and interlinked series of mechanical failures, human judgements, engineering design, operational implementation and team interfaces came together to allow the initiation and escalation of the accident. Multiple companies, work teams and circumstances were involved over time" (BP, 2010, p. 11). This shows how difficult it may be to pinpoint responsibility and blame in cases in which many actors are involved. Furthermore, all three companies, BP, Halliburton, and Transocean, blame each other for various events that cause the blowout.

By May 2010, over 100 lawsuits relating to the spill had been filed against one or more of BP, Transocean, and Halliburton Energy Services. In the civil trial on September 4, 2014, US District Judge Carl Barbier ruled BP was guilty of gross negligence and willful misconduct, and apportioned 67% of the blame for the spill to BP. Transocean was held 30 percent liable and Halliburton 3 percent liable; both companies were deemed "negligent." BP agreed to settle claims made by the plaintiffs' steering committee, the consolidated representative body for many of the individual victims of the spill, for at least $7.8 billion. BP reached an agreement with the US Department of Justice to plead guilty to 14 criminal charges, among them 11 counts of felony manslaughter, and violations of the Clean Water and Migratory Bird Treaty Acts. The agreement carried penalties and fines amounting to more than $4.5 billion, the largest of its kind in US history. Transocean agreed to a $1 billion civil penalty under the Clean Water Act, resolved claims made by the plaintiffs' steering committee for some $211.7 million, and also pled guilty to criminal violations of the Clean Water Act, resulting in a $400 million criminal penalty. Halliburton agreed to pay a $200,000 penalty after pleading guilty to criminal charges that its employees had destroyed evidence related to the spill, and settled claims with the plaintiffs' steering committee for some $1.1 billion.

Five individuals were charged with federal crimes, including the two rig supervisors Bob Kazula and Donald Vidrine who were prosecuted for involuntary manslaughter, however, the federal prosecutors had to drop the manslaughter charges because they could not meet the legal standard for instituting these charges. None of the charges against the five individuals resulted in any prison time, only probations and community services.

Source: Based on Van de Poel (2015), and https://www.britannica.com/event/Deepwater-Horizon-oil-spill/last, accessed November 17, 2022.

This case illustrates a number of issues with respect to responsibility in engineering. First, it shows how difficult it may be to pinpoint responsibility and blame in cases in which many people are involved in an activity and in which many causes contributed to a disaster. This is known as the problem of many hands. Second, it shows that even if we may have good reasons to hold someone morally responsible (blameworthy) for their actions that person might not be legally guilty or liable. So there is a difference between moral responsibility and legal liability. Third, the case raises the question of how we can best organize active responsibility in complex organizations in order to avoid disasters as with the *Deepwater Horizon* drilling rig.

In this chapter, we will first discuss the so-called problem of many hands in Section 7.2. Dealing with the problem of many hands requires attention for the distribution of responsibility in engineering. In this chapter, we will discuss three ways in which responsibilities are actually distributed in engineering, that is, through (1) the law (2) organizational models for responsibility, and (3) technological design. We will argue that in each case the resulting responsibility distribution can be evaluated in terms of moral fairness (are the appropriate persons held responsible?) and in terms of effectiveness (does the responsibility distribution contribute to avoiding harm and to achieving beneficial results?).

7.2 The Problem of Many Hands

Up until this point, we have focused on how individual engineers can behave responsibly. The social consequences of technology are, however, the result of the interaction between the actions of many different actors. Apart from engineers, this includes users, governments, companies, managers, and the like. One might assume that if all of the actors would behave individually responsibly, the overall result would be beneficial for society. This assumption does not always hold water: the fragmentation of decision-making may lead to different parts of an organization focusing purely on their areas of responsibility, and thus not feeling responsible for safety as a whole. For example, in the Deepwater Horizon case, BP did not adequately supervise the work of its contractors, and Halliburton appears to have done little to supervise the work of its key cementing personnel. So, different organizations were responsible for the supervision, but no one had overall authority. Thompson (2014) concludes from this that the responsibility design was diffuse, which probably contributed to the disaster. In addition, the principal agency for regulating the drilling, the federal Minerals Management Service (MMS), granted exceptions, such as the "special exception" in 2009 exempting *Deepwater Horizon* from the obligation to comply with the National Environment Policy Act. MMS also had never achieved the reform of its regulatory oversight of drilling safety consonant with practices that most countries had embraced decades earlier (National Commission on the BP Deepwater Horizon Oil Spill and Offshore Drilling, 2011b, p. 71). Furthermore, MMS conducted since 2005 at least 16 fewer inspections aboard the Deepwater Horizon than it should have under its policy.[1]

The problem of many hands describes the problem where a lot of people are involved in an activity, like a complex engineering project, therefore making it difficult to identify where the responsibility for a particular outcome lies (Thompson, 1980; Bovens, 1998). In part, this is a practical problem. It is often difficult in complex organizations or engineering projects to identify and prove who was responsible for what. Especially for outsiders, it is usually very difficult, if not impossible, to know who contributed to, or could have prevented a certain action, who knew or could have known what, et cetera. This is especially problematic if one wants responsibility to have juridical implications, because the law requires evidence of irresponsible behavior and this evidence has to meet a certain standard of proof.

The problem of many hands is also a moral problem. This is so because it may turn out that nobody can reasonably be held morally responsible for an engineering disaster. This is morally problematic for at least two independent reasons. The first is that many people, including victims, members of the public, and also the engineering community, may find it morally unsatisfactory that if an engineering disaster occurs nobody can be held responsible. Of course, the search for somebody to blame may be misunderstood, but at least in some situations, it seems reasonable to say that someone should bear responsibility. In fact, some philosophers have introduced the notion of **collective responsibility** to deal with the intuition that there is more to responsibility in complex cases than just the sum of the responsibilities of the individuals considered in isolation. Intuitively, we may say that a collective is responsible in cases where, had it been an action performed by one person, they would have been held responsible. This addresses the intuition that for outsiders it should not make a difference whether a complex engineering project was undertaken by one person or by a large number of persons in a division of labor. The second reason for attributing responsibility is the desire to learn from mistakes, and to do better in the future (Fahlquist, 2006a, 2006b). If nobody is held morally responsible for a disaster, this may not happen.[2]

> **Collective responsibility** The responsibility of a collective of people.

We can now characterize the **problem of many hands** as the occurrence of the situation in which the collective can reasonably be held morally responsible for an outcome, while none of the individuals can reasonably be held responsible for that outcome. In this definition, the case of the *Deepwater Horizon* is probably not a problem of many hands because it is likely that at least some of the individuals involved meet the conditions for individual moral responsibility, although it may be difficult to distribute the share of moral responsibility in a fair way. To illustrate the problem of many hands we will therefore look at another example.

> **Problem of many hands** The occurrence of the situation in which the collective can reasonably be held morally responsible for an outcome, while none of the individuals can be reasonably held responsible for that outcome.

7.2.1   The CitiCorp building

To illustrate the problem of many hands in engineering we will return to a case described in Subsection 3.9.3: the design and construction of the Citibank Headquarters in midtown New York. As we saw there this 59-story building was completed in 1977 and was designed by LeMessurier, a renowned structural engineer. In 1978, LeMessurier learned due to a series of serendipitous events that the tower's steel frame was structurally deficient. In Chapter 3 we focused on LeMessurier's behavior after this discovery. Here we focus on the situation in 1977 before LeMessurier discovered the flaw in the building. Apparently, the building was structurally deficient at that time, although nobody knew that. Who is to be held responsible for this structural deficiency?

To answer this question, we start by briefly sketching the main causes of the structural deficiency of the building. We then focus on the three main actors that causally contributed to this structural deficiency: (1) LeMessurier who designed the building; (2) the contractor who during construction decided to replace the welded joints with bolted joints; and (3) the employee at LeMessurier's firm who approved this change but did not inform LeMessurier about it (the "approver"). We will argue that none of these actors can reasonably be held responsible for the building being structurally deficient in 1977 and that this leads to a problem of many hands.

The structural deficiency of the CitiCorp building was mainly caused by a combination of two facts.[3] One was the peculiar design of the building, the other the change from welded to bolded joints. It was the combination of these two facts that made the building structurally deficient. Each of these facts considered in isolation did not jeopardize the structural strength of the building. The design was peculiar because the first floor was several stories above ground, with the ground support of the building being four pillars placed in between the four corners of the structure rather than at the corners themselves. The reason for this construction was that there had been a church on the building site and it had been agreed that this church would be reconstructed beneath the building after its completion. However, as LeMessurier found out in 1978 the combination of the peculiar design and the bolded connections made the structure vulnerable to high winds that strike the building diagonally at a 45-degree angle. Based on the New York weather records, a storm with a probability of occurrence once every 16 years (a so-called 16-year storm) would be sufficient to cause total structural failure.[4]

Now that we have some insight in the causes of the structural deficiency of the Citicorp building, let us see whether each of the three mentioned actors can reasonably be held responsible: LeMessurier, the contractor, and the approver. In doing so, we will apply the conditions for individual moral responsibility that were presented in Section 1.3. An individual is thus morally responsible for the structural deficiency if:

1 They actually caused that (i.e., the structural deficiency) for which they are being held responsible;
2 They did something wrong;
3 They could reasonably have known that the building was structurally deficient; and
4 They acted freely.

The fourth condition is fulfilled for all three persons: they were not forced to act in a certain way. The first condition is also met: each of them made a causal contribution to the structural deficiency, for example, by changing the design from welded to bolded joints (the contractor), by approving the design change (the approver) and by choosing this particular design (LeMessurier).

The crucial responsibility conditions here are, therefore, the knowledge condition (the person could have known of the deficiency) and the wrong-doing condition. If we apply these conditions, the following picture arises. LeMessurier cannot reasonably be held responsible in 1977, because he then did not know of the change from welded to bolted joints, which was crucial to foresee the structural deficiency of the building. The contractor, of course, knew about the change and probably also about the peculiar design but it seems reasonable to say that the contractor could not have known that the combination of these two factors would lead to structural deficiency. There are two reasons why the contractor could or should not have known this. First, in normal circumstances it would not have been a problem to change from welded to bolted joints. Second, the contractor, not being a structural engineer like LeMessurier, lacked the knowledge and expertise that was required to foresee this particular structural deficiency.[5] Moreover, the contractor asked for and received approval for the change from the approver and, therefore, was not at fault. Consequently, the contractor cannot reasonably be held responsible.

What about the approver? Could or should they have foreseen the structural deficiency before approving the change? And if so, did they act wrongly in approving the change? Here are some reasons why the approver cannot reasonably be held responsible. According to Morgenstern, LeMessurier argued that the

> choice of bolted joints was technically sound and professionally correct. Even the failure … to flag him [LeMessurier] on the design change was justifiable; had every decision on the site in Manhattan waited for approval from Cambridge, the building would never have been finished. Most important, modern skyscrapers are so strong that catastrophic collapse is not considered a realistic prospect; when engineers seek to limit a building's sway [the purpose for having welded joints], they do so for the tenants' comfort. (Morgenstern, 1995)

Furthermore, it even took LeMessurier several weeks in 1978 after hearing about the change in joints and being asked by a student about the structural strength to find out the vulnerability to 45-degree winds. Even if LeMessurier could, and possibly should, have foreseen the structural deficiency if he had known about the change from bolted to welded joints (which he did not), it seems reasonable to assume that the approver could not have foreseen the structural deficiency. The reason for that is that the approver is likely to have had considerably less experience and knowledge about the rationale for the design compared to LeMessurier. Hence, it is not reasonable to hold the approver responsible. It then turns out that none of the actors can reasonably be held responsible. To show that this is a problem of many hands, we also need to show that the collective can reasonably be held responsible in this case, a task to which we turn now.

In the CitiCorp case, we can define the collective as LeMessurier, the contractor, and the approver together. We assume that these three people can cooperate and share information. To attribute responsibility reasonably to the collective some conditions need to apply. For the moment, we will assume that these conditions are similar to the ones applying to individuals, that is, the collective acted freely, made a causal contribution, could have known it and was doing wrong.[6] It seems obvious that the collective acted freely, as each of the individuals acted freely. The collective also made a causal contribution to the structural deficiency. It is less clear whether the collective also meets the knowledge and wrong-doing conditions.

An important argument why the collective meets the knowledge condition is that if they had shared their knowledge and expertise they could have known that the building was structurally deficient. LeMessurier in fact drew this conclusion in 1978 after being informed about the change from welded to bolded joints.

Is the wrong-doing condition also met? From a consequentialist point of view, it obviously is: structural failure once in 16 years is unacceptably high; no engineer would contest that. One possible counter-argument is that the building still met the New York City building code because that code only requires taking into account 90-degree winds and not 45-degree winds and the building was only structurally deficient for the latter. Nevertheless, the effect of quartering winds was known long before the 1970s – the city's building code of 1899 already required to take all possible directions into account, although some later codes did not (Kremer, 2002). Moreover, wrong-doing is not confined to breaching the code. Engineers are expected to live up to a standard of reasonable care (like anyone else) as well as to act on state-of-the-art knowledge. According to Pritchard, "What counts as reasonable care is a function of both what the public can reasonably expect and what experienced, competent engineers regard as acceptable practice" (Pritchard, 2009). In this case, given the innovative design of the structure, it seems that engineers were required to take into account 45-degree winds (cf. Kremer, 2002). It thus seems reasonable to hold the collective in 1977 responsible for the structural deficiency of the CitiCorp building.

7.2.2 Causes of the problem of many hands

In the Citicorp case, the problem of many hands is primarily due to the distribution of information among the various actors. Due to the way information was distributed, neither LeMessurier nor the contractor nor the approver could reasonably have known that the actually built construction was structurally deficient. Still, at the collective level, the structural deficiency could reasonably have been foreseen (and the other responsibility conditions are also met). This reveals a more general cause of the problem of many hands: the distribution of information. The crucial point is that applying the knowledge condition to each of the individuals in isolation might yield a different result than applying it to the entire group of actors at once. This is why we might sometimes judge that none of the individuals could reasonably foresee a certain harm, while at the collective level that same harm is foreseeable.

The conflict between applying the responsibility conditions to the individuals and to the collective can also occur for other conditions, like the wrong-doing condition. An example is the responsibility of individual car drivers for the greenhouse effect. Individual car drivers by using their car to emit concentrations of greenhouse gases that are – considered in isolation – completely harmless (assuming that there is a level below which greenhouse gas emissions have no effect); all car drivers together, however, introduce a considerable risk for future generations. What is essential about this example is that while none of the individuals is doing something wrong or is at fault, at the collective level there is obviously harm done, so it would be natural to assume that there is also wrong-doing.

The problem of many hands can also arise due to a combination of conditions for responsibility. For example, an employee of a company who knows of a defect in a product may – due to the hierarchical nature of the organization and the specific procedures within the organization – lack the freedom to repair the defect or to warn customers about it. Their superior may have the freedom to act but maybe could not have known about the defect. The above suggests that the CitiCorp example is not an exception but that the problem of many hands is likely to occur regularly in engineering (and elsewhere).

7.2.3 Distributing responsibility

The notion of collective responsibility is helpful to articulate the moral intuition that under certain conditions people should be held responsible for disasters in complex engineering projects even if none of the individuals meet all the conditions for blameworthiness. However, it is not immediately clear what ascribing responsibility to the collective implies for the individuals who together form that collective. This requires attention to the **distribution of responsibility** among the members of a collective. But how should we distribute responsibility? In answering this question, we should keep

> **Distribution of responsibility** The ascription or apportioning of (individual) responsibilities to various actors.

Moral fairness requirement   The requirement that a distribution of responsibility should be fair (just). In case of passive responsibility, this can be interpreted as that a person should only be held responsible if that person can be reasonably held responsible according the following conditions: wrong-doing; causal contribution; foreseeability; and freedom of action. In terms of active responsibility it can be interpreted as implying that persons should only be allocated responsibilities that they can live by.

in mind that there are at least two reasons for ascribing responsibility. One is that we consider it morally important to hold people responsible for their actions and the consequences of these actions if certain conditions are met. In Section 1.3, we discussed the conditions that need to apply for holding people fairly responsible: wrong-doing, causal contribution, foreseeability, and freedom of action. We will call this the **moral fairness requirement**. The moral fairness requirement can also be applied to active responsibility: in that case, we will take it to mean that people should only be ascribed a certain active responsibility if they are able to live up to that responsibility. This, among other things, means that they should have the means and authority to fulfill their active responsibility.

The other reason why we ascribe responsibility is that we want to avoid harm and stimulate desirable outcomes. For utilitarians, this aim is the only aim of responsibility ascrip-

Effectiveness requirement   The moral requirement that states that responsibility should be so distributed that the distribution has the best consequences, that is, is effective in preventing harm (and in achieving positive consequences).

tions. The distribution of responsibility that has the best consequences, that is, is effective in preventing harm, is the morally required distribution of responsibility. We will call this the **effectiveness requirement**. The effectiveness requirement seems especially relevant for active responsibility because then nothing has gone wrong yet, but it is also relevant for passive responsibility because it is desirable that people learn from their mistakes and are deterred from doing certain things and both aims presuppose assuming responsibility for what went wrong.

We will assume that an ideal distribution of responsibility is both morally fair and effective. The problem of many hands shows that it is sometimes hard to meet both requirements at once. In cases like the CitiCorp case, it seems morally unfair to hold one of the actors responsible for the structural deficiency. Yet this distribution of responsibility, or rather the absence of it, does not seem very effective in avoiding harm. How can we reconcile the requirements of fairness and effectiveness? We do not have a clear-cut answer to this question. Instead, we will discuss below a number of mechanisms for distributing responsibility and their moral fairness and effectiveness. These mechanisms are the law (Section 7.3), organizational models for distributing responsibility (Section 7.4) and technological designs (Section 7.5).

7.3   Responsibility and the Law

Responsibility is not only a moral concept, but also a legal concept. The way the notion of responsibility is used in the law is however different from how it is used in ethics.

Liability   Legal responsibility: backward-looking responsibility according to the law. Usually related to the obligation to pay a fine or repair or repay damages.

We will therefore use the term **liability** to refer to legal responsibility. In what respects is liability different from moral responsibility? First, the conditions or basis by which someone is held liable are often different from the conditions by which someone is held morally responsible. The conditions for liability are laid down in the law and may differ for different types of actions, for different types of consequences and in different countries. For moral responsibility, usually the conditions set out in Section 1.3 are used. This difference means that it may well be possible for a person to be morally responsible for an action while they are not liable as we saw in the *Deepwater Horizon* case. Also, the opposite may occur: a person may be liable without being morally responsible. Secondly, liability is established in an official and well-regulated procedure in court. It requires a verdict by a judge or a jury and the liability conditions must be proven to apply in a formal juridical sense. Moral responsibility can be established more informally. Third, liability usually implies the obligation to pay a fine or to repay damages, while this is not necessarily an implication of moral responsibility. Fourth, liability always applies after something undesirable has occurred, while responsibility is relevant both after the fact as well as before something undesirable has occurred (active responsibility; see Section 1.4).[7] In the *Deepwater Horizon* case, it could, for example, be argued that the Halliburton's engineers, who knew about the poorly designed cement slurry, had an active responsibility to internally blow the whistle, even if they are not legally liable when they do not do this. The key differences between moral and legal responsibility are summarized in Table 7.1.

Even if liability and moral responsibility are different notions, one might make an attempt to make them as similar as is feasible. One could for example base liability on the same conditions as passive moral responsibility. An argument in favor of translating moral responsibility into liability is that if morally irresponsible behavior never leads to punishment on the grounds of legal liability then there remains little incentive to act morally so that few people will be encouraged to do that. However, one may doubt the assumption that people are inclined to behave immorally unless they are punished. An argument in favor of basing liability upon moral responsibility is the consideration that it would be undesirable to have immoral laws. However, laws that deviate in some respects from morality are not necessarily immoral. One reason for this is that not everything that is legally allowed is also morally allowed. In most countries, adultery is not forbidden by the law but that does not make adultery morally allowed in these countries. Rather the law is silent on it.

Both arguments thus do not necessarily lead to the conclusion that moral responsibility and liability *must* coincide. One might equally well say that the law, and therefore also liability, could never apply to all cases of moral responsibility. Therefore, even if one tries to translate moral responsibility into liability, this will never completely succeed. In some cases, it may even be desirable to make someone liable even if their moral responsibility is debatable. Holding people liable, even if they cannot reasonably be held morally

Table 7.1 Key differences between moral responsibility and legal liability

Moral responsibility	Legal liability
Moral blameworthiness based on conditions of wrong-doing, causality, freedom, and foreseeability.	Based on conditions formulated in the law
Can be established more informally; you can also consider whether you are yourself responsible	Established in well-regulated procedure in court; juridical proof of conditions required
Not necessarily connected to punishment or compensation	Usually implies the obligation to pay a fine or to repay damages
Backward-looking and forward-looking	Backward-looking

responsible, may make them more cautious, and may prevent negative effects and so help to solve the problem of many hands. In other words, the ascribing of liability can be based on considerations of effectiveness rather than on considerations of moral fairness. Below we will discuss some possibilities and limitations of liability as a tool for preventing undesirable consequences of technology.

7.3.1   Liability versus regulation

Liability is one of the legal tools that can be used to deal with the social consequences of technology. It is, however, not the only possible tool. The other main legal tool is **regulation**. Regulation can forbid the development, production, or use of certain technological products, but more often it formulates a set of the boundary conditions for the design, production, and use of technologies. However, such regulation is usually absent in the case of innovative designs. One reason for this is that regulation is usually based on our current knowledge of a technology and its consequences and on past experiences with that technology. Regulation is therefore often not able to deal with innovation. As a consequence, regulation will either have to forbid certain innovations or will lag behind the technological developments.

> **Regulation**   A legal tool that can forbid the development, production, or use of certain technological products, but more often it formulates a set of the boundary conditions for the design, production, and use of technologies.

Given the large economic and social benefits of innovation, most of today's governments refrain from regulation that forbids certain innovations outright. The consequence is that regulation tends to lag behind technological development and its consequences. This is primarily a problem of lack of knowledge and experience, but it is further aggravated by the fact that legal regulation is a long and cumbersome process. Even if certain negative consequences of a technology are discovered it may take years before they are adequately addressed in new regulations. In such situations, liability may provide an attractive alternative legal framework for dealing with the social consequences of technology.

Liability does not require the government to foresee the consequences of new technology but rather makes the ones developing those technologies, usually companies and the engineers employed by them, legally liable for those consequences under certain conditions. One might argue that this places the responsibility where it can be met best: in the hands of the ones developing technology. They have the best knowledge of new innovations and their possible effects and are in the best position to avoid certain disadvantages. Moreover, a scheme of liability would stimulate them to employ their (active) moral responsibility. A next question then is: what is the best form of liability to stimulate this?

7.3.2   Negligence versus strict liability

> **Negligence**   Not living by certain duties. Negligence is often a main condition for legal liability. In order to show negligence for the law, usually proof must be given of a duty owed, a breach of that duty, an injury or damage, and a causal connection between the breach and the injury or damage.

The conditions that must be met in order for a person to be liable depend on the law and may, therefore, differ from country to country. Nevertheless, in large parts of the Western world, the main condition for liability is **negligence**. In order to claim negligence, proof must be given of:

1   A duty owed. This is usually a **duty of care**, which is the legal obligation that individuals adhere to a reasonable standard of care while performing any acts that could foreseeably harm others. Duties of care typically arise in particular relationships such as between parent and child or between landlord and

> **Duty of care**   The legal obligation to adhere to a reasonable standard of care when performing any acts that could foreseeably harm others.

tenant. Also the relation between engineer and the public defines such a duty of care. For engineers, the standard of care both depends on what the public can reasonably expect from engineers and what is common practice in engineering. The duty of care is thus based on the sometimes implicit moral responsibilities of engineers and need not be made explicit in the law;

2   A breach of that duty;

3   An injury or damage; and

4   A direct causal connection between the breach and the injury or damage. Usually it is not enough for liability that the breach is part of the causal chain, but it has to be a proximate cause of the injury or damage.

Negligence does not require that the defendant actually foresaw the damage but that a reasonable person in the position of the defendant could have foreseen the damage.

In contrast to negligence, **strict liability** does not require the defendant to be negligent in order to be liable. It is usually enough that the defendant engaged in a risky activity and that this activity caused the damage done. Technological innovation is obviously a risky activity in the sense that it might produce unknown hazards to society. So innovation is a possible candidate for strict liability. In fact, the US and the countries of the European Union recognize **product liability**, which makes a manufacturer liable for defects

> **Strict liability**   A form of liability that does not require the defendant to be negligent.
>
> **Product liability**   Liability of manufacturers for defects in a product, without the need to prove that those manufacturers acted negligently.

in a product, without the need to prove that that manufacturer acted negligently.

EU Council Directive 85/374/EEC for Product Liability

Article 1   The producer shall be liable for damage caused by a defect in his product.

Article 4   The injured person shall be required to prove the damage, the defect and the causal relationship between defect and damage.

Article 6.1   A product is defective when it does not provide the safety which a person is entitled to expect, taking all circumstances into account ...

Article 7   The producer shall not be liable as a result of this Directive if he proves: ... (e) that the state of scientific and technical knowledge at the time when he put the product into circulation was not such as to enable the existence of the defect to be discovered ...

Article 15.1 (b)   Each Member State may by way of derogation from Article 7 (e), maintain or ... provide in this legislation that the producer shall be liable even if he proves that the state of scientific and technical knowledge at the time when he put the product into circulation was not such as to enable the existence of a defect to be discovered.

One reason for applying strict liability to technological products is that it motivates engineers and the other people involved in innovation to be very careful, for example, by investigating possible hazards and taking precautions. Strict liability will therefore probably result in a higher level of safety. It may in fact be the only way to meet Mill's freedom principle (Subsection 3.7.2) or the principle of informed consent (Subsection 6.5.1) for technological risks (see Zandvoort, 2000). Both principles forbid subjecting people to (unknown) risks unless they consented to the risks or the hazardous activity.

Strict liability, however, also has disadvantages. First, strict liability may well slow down the pace of innovation. This is often considered undesirable because innovation is an important source of social and technological progress in today's society. Against this, it may be argued that strict liability does not outlaw innovation, but only requires careful innovation. Second, it seems morally unfair to hold people liable when they are not at fault or could not have foreseen the damage. If strict liability is applied, engineers or the corporations for which they work may well be liable for the hazards of a technology, while they are not morally responsible (if moral responsibility is understood in terms of the conditions discussed in Chapter 1). On the other hand, it also seems unfair to the potential victims that they have to bear the damage: they could have done even less than the engineers to prevent the damage. In fact, one of the motivations for the European Union to introduce product liability is that strict liability (also called "liability without fault") in their view would result in a fairer distribution of the risks and benefits of technological innovation:

> liability without fault on the part of the producer is the sole means of adequately solving the problem, peculiar to our age of increasing technicality, of a fair apportionment of the risks inherent in modern technological production (Council Directive 85/374/EEC).[8]

Development risks In the context of product liability: Risks that could not have been foreseen given the state of scientific and technical knowledge at the time the product was put into circulation.

Nevertheless, the EU directive for product liability makes an exception for defects that could not have been foreseen given the state of scientific and technical knowledge at the time the product was put into circulation (see box).[9] Such unforeseeable risks are called **developments risks** and most schemes of product liability make an exception for development risks.

The exception for development risks is based on two considerations. One is that otherwise innovation would be too much hampered. The other is the concern that it would be morally unfair to make manufacturers responsible for damage they cannot foresee. Although both considerations are reasonable, it is questionable whether they provide a conclusive argument for excluding development risk from liability. First, excluding development risks from liability seems to put a bonus on not developing scientific and technological knowledge about the potential harms of new products. As soon as such knowledge is available, liability may apply. This is obviously an undesirable effect from the point of view of effectiveness because it might mean that society is unnecessary subject to certain risks. Second, even if it may be impossible to predict all the dangers of a new technology beforehand – and there are good reasons to suppose so as we saw in Chapter 6 – it is not obvious that the victims of the yet unknown risks should bear the damage rather than the ones having introduced the product or society as a whole. The fairness argument of the EU for product liability seems to apply to the development risk as well.

7.3.3   Corporate liability

Not only individual persons, but also corporations can be held legally liable. In such cases, the corporation is treated as a legal person. This is called **corporate liability**. An example is found in the *Deepwater Horizon* case in which the companies BP, Transocean, and Halliburton Energy Services were liable for the oil spill. A main advantage of corporate liability is that victims or the government do not need to find out which individuals in a company were responsible for, for example, a defect in a product but that they can simply sue the company as a legal person. If the company as a legal person is convicted, it is bound to pay a fine or compensate for the damage done. Despite this advantage, corporate liability also has a number of disadvantages and limitations. First, corporations, unlike natural persons, do not possess a conscience. They have "no soul to damn and no body to kick."[10] They cannot be put in prison for example. Therefore, legal instruments that are reasonably effective when applied to natural persons are not necessarily effective when it comes to corporations.

> **Corporate liability** Liability of a company (corporation) when it is treated as a legal person.

Second, most modern corporations are characterized by **limited liability** (Kraakman et al., 2004, pp. 8–9). This means that the shareholders are liable for the corporation's debts and obligations up to the value of their shares. Corporations may, however, well inflict more damage than the total value of their shares. This additional damage is then to be borne by the victims. This point is further aggravated by the fact that companies, unlike natural persons, can disappear by division, merger, or bankruptcy. Bankruptcy, mergers, or splitting up are sometimes deliberate strategies employed by corporate officials to avoid or limit liability claims.

> **Limited liability** The principle that the liability of shareholders for the corporation's debts and obligations is limited to the value of their shares.

Third, both the moral fairness and the effectiveness of corporate liability to a large extent depend on how the liability of the corporation is translated to individuals within the organization. A liability claim on a company may, for example, result in the dismissal of employees who did not partake in the damage for which compensation is claimed. On the other hand, managers who played a major role in the damage done may emerge unscathed. This is especially the case if their behavior was not illegal and if no individual liability on their part can be shown. This point draws attention to the allocation of responsibility in organizations, a theme that will be discussed next.

7.4   Responsibility in Organizations[11]

Most modern organizations are characterized by a division of tasks and roles. This has implications for who can be held responsible for what in organizations. In this section, we will discuss three different models for distributing responsibility in organizations:

1 The hierarchical model where those highest in the organization's hierarchy are held responsible;
2 The collective model in which each member of the organization is held to be jointly and severally responsible for the acts of the organization as a whole; and
3 The individual model in which each member of the organization is held responsible in relation to their contribution.

The models are in the first place intended to establish who is passively responsible for undesirable consequences. Which model is actually applicable depends to an important degree on the formal and actual organizational form of an organization. In the *Deepwater Horizon* case, some individuals – like the rig supervisors Vidrine and Kaluza, – were prosecuted for manslaughter (individual responsibility model), others – like David Rainey, BP's former vice-president – were primarily prosecuted because they were highest in hierarchy (hierarchical responsibility model). We might also pose the normative question of which model could "best" be applied, not only for allocating passive responsibility (after something has happened) but also for the distribution of active responsibility. Two considerations are then, again, important. First, whether the model is morally fair in how it allocates responsibility and, second, whether, it is effective in avoiding undesirable behavior. Below, we will discuss these issues for the three models.

Hierarchical responsibility

Hierarchical responsibility model The model in which only the organization's top level of personnel is held responsible for the actions of (people in) the organization.

In the case of **hierarchical responsibility model** it is only the organization's top level of personnel that is responsible for the actions of the organization. The hierarchical model is attractive because of its relative simplicity and clarity. In present-day practice, though, the hierarchical responsibility model is not always effective in preventing undesirable consequences. A main reason for this is the fact that the managers of organizations may be, to an extent, outsiders within their own organization. In practice, it is often very difficult for executives within an organization to get hold of the right information in time or to effectively steer the behavior of lower organizational units. Nevertheless, the knowledge that they will later have to account for damage done by the organization may motivate managers to gather the necessary knowledge, to better steer the organization, and to create a safety culture. For example, in the Deepwater Horizon case, BP's corporate culture, a cost-cutting culture for compromising safety, was identified as a factor leading to the disaster.

The hierarchical responsibility model also may seem somewhat morally unfair. If managers are not well informed about what is going on within their organization and are only able to steer to a limited degree, how can they then fairly be held responsible for undesirable activities? Therefore, the allocation of responsibility along strict hierarchical lines may lead to moral objections.

Collective responsibility

Collective responsibility model The model in which every member of a collective body is held responsible for the actions of the other members of that same collective body (and for the responsibility of the collective).

With the **collective responsibility model** every member of a collective body is responsible for the actions of the other members of that same collective body. The collective responsibility model is not attractive to large organizations, because it is not possible to allocate responsibility in differing degrees to individual members of the collective. Everyone is responsible to an equal degree for the actions of the collective body. Individual differences in being at fault or being able to prevent certain damage cannot be accounted for in this model. This is often seen as morally unacceptable. Another disadvantage of the collective responsibility model is that no one in particular tends to feel morally responsible for the consequences of the activities of the organization as a whole. In fact, everyone is held equally responsible for the actions of the whole, whether people as individuals have contributed to that or not. In such a situation it quickly becomes attractive to let someone else burn their fingers, especially in large organizations. The collective responsibility model would only seem to be applicable in a number of more exceptional cases. One important condition when it comes to introducing the model is that the members of the collective must be able to effectively influence each other.[12] This demands small-scaleness and equality between the members of the collective. In large organizations this condition is usually not met. One possibility would be to strive toward achieving smaller organizations with greater solidarity.

Individual responsibility

In the **individual responsibility model** each individual is held responsible insofar as they meet the conditions for individual responsibility as discussed in Section 1.3. A main advantage of this model is that it is morally fair. The model also might seem effective because it encourages individuals to behave responsi-

> **Individual responsibility model** The model in which each individual is held responsible insofar as they meet the conditions for individual responsibility.

bly. Nevertheless, as we have seen, individual moral responsibility may lead to the problems of many hands: the organization may collectively bring about undesirable consequences for which no individual can be held responsible. This is clearly a disadvantage of this model. Another disadvantage of this model is that companies may use their employees as scapegoats. In the Deepwater Horizon case, for example, BP has been accused that their findings about the disaster were not meant to determine what had gone wrong but rather who should take the fall, and the document was pointing a finger at Kaluza and Vidrine "who were too far down the corporate hierarchy to be compelling defendants in a case that involved a corporate culture run amok," according to the former chief of the Justice Department's environmental crimes section David Uhlmann.[13] One needs, however, not conclude that the individual model should never be applied. One also could try to avoid the problem of many hands through certain organizational measures, like the better sharing of information or empowering individual employees so that they can live by their individual responsibility. The latter can for example be achieved by a good company policy for internal whistle-blowing so that employees can raise issues without the fear of being dismissed (see also Subsection 2.3.4).

Conclusion

Obviously, none of the models discussed is ideal in terms of moral fairness and effectiveness. There is no single answer to the question of how responsibility can best be distributed in organizations. It should be noted that which model can be best applied in a particular case partly depends on how the organization in question is actually organized (and on the legal status of the organization). If an organization is, for example, organized along strictly hierarchical lines, it will be both morally unfair and ineffective to apply the collective responsibility model. The relation, however, also works in the other direction: if a certain responsibility model is judged most desirable in a particular situation or for a particular task, attempts can be made to make the organization fit the responsibility model. The box discusses a situation in which a design team has to be set up and a choice has to be made for a certain responsibility model and hence for certain organizational set-up of the design team.

How Should the Responsibility for Safety be Distributed?

Suppose that you are working as an engineer for a company producing cars. The head of the R&D department asks you to set up a design team for the design of a new type of truck. One of the issues you will have to take into account when setting up the design team is how to distribute the responsibility for the safety of the truck. In this case, the question is not how to allocate responsibility for safety given a certain organizational set-up, but rather how to best allocate this responsibility and of, next, finding an organizational set-up that matches that allocation. (Obviously, in your considerations other concerns than the responsibility for safety will play a role, but we will leave aside such concerns for the moment.) In this case, you could start with formulating the desiderata for an allocation of responsibility for safety. One could think, for example, of the following desiderata:

1 All individuals (or groups) to which a certain responsibility is allocated should be able to live up to those responsibilities.
2 For outsiders, that is, people who are not members of the design team, it should be clear whom to address if there is a concern or question about the safety of the truck.
3 The distribution of responsibilities should be effective in the sense of resulting in a safe truck.

Different models or combinations of models for allocating responsibility could be considered for achieving this. For example:

1 The hierarchical model, in which the leader of the design team is responsible for safety;
2 The collective model, in which all design team members are equally responsible for the safety of the truck;
3 The individual model in which each member of the design team is responsible relative to their individual contribution;
4 The responsibility for safety can also be allocated to a special safety official. Many organizations, for example, have special HSE (Health, Safety, and Environmental) officers for this type of concern.

When we evaluate these models with the mentioned desiderata, each seems to have specific pros and cons. The hierarchical model may do good on desideratum 2 (a clear address for outsiders), but it might be questionable whether the design team leader is able to oversee and steer all safety-related decisions, so that desiderata 1 and 3 are not fully met. The individual model will motivate all design team members to take safety seriously. However, guaranteeing the safety of the entire truck may require an integral approach that takes into account the interaction between different parts of the truck that are designed by individual engineers (or sub teams). On the individual model, nobody may be responsible for this interaction, which may have disastrous results for safety, for which it is very difficult to hold any of the individual design team members responsible; hence the individual model may not meet desiderata 2 (address) and 3 (effectiveness). On the collective model, the whole team (each individual member) can be addressed for failure to approach safety in an integral way. It is, however, questionable whether this model is effective (desideratum 3) and fair (desideratum 1) because the model presupposes that all design team members know and understand what the other members are doing and are able to influence that, which may be very difficult to attain in practice. For such reasons, one could choose for appointing a safety official who is responsible for an integral approach (the fourth model), possibly in combination with the individual model, so that each team member is also responsible for their own contribution.

In the philosophical literature on responsibility, various authors have pleaded for reinforcement of the individual model (Bovens, 1998). It should be noted that if the individual model is chosen, this would require changes in the way most organizations are currently organized. The individual model requires that within organizations people have the freedom to operate in actively responsible ways.[14] In some organizations this freedom may be limited. Moreover, the law often allocates liability to the organization as a whole (corporate liability) or to the owners or managers, rather than to individual employees. Again, this is a reflection of the way most organizations are currently organized and of the legal status of employees. As we have seen in Subsection 2.3.4, freedom of speech is, for example, not guaranteed within organizations to the same degree as it is in the relation between individuals and the state. Reinforcing the individual model may thus require major organizational and legal changes. In addition, the individual model has the great disadvantage that no one in particular is responsible for the collective consequences of individual actions. In all the other models some kind of provision is made for this, even though these other models have their own drawbacks. We may then conclude that a combination of the models, tailored to the specific requirements of the situation, will often be the best option.

7.5 Responsibility Distributions and Technological Designs

Not only do the law and organizations influence the distribution of responsibility, but engineering design does too. An example is the automatic pilot in an airplane. The automatic pilot takes over a number of actions from the pilots and consequently also takes over parts of their task. Design decisions about which tasks to allocate to the automatic pilot and which to human pilots are usually made with an eye to effectiveness in terms of safety and costs. What is less often taken into account, is that such design decisions also affect the (passive) responsibility for errors, for example in case of an accident. For example, if the automatic pilot is designed in a way that it can only be turned on and off during take-off and landing, the human pilots no longer have the freedom to correct the plane in case of a calamity during flight and can no longer be held responsible if such a calamity results in a disaster because freedom to act is one of the conditions for responsibility. In such cases, the designers (or producers) of the automatic pilot may rather be the ones that are to be held responsible. Sometimes, however, they may also not meet the responsibility conditions, so that nobody can be held responsible, and a problem of many hands occurs. This may be considered as an undesirable effect of the way the automatic pilot was designed in the first place, even if the most effective design, that is, the design that results, for example, in the lowest number of accidents, was chosen.

Another important, though often overlooked, issue is that if certain tasks are allocated to humans through design decisions, it should be ascertained that the conditions exist or can be created under which those individuals can responsibly carry out those tasks. Human pilots, for example, need information to be able to steer a plane. (The knowledge condition is one of the conditions for responsibility.) The system thus should provide them the right information on time. Note that even if this is the case, the pilots are dependent on the system, and on the system designers, for getting reliable information; pilots cannot simply look out of the window of the plane and estimate the flight altitude.

Technological design may not only allocate responsibilities to individuals, as in the case of the automatic pilot, but may also imply more complex divisions of labor and responsibility. Some technologies, for example, require a certain social structure to function properly. The Greek philosopher Plato in the *Republic* already argued that to navigate a ship one needs one and only one captain and that the crew needs to obey the captain. In other words, navigating a ship requires a hierarchical social structure. In a similar fashion, the contemporary philosopher of technology Langdon Winner has argued that the atomic bomb requires an authoritarian social structure: "[T]he atom bomb is an inherently political artefact. As long as it exists at all, its lethal properties demand that it can be controlled by a centralized, rigidly hierarchical chain of command closed to all influences that might make its working unpredictable. The internal social system of the bomb must be authoritarian; there is no other way" (Winner, 1980, p. 131). Winner might exaggerate the extent to which an authoritarian structure is required, but that the atomic bomb requires some social structure of control to prevent certain misuse seems undeniable. Technologies might therefore further or require an allocating of tasks and responsibilities along hierarchical lines. They may also further other, more complex, divisions of labor and responsibility as in the case of the V-chip (see box).

Case   V-chip

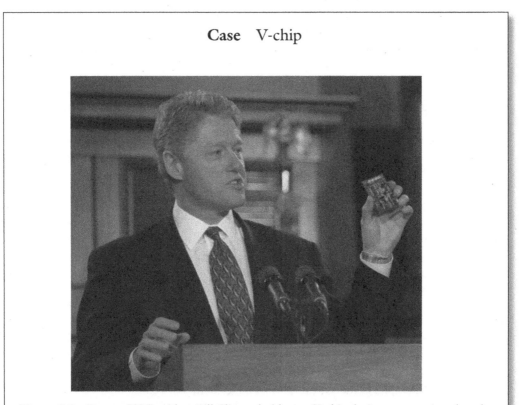

Figure 7.2 Former US President Bill Clinton holds up a V-chip during ceremonies where he signed the Telecommunications Reform Act at the Library of Congress in Washington, DC February 8, 1996. Photo: U.S. Federal Government/Wikimedia Commons/Public Domain.

The V-chip is an electronic device that can be built into televisions to block television program that are violent or otherwise unsuitable to children (FCC, 2009). In the USA, the V-chip is required for all televisions of 13 inch and larger since January 2000. The V-chip functions as follows. The television stations broadcast a rating as part of the program. The parents program the V-chip by setting a threshold rating. All programs above the rating are then blocked by the V-chip.

In order for the V-chip to function properly, it requires a uniform rating system and organizations doing the rating. The latter can be the TV station, the program makers, the government, or an independent review board. In the USA, the National Association of Broadcasters, the National Cable Television Association, and the Motion Picture Association of America have established a ratings system known as "TV Parental Guidelines." The program makers or television stations give ratings to the programs. A TV Parental Guidelines Monitoring Board monitors the application of the rating system and deals with complaints.

If one compares the V-chip with the traditional situation (see Figure 7.3) some interesting shifts in the division of tasks become clear. In the traditional situation, the parents decided directly what their children saw on television. They did so presumably by switching on or off the TV. There were in most countries no formal or legal restrictions for TV programs although there were some moral and aesthetic constraints, for example of "good taste."

The second situation is one in which the programs contain a rating (see Figure 7.4), which may be helpful for parents to decide which programs their children are allowed to see and which not. This system still operates in many countries outside the USA. In this system the parents still mainly decide what their children watch, although the judging of the programs have partly been taken over by others, who are applying the ratings to programs.

With the V-chip the actual role of the parents has further diminished (see Figure 7.5). If they choose to use the V-chip, they only have to set the rating they find

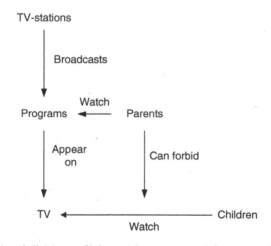

Figure 7.3 Traditional divisions of labor with respect to violence on television.

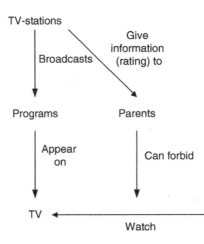

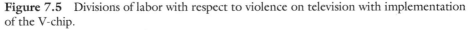

Figure 7.4 Divisions of labor with respect to violence on television in countries with a rating system.

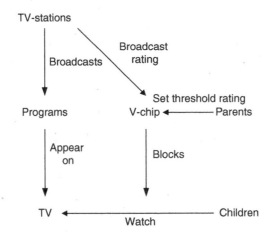

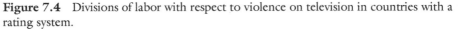

Figure 7.5 Divisions of labor with respect to violence on television with implementation of the V-chip.

acceptable for their children. Note, however, that in doing so they presume – at least tacitly – not only that the rating system is applied properly but also that the rating scheme coincides with their own norms and values. Actually, it might be the case that parents would judge some programs as unacceptable which are still rated as acceptable and, equally, they might consider some programs acceptable or even useful which are rated unacceptable.

Although the V-chip is still required in the USA, it has lost most of its relevance, since children hardly ever watch cable TV due to streaming platforms and social media. These streaming platforms also offer all kinds of parental control. For example, Netflix has created a system in partnership with the British Board of Film Classification (BBFC) to automatically age rate content for everything on its platform using an algorithm.[15] Parents can block specific shows, or set a profile so that their children can only view specific content by setting up a control on Netflix which can move the level from TV-Y, that is, appropriate for all children, all the way up to NC-17 ("No Children under 17 admitted"), or anywhere in-between. Also YouTube has created some tools to block content on YouTube, such as YouTube's Restricted Mode and YouTube Kids App.[16] A disadvantage of these video sharing websites is that their protection is at the browser level, so that parents have to set it up for every device their children use.

These parental controls allocate a number of tasks and responsibilities to a variety of actors. We could judge this resulting responsibility distribution in terms of effectiveness and moral fairness. For effectiveness, a major question is what the consequences of the responsibility distribution are. For example, do fewer children watch violent programs than without these controls? How do these parental controls affect the content that is broadcasted or streamed? It seems that TV stations, program makers, and streaming platforms get an incentive to make responsible content but the system might also be an excuse to make tasteless movies and shows because everybody can block these movies and shows they don't like with the parental controls.

With respect to moral fairness, one question is whether the actors involved can reasonably live up to the responsibilities that are allocated to them. Can parents be expected to program the parental controls? Can we expect TV Stations and steaming platforms rate videos and shows? Can we leave it to algorithms to rate videos and shows? A further question is whether these parties are the appropriate parties to assume these responsibilities. For example, if one would argue that parents should be the ones who are primarily responsible for what their children watch a crucial question is: Do the parental controls diminish the responsibility of parents (because other parties assume a role too) or do they enable the parents to assume more responsibility than before (because they now have better means to control what their children watch)? One could also wonder whether it is appropriate that TV stations and streaming platforms, or even algorithms, rate programs, and so influence what children watch. Some people would probably argue that this is not a proper task of TV stations and streaming platforms in a liberal society and that these decisions should be entirely left to the parents. And if algorithms are used to do the rating, can we sufficiently trust them and who is responsible if they rate in improper ways? (Perhaps the designers of the algorithm? But can they really foresee how the algorithm will behave?) Some authors even state that there should be no parental control at all, because "children have a right to privacy from their parents, because such a right respects their current capacities and fosters their future capacities for autonomy and relationships" (Mathiesen, 2013, p. 263).

Designing for responsibility

Technologies can be designed in such a way that they further the responsibility of users and other stakeholders. Fahlquist et al. (2015) have suggested the following heuristics when designing for responsibility:

1 *Moral agency*: Technological designs should not diminish the moral agency of users, operators and other stakeholders.

2 *Voluntariness or freedom*: Technological designs should respect or improve the voluntariness of actions, for example, by increasing the options of actions for users, operators, and other stakeholders.

3 *Knowledge*: Technological designs should provide the right knowledge in the right form for acting responsibly.

4 *Causality*: Technological designs should increase the control over relevant outcomes.

5 *Transgression of a norm*: Technological designs should make people aware of relevant moral norms and potential transgressions of them.

6 *Behavior*: Technological designs should encourage morally desirable behavior of users. It should, however, do so in a way that respects design heuristic 1, 7, and 8.

7 *Capacity*: Technological designs should encourage the capacity of users, operators, and other stakeholders to assume active responsibility, that is, their ability to reflect on their actions, and their ability to behave responsibly.

8 *Virtue*: Technological designs should foster virtues in users, operators, and other stakeholders.

9 *Completeness*: Technological designs should distribute responsibility in such a way that for each relevant issue at least one individual is responsible.

10 *Fairness*: Technological designs should distribute responsibilities in a fair way over individuals.

11 *Effectiveness*: Technological designs should distribute responsibility in such a way that harm is minimized and that goods are achieved as much as possible.

12 *Cultural appropriateness*: Technological designs should strike the balance between individual and collective responsibility in a way that is culturally appropriate.

7.6   Chapter Summary

Even if all the individuals involved in technological development act responsibly, the overall effect of their actions is not necessarily benign. A main reason for this is the problem of many hands. This is the situation whereby a collective can reasonably be held responsible for an outcome (such as the negative consequences of technological development) while none of the individuals involved meet the conditions for responsibility discussed in Section 1.3 (wrong-doing, causal contribution, foreseeability, and freedom of action). Overcoming the problem of many hands requires paying attention to the distribution of responsibility. In this chapter, we have discussed three mechanisms for distributing responsibility: (1) the law; (2) organizational models for allocating responsibility; and (3) technological design. In all these cases, the resulting responsibility distribution can be assessed in terms of moral fairness (are the appropriate persons held responsible?) and effectiveness (does the responsibility distribution contribute to avoiding harm and to achieving desirable effects?).

   The law makes certain people liable for certain actions and outcomes. Liability is the legal counterpart of moral responsibility. One can, however, be morally responsible without being legally liable and the other way around. Liability can also be used as a tool to prevent possible negative consequences of technology. It is then an alternative to regulation. Unlike regulation, liability does not require that the government foresees the consequences of new technology. Rather it makes the ones developing those technologies, usually companies and the engineers employed by them, legally liable for those consequences under certain conditions. What these conditions comprise of depends, among others things, on whether liability is based on negligence or is strict. Negligence requires a causal connection between the breach of a duty owed and injury or damage. Strict liability does not require the breach of a duty; it is enough to show that the defendant undertook a risky activity that caused the damage. An example of strict liability that is relevant for engineering is product liability. Product liability, however, usually excludes developments risks, that is, risks that could not have been reasonably foreseen at the time the technology was developed.

We have discussed three models for distributing responsibility in organizations: the hierarchical; the collective; and the individual. Each of the models has its own particular advantages and disadvantages in terms of fairness and effectiveness. None of the models provides a general solution to the problem of many hands. In peculiar circumstances, the models might, however, be usefully applied or combined. Also, engineering design influences the allocation of responsibility in technology. Differently designed technologies may provide users with different degrees of freedom and knowledge and may, hence, influence their responsibility because freedom of action and foreseeability are preconditions for responsibility. Technologies may also allocate certain tasks to certain actors and so influence the allocation of active responsibility.

Even if there is often not one responsibility distribution that is obviously the most attractive, additional attention on how to distribute responsibility might help to avoid or at least soften the problem of many hands. In the end, paying attention to the question of how to distribute responsibility is a responsibility to be taken up by individuals, but it is a responsibility that is easily overlooked if one focuses on individual responsibility only.

Study Questions

1 Explain why the problem of many hands is a moral problem.
2 In what ways do you think corporations may be moral agents? How do they differ from human agents?
3 What are the disadvantages of corporate liability?
4 a. What is strict liability?
   b. On which ethical principle(s) is strict liability based?
   c. Do you think strict liability is ever justified?
5 What is the difference between responsibility and liability?
6 What are the three models for allocating responsibility in organizations? Describe these models.
7 Looking back, which model for allocating responsibility could be best applied in the LeMessurier case for safety?
8 Explain how technological design can influence the allocation of responsibility.
9 Do you think that TV stations have the responsibility to rate programs with respect to the V-chip? Explain your answer.
10 Read the case of the *Herald of Free Enterprise* (see Appendix VI). Consider the following additional information on the safety of roll-on/roll-off ferries:[17]

   When it comes to formulating legal safety requirements, it is the *International Maritime Organization (IMO)* that has an important part to play. This international organization is responsible for adopting legislation for ships. IMO knew as early as 1981 that if water entered the car decks of roll-on/roll-off ships, they could be lost in a rapid capsize (Van Poortvliet, 1999, p. 52). The IMO did not adjust its regulations at the time to solve this problem, while a simple solution was available.

Because legislation adopted by the IMO needs to be implemented by *governments*, only governments accepting an IMO convention have to implement it. When making a convention it is, therefore, important to make it acceptable for as many governments as possible, otherwise only a small percentage of all fleets will be obliged to abide by the convention. A shipping company can decide to sail under the flag of another country which has not ratified an IMO convention, if complying with the convention costs a lot of money. So there is a certain amount of pressure on the IMO not to issue safety requirements that are too tight.

Apart from the IMO, *insurance and classification companies* also have a part to play in the formulation of safety requirements. For hull insurance bought by operating companies from insurance companies such as Lloyd's of London, a ship needs to be classified. Classification organizations are private organizations that have to monitor compliance with legislation during construction and the certification of sea worthiness during a ship's lifetime. Only the equipment and the construction are taken into account by the classification organizations, not passenger safety.

There is little incentive for *shipping companies* to ask for, or for *shipyards* to design ships, that are even safer than required by IMO conventions and hull insurance regulations. When disasters occur, the investigation that follows usually concludes that it was a human error that led to the disaster. Little attention is given to the design of the ship as long as on completion it complies with regulations.

a. Discuss for each of the following actors whether they are responsible (blameworthy) for the inherent instability of roll-on/roll-off ferries. Use the conditions for responsibility discussed in Section 1.3.
   • Maritime engineers designing these ferries
   • The IMO
   • Governments
   • Insurance companies
   • Classification organizations
   • Shipping companies
   • Shipyards
b. Is this a problem of many hands?
c. How could the active responsibility for increasing the safety of roll-on/roll-off ferries best be allocated?

Discussion Questions

1 In the aftermath of technological disasters like that of the *Deepwater Horizon* drilling rig, there is often a lot of attention on who is to blame. It could, however, be argued that for engineers the main concern is not blame but how to prevent such disasters in the future. In other words: one should not focus on backward-looking responsibility and blameworthiness but rather on forward-looking responsibility or active responsibility. Do you agree? Do the two perspectives exclude each other or are they somehow connected? Would the problem of many hands still be a problem if one focuses on forward-looking responsibility?

2 The text mentions two requirements for distributing responsibility: moral fairness and effectiveness. Do you consider one of these requirements more important than the other? How should conflicts between both requirements be dealt with?

3 Should legal liability in your view be based on moral responsibility as much as possible or not? What other considerations may be relevant for legal liability apart from moral responsibility (if any)? Would your view have consequences for currently existing forms of legal liability as they are discussed in this chapter?

4 Engineers often try to increase the safety of technological systems by technological devices (for example, automatic pilot, automatic shut-down of system, completely automated process control). What does this imply for the responsibility of the operators of these systems? Do you think that this makes those systems safer overall? Do you consider increasing safety by safety devices a desirable development or should safety be dealt with in another way?

Notes

1 https://www.huffpost.com/entry/deepwater-horizon-inspect_n_578079, accessed May 22, 2022.

2 It is sometimes argued that pinpointing responsibility may hamper learning and openness about incidents and near-accidents, because the focus is on blame instead of on openness and on learning. This may be true if the focus is on juridical responsibility, the paying of damage and/or the public blaming of the culprit. However, willingness to learn from accidents seems to imply the acceptance of at least some moral responsibility.

3 There was in fact a third factor: people from LeMessurier's team had defined the diagonal wind braces as trusses instead of columns so that no safety factor applied. The result was a smaller number of joints, which increased the structural deficiency. We leave this out because the building would also have been structurally deficient without this mistake; although the probability of failure would probably have been lower than once every 16 years, it would still have been unacceptably high.

4 The building was designed with an electric damper that, if functioning, would reduce the probability of failure to once every 55 years. That damper might however fail due to a power failure during a heavy storm.

5 One might argue that the contractor then should have consulted an expert about the change, but given that the change was approved by the approver, it would seem reasonable for the contractor to assume that this approval was based on the required expertise.

6 Applying these conditions to the collective raises the important philosophical question whether the collective can act and be held responsible. We here side-step this problem.

7 The law also implies sometimes forward-looking responsibility (e.g., the responsibility of parents for their children), but here we focus on liability, which seems to occur after the fact.

8 One may doubt whether it is the *sole* means for (fairly) apportioning responsibility for risks. Another possibility might be insurance.

9 Countries are free not to incorporate this exception in their national law, but most EU countries have followed the directive in this.

10 Attributed to Edward Thurlow, 1st Baron Thurlow (1731–1806).

11 This section is based on Bovens (1998).

12 Bovens gives four conditions under which the collective responsibility model can be usefully applied (1998, p. 103):
   - The collective must be characterized by a high degree of de facto solidarity.
   - Efficient, professional supervision from outside is not feasible.
   - Those who are held responsible for the conduct of other members or the collective should be aware beforehand that such a model of responsibility will be employed.

   Those who are held responsible should have the chance to exercise a certain degree of influence on the eventual outcome.

13 https://www.texasmonthly.com/articles/deepwater-horizon-prosecution, accessed May 30, 2022.

14 Here it has been presumed that the model can only be introduced if the opportunities are created for individuals within the organization to operate in responsible ways.

15 https://www.makeuseof.com/netflix-age-rating-algorithm, accessed May 31, 2022.

16 https://www.businessinsider.com/youtube-parental-controls?international=true&r=US&IR=T, accessed May 31, 2022.

17 Text is based on and partly drawn from Van Gorp and Van de Poel (2001).

8

Sustainability, Ethics, and Technology

Having read this chapter and completed its associated questions, readers should be able to:

- Distinguish between anthropocentrism and biocentrism;
- Describe what environmental problems are;
- Discuss the special responsibility of engineers regarding the environment;
- Reflect on the notion of "sustainable development" and how it can be justified and operationalized;
- Integrate considerations of sustainability in the design process, and think about how to deal with value conflicts in design for sustainability.

Contents

8.1 Introduction 242

8.2 Environmental Ethics and the Responsibility of Engineers 244

8.3 Sustainable Development 246

    8.3.1 The Brundtland definition 246
    8.3.2 Moral justification 248

8.4 Engineers and Sustainability 252

    8.4.1 Design for sustainability 252
    8.4.2 Specifying sustainability 253
    8.4.3 Life cycle analysis (LCA) 256
    8.4.4 Value conflicts in design for sustainability 258

8.5 Chapter Summary 262

Study Questions 263

Discussion Questions 264

Ethics, Technology, and Engineering: An Introduction, Second Edition. Ibo van de Poel and Lambèr Royakkers.
© 2023 John Wiley & Sons Ltd. Published 2023 by John Wiley & Sons Ltd.

8.1 Introduction

Case   Biofuels

Figure 8.1   Biofuel. Photo: Hajohoos/E+/Getty Images.

How can we continue to provide the energy for our transport needs? According to a theory known as "Peak Oil," the production of oil will decline in the near future. In the meantime, our transportation needs seem to continue growing. This scenario implies that we cannot continue to rely on oil for transportation. An additional reason for aiming at a reduced or no reliance on oil for transportation is the substantial contribution that the use of oil makes to the greenhouse effect through car emissions.

One proposed approach to this problem is to develop and use biofuels. Biofuel is solid, liquid, or gaseous fuel obtained from relatively recently lifeless or living biological material. The main difference with fossil fuels is that the latter is derived from long-dead biological material. The use of biofuel appears to have two obvious advantages. First, we will not run out of fuels since we can always grow more. Second, the plants that are used to make biofuels extract the green-house gas CO_2 from the atmosphere, so reducing the greenhouse effect. The proverbial two birds with one stone.

But as with all things, if something sounds too good to be true, it usually is. To start with its contribution to global warming. The reasoning stated earlier ignores the fact that farming costs energy because of the machinery used, and because of the fertilizers that need to be produced. Turning crops into biofuels also consumes energy. Given that this is energy spent "outside" the cycle of letting the plants grow and using the end products as fuel, it is obvious that the contribution to global warming is not as minimal as it would first seem. Perhaps more important are the unwanted side effects of using crops for producing fuel. Higher demand generally means higher prices. Crops such as sugarcane, wheat, and corn are likely to become more expensive which could contribute to hunger and poverty, especially in the Third World.

Sometimes it is claimed that second-generation biofuels solve these problems. Second-generation biofuels are based on non-food crops. However, producing such crops still requires land, water, and fertilizer, which may become short in supply or the prices of which will drastically increase. Therefore, second-generation biofuels are likely to contribute to an indirect increase in food prices. Moreover, the drive to produce large amounts of such crops is likely to reinforce already existing negative trends such as deforestation (to obtain arable land) and reliance on monocultures in agriculture.

Third and fourth-generation biofuels – producing fuels by means of bacteria or algae, respectively, producing fuels by means of genetically modified algae – are being researched and developed in answer to these concerns. Large-scale commercialized production of third and fourth-generation biofuels is not carried out yet due to the insufficient biomass production, high production costs as well as environmental and health concerns, and in the case of the fourth generation, the associated stringent regulations with respect to genetical modifications are considered a main barrier.

Source: Based on Naylor et al. (2007), Inderwildi and King (2009), Zah et al. (2007), and Abdullah et al. (2019).

A number of questions emerge from the biofuel case. Do engineers have an obligation to prevent future environmental problems, even if they mainly affect future generations? Isn't it better to leave every generation to solve the problems that are relevant to their lifetime? If engineers do have such an obligation, then why is that?

Environmental problems also have an important social dimension, since some problems hit those lower on the social ladder harder or first. For example, housing near sources of pollution is cheaper, such as housing near highways, municipal dumps, major industrial installations, etc. In an international context, economically weaker countries are willing to allow investments or encourage industry at the expense of strict environmental legislation. One example is the demolition of large ships including oil tankers. At high tide, these tankers are driven onto the beach (in Bangladesh among other places) at full speed. Following that they are demolished in a primitive way without any regard for the environment. One could say that environmental problems hit those lower on the social ladder harder, but also that the existence of inequalities (and their persistence) supports environmental problems.

Moreover, the discussion about biofuels clearly shows that there is a trade-off between our obligations to future generations and the present one: to what extent would we be justified in endangering the food provision to the poorest people alive now, in our efforts to minimize the impact to future generations of the fuel used for transportation? These are some of the questions that we shall address in this chapter.

Environmental problems

Three kinds of environmental problems can be distinguished:

1 We speak of *pollution* if something is added to the environment. Examples of this are the various ways air, water, and soil absorb "foreign" matter making them less suitable for supporting life.

2 If something is taken away from the environment, we speak of *exhaustion*. An example is the use of *non-renewable resources* like fossil fuels, ore, and materials like tropical wood. We call a resource non-renewable if it can only be consumed without the possibility of producing more of it now or in the future by for example growing plants. *Renewable resources* are resources of which more can be produced. So while fossil fuel is a non-renewable source, biofuel is a renewable source. Sunlight and wind, for example, are renewable sources that are constantly being replenished. A resource can also be renewable but easily depleted, as is the case with overfishing. Once too much fish of a certain kind has been caught it may take years, or even become impossible, to restore the fish population.

3 We refer to *degradation* if there are structural changes to the environment. Examples of this are soil erosion, a decline in biodiversity, the buildup of greenhouse gases, and the hole in the ozone layer. Note that degradation often occurs in conjunction with adding to or taking away from the environment, so the categories are not mutually exclusive.

Climate change, deforestation, air pollution, soil contamination, the availability of clean water, and overfishing are just a few examples of environmental problems that have increasingly drawn the attention and concern of both governments and individuals over the past few years. In many cases, technology is part of the cause of these problems but often it is also potentially a part of the solution. In philosophical ethics, these problems have led to the attention of what is called "environmental ethics." We will, therefore, begin this chapter with a short discussion of what we are to understand by "environmental ethics" (Section 8.2). In Section 8.3, we shall consider the term sustainability, how we can justify sustainable development from a moral standpoint, and problems related to the operationalization of sustainability. In Section 8.4, we shall look specifically at the role engineers can play to design for sustainability, specifically, the life cycle analysis, a useful instrument in design for sustainability, and how engineers can deal with value conflicts in design for sustainability.

8.2 Environmental Ethics and the Responsibility of Engineers

Ethics is traditionally involved with interaction between people. So even though you may have come across the notion that people have a certain responsibility regarding the environment, is this morally justifiable?

In the first instance, two different answers can be given to this. First, you could say that responsibility for the environment is derived from responsibility for humankind and society. For if we pollute the environment at will and exhaust our resources – burn up all our fossil fuels now – there may be severe consequences for people now and in the future. In environmental ethics this kind of justification is known as **anthropocentrism** (Achterberg, 1994); it is a position that states that the environment only has an instrumental value, that is, the value of its use by us (Baxter, 1974). Note that within this position it is possible to argue that for reasons of self-interest we must take far-reaching measures in the long run to protect the environment. A sustainable environment is necessary to provide the essential basic needs, such as food,

> **Anthropocentrism** The philosophical view that the environment has only instrumental value, that is, only value for humans and not in itself.
>
> **Biocentrism** The viewpoint that the environment has intrinsic value (value of its own).

clean air, and clean water. The second answer argues that the environment has a value of its own (an intrinsic value) and, therefore, should be considered in moral arguments. This is known as **biocentrism**. So even if the environment (or part of it) does not have any use value, it still has moral value. The notion of intrinsic value stems from environmental ethics and can be found in many policy documents on the protection of the environment. This does justice to the moral intuition that being worthy of protection does not depend on use value. We find this perfectly plausible for works of art, so why not for the environment?

If the environment has moral value, be it instrumental or intrinsic, it deserves to be protected. From this, it follows that people in general are responsible for the environment. Moreover, engineers have a special responsibility, because technology affects the environment in a number of ways. Some technologies contribute to the pollution of the environment; others lead to the use and possible exhaustion of nonrenewable natural resources, like coal or uranium. Technology also contributes to environmental problems like the greenhouse effect, overfishing, and loss of ecosystems. In the meantime, technology can contribute to the solution or prevention of many environmental problems. Innovative technologies may result in a reduction of energy consumption; CO_2 capture and storage may contribute to avoiding or at least reducing the greenhouse effect; and technologies can contribute to cleaning up environmental pollution.

Whether a particular kind of technology increases or solves environmental problems partly depends on the type of technology at stake. The environmental impact of a given kind of technology also depends on its design. Some refrigerators, for example, consume less energy than others. Copying machines that have double-sided printing as the default option, rather than single-sided printing, are likely to lead to reduced consumption of paper. How a particular technology is designed thus matters for its environmental impact, and this gives engineers a particular responsibility concerning the environment.

There is an increasing consensus in society concerning the existence of environmental problems and that environmental care is important, and moreover, that technology can play an important role in this. Engineers recognize the special responsibility for the environment, as we can see in the professional codes of conduct (see Chapter 2). The Code of conduct of the NSPE (National Society of Professional Engineers, see Chapter 2), for example, states that "Engineers are encouraged to adhere to the principles of sustainable development in order to protect the environment for future generations," and the Code of Conduct of ENGINEERS EUROPE states that "Engineers shall carry out their tasks so as to prevent … avoidable adverse impact on the environment." These broadly formulated responsibilities also imply that engineers have a responsibility to design technological products for sustainability.

8.3   Sustainable Development

8.3.1   The Brundtland definition

> **Sustainable development**   Development that meets the needs of the present without compromising the ability of future generations to meet their own needs (Brundtland definition).

Nowadays, sustainability is often conceived as the result of a societal and economic process of development, and that is why discussions about sustainability are often conducted in terms of sustainable development. The best-known and most influential definition of **sustainable development** originates from the Brundtland report:

Sustainable development is development that meets the needs of the present without compromising the ability of future generations to meet their own needs. It contains within it two key concepts:

- the concept of 'needs,' in particular the essential needs of the world's poor, to which overriding priority should be given; and
- the idea of limitations imposed by the state of technology and social organization on the environment's ability to meet present and future needs. (World Commission on Environment and Development, 1987)

The Brundtland Commission or the World Commission on Environment and Development (WCED) proposed the concept of sustainable development as an ideal for the global economy and corporations. Sustainable development is the foundation for today's leading global framework for international cooperation: the 2030 Agenda for Sustainable Development and its Sustainable Development Goals (SDGs) (see also Subsection 2.2.2). The SDGs form the framework for improving the lives of populations around the world and mitigating the hazardous man-made effects of environmental problems:

- SDG 13: Climate Action: Take urgent action to combat climate change and its impacts;
- SDG 14: Life Below Water: Conserve and sustainably use the oceans, seas, and marine resources for sustainable development;
- SDG 15: Life on Land: Protect, restore, and promote sustainable use of terrestrial ecosystems, sustainably manage forests, combat desertification, and halt and reverse land degradation and halt biodiversity loss.[1]

Sustainable development is not only related to the natural living environment, but requires an integrated approach that takes into consideration environmental concerns along with economic and social development. This means that certain considerations have to be taken into account: ecological values cannot unrestrictedly be given precedence over social justice or economic feasibility. For that matter, you might wonder whether these three aspects do not always clash. However, it is well possible that a measure or design choice can be positive for more than one field at the same time. Take, for example, a design in which you manage to reduce the amount of material required. This is favorable both from an economic and environmental point of view.

Points for discussion

What are the most important questions that the Brundtland definition raises?

- *Needs.* Which needs are we talking about here? Are none of the present needs disputable? Is any development that places limitations on people by definition non-sustainable? We are all becoming increasingly mobile, for example, we have at least one car, we go on holidays several times per year (using a plane), we desire spacious, comfortable, and properly heated living space, we want the latest model of mobile phone, and want to eat vegetables that come from abroad the whole year round. Should all this simply be allowed? If not, which needs are legitimate ones? Are they only the basic ones like food and shelter? And if we feel they go further than basic needs, how do we justify them?
- *Present needs.* Whose are they? The less quoted second part of the definition clearly shows that the Brundtland committee wished to have the basic needs of the Third World addressed first. As a result, we can deduce that the needs in the more prosperous parts of the world can only be fulfilled if the environment allows space for them.
- *Needs of future generations.* How do we know what these are? Superficially this point could lead to a statement like, "Of course we don't know what the needs will be in 50 or a 100 years, because 50 to a 100 years ago people did not know what we feel to be obvious now!" This interpretation does not suffice due to the above point of the legitimacy of needs: "obvious" needs are not necessarily legitimate. Nevertheless, this is an important point. Economic and technological developments may turn needs that are legitimate now into illegitimate ones and vice versa.
- *"Without compromising the ability of future generations to meet their own needs."* Today, there are almost 140 developing countries in the world seeking ways of meeting their development needs, but with the increasing threat of climate change, concrete efforts must be made to ensure development today does not negatively affect future generations.[2] Should we interpret this in such a way that we may not reduce the options of future generations at all, or should we interpret it as giving us the room to leave future generations with some ways of fulfilling their needs? If we take the second interpretation, based on the notion that technology will continue to develop, then to what extent should we assume that future technology will be able to meet needs? As the future level of technology is uncertain, we are running a risk that may be felt to be irresponsible. Moreover, there is another catch: technology does not develop on its own – it is something the present generation has to work on. That is why any argument that states we do not need to do much about sustainability because future technology will be capable of providing future needs more efficiently is suspicious to say the least. The first interpretation – in which we may not reduce the options for future generations in any way – is also known as the **standstill principle**. Essentially it states we must not pass on a poorer environment to the next generation than the one we received from the previous generation. The idea is also expressed in the notion that we must not rely on environmental loans, that is, do not create problems that we trust future generations will solve. There is at least one main disadvantage to the above definition: what do we mean by poorer? Does it

> **Standstill principle** The principle that we must not pass on a poorer environment to the next generation than the one we received from the previous generation.

mean that everything has to stay the same, or could the worsening of one environmental aspect be compensated by the improvement of another? The problem here is that the comparison soon becomes very hard. Can the storage of radioactive waste with a very long half-life be compensated by lower CO_2 production as a result of using electricity produced by nuclear power plants? Again it would be wrong to think that all answers to this question are equally good. Some answers are more ethical than others, that is, they lead to a more defendable justification for sustainable development.

8.3.2 Moral justification

The heart of sustainable development lies in two kinds of justice. The first kind of justice relates to the division of resources between our own generation and future generations: **intergenerational justice**, "without compromising the ability of future generations to meet their own needs." The question, "can we continue to use fossil fuels simply until we run out and let the next generation find alternatives?" falls under this heading. Next to that, sustainable development requires a just division of resources within our own generation (compare the First and the Third World): **intragenerational justice**, "the essential needs of the world's poor, to which over-riding priority should be given." The question, "can we use agricultural resources for producing biofuels even if that makes food more expensive?" falls under this heading. These two types of justice can be justified and described in various ways. We shall discuss three of the theoretical backgrounds. It will make little difference whether we refer to intergenerational or intra-generational justice; the only difference is that the first group is separated by time while the latter is separated by place.

> **Intergenerational justice** Justice that relates to the just distribution of resources between different generations.
>
> **Intragenerational justice** Justice that relates to the just distribution of resources within a generation.

Property rights

The first possible foundation for sustainable development stems from the historical principle attached to the justification of **property rights**.[3] The traditional question asked about property is whether we can justify that some matters belong to individuals. If we apply this historical principle to the environment and sustainable development, we get the following statement: the environment belongs to all of humanity, but what people produce with their own labor using natural resources belongs to them.[4] However, this only works as long as there are enough natural resources of the same quality left over for others. This is exactly the point that links up the formulation "without compromising the ability of future generations to meet their own needs" from the Brundtland definition. Only to the extent that we leave enough for future generations can we consider the environment and its resources as the property of this generation, and may we use it as we see fit. This induces that the present generation should take precautionary measures against natural resource abuse and pollution, and creates institutions to assure restitution when the environment is damaged (Brätland, 2006). This relates to the precautionary principle and the

> **Property right** The right to ownership of a specific matter or resource like money, land, or an environmental resource (like clean air).
>
> **Polluter pays principle** The principle that damage to the environment must be repaired by the party responsible for the damage.

polluter pays principle. The polluter pays principle is the principle that damage to the environment must be repaired by the party responsible for the damage.

The first mention of this principle at the international level is to be found in the 1972 Recommendation by the Organization for Economic Co-operation and Development (OECD) Council on Guiding Principles concerning International Economic Aspects of Environmental Policies, where it stated that: "The principle to be used for allocating costs of pollution prevention and control measures to encourage rational use of scarce environmental resources and to avoid distortions in international trade and investment is the Polluter-Pays Principle." It is still regarded as a regional custom because of the strong support it has received in most OECD and European Community countries.

The precautionary principle, which we introduced in Subsection 6.7.1, is often mentioned when we consider which risks of irreversible environmental damage are acceptable. As we have seen in the Rio Declaration the precautionary principle was defined as follows: "Where there are threats of serious or irreversible damage, lack of full scientific certainty shall not be used as a reason for postponing cost-effective measures to prevent environmental degradation." States should not wait to take preventive measures in case there is no clear or convincing scientific evidence of a cause-result relationship between an activity and its effects on the environment (or health) or between the input of substances and energy in the environment and their effects.

The important connection between sustainable development and the precautionary principle lies in the notion that we should not leave environmental loans to the coming generations. In other words, we should not give future generations problems that we allow to continue because we cannot agree on the question of whether there are serious environmental effects. In this way, we are not guilty of falling into the trap of a wait-and-see policy. The main problem with the precautionary principle is that it seems to forbid too much. Opponents point out that a number of important technical innovations that we now consider to be desirable would not have been implemented if we adhered to the precautionary principle. The high speed of train traffic was thought to have all sorts of negative effects, and if the suspicions had been taken seriously at the time then train traffic could never have become so important (see Schivelbusch, 1986). According to them, the precautionary principle places too high demands – in fact they are nonsensical demands since you can never prove that something will not cause damage. It seems more justifiable, though, to demand that we have reasonable grounds rather than incontrovertible proof that there will be no damage. But what is reasonable? Another problem lies in the cost-effectiveness. If we are ignorant of which damage will occur and especially what the chances are of such damage occurring, it is particularly difficult to say something about when a measure will be cost-effective.

Utilitarianism

The utilitarian approach defends intergenerational justice in the following way. Say that we develop in a way that does not allow us to continue our lifestyle. It would mean that sometime in the future the total utility for all would diminish. In turn, the aim of utilitarianism – the greatest happiness for the greatest number – would not be achieved. Development along non-sustainable lines, therefore, is undesirable. Two points can be mentioned in connection with the argument. First, the total utility must be maximized over an extremely extended period: maximization of utility in the short term will not lead to sustainability. Second, one could dispute this utilitarian reasoning by arguing that the expected aggregated total utility across the entire period would be greatest if we use a lot now. From an economic point of view, resources can often be better used now: due to, for example, inflation, one euro spent now is worth more than one euro spent in ten years. (Moreover resources well spent now will give a return on investment in the future). To correct this, economic analyses usually use a discount rate for future costs and benefits. High discount rates will make benefits and harms after 20 to 30 years almost negligible. While the use of discount rates is perfectly acceptable for traditional economic investments, its applicability to environmental problems is controversial (e.g., Hellweg et al., 2003). One might argue that we should not discount the rights of future generations, and moreover that some environmental problems, like, for example, climate change, should not be conceived as economic risks that can be compensated by higher benefits now, but rather as an existential threat. A discount rate of zero (or close to zero) might therefore be more appropriate. Such arguments, however, seem to extend beyond utilitarianism, which is insensitive to questions of distribution, both between different groups of people or across time.

Duty ethics

Intergenerational justice can be defended on Kant's categorical imperative (see Section 3.8). For example, one can also reason along the lines of the second formulation of the categorical imperative, the reciprocity principle: "Act as to treat humanity, whether in your own person or in that of any other, in every case as an end, never as means only." Future generations should be treated the same as all groups of people; they are not only a means but also an end in themselves. If we were to live a life of abandonment so that future generations cannot provide for their own needs, then we have used them as a means to achieve our own ends. Without their permission, we deny them the opportunity to strive for their own ends in the way that we did. However, if we strive to maintain the ability to fulfill needs in the future too then we do justice to the fact that future generations will have their own ends that they wish to strive for in a rational manner.

Often two arguments are mentioned against sustainable development. These can be set aside using duty ethics theories. The first counterargument is whether previous generations made an effort for us too. Apart from a high standard of living, we have also inherited substantial environmental problems. Given that we must use our ingenuity to cope with this, why would it be unjust to desire the same thing from future generations? According to duty ethics, however, it is not relevant how others factually treat us, but rather how we would prefer them to treat us. The fact someone steals my bicycle does not justify my decision to steal somebody else's bicycle.

The second counterargument is that the largest environmental problems are caused by population and consumptive growth in developing countries. This can lead to the question of why we should make efforts to solve a problem that has its origins elsewhere. However, we need to realize that as far as consumptive growth is concerned these countries still have some catching up to do. In other words, it would be odd to say that the problem does not lie with the developed world. If their own welfare could only be maintained by denying that welfare to others, then we can hardly call the maintenance of such welfare just. We would be using the others as a means and no longer view them as an end.

The three ethical foundations may defend intergenerational and intrageneration justice, however, they cannot solve the tension between these two types of justice which remains a serious and difficult issue. This tension also reveals itself in the biofuel example at the beginning of the chapter. Biofuels may contribute to intergenerational justice, by decreasing problems like oil shortages and the greenhouse effect for future generations, but at the same time, they may very well decrease intragenerational justice, because they will likely result in higher food prices that especially damage the already poor and hungry in the Third World.

Questions of justice are more prominent for some environmental problems than for others. One area in which they are particularly prominent is that of global warming, which is without doubt one of the main environmental issues of the present. In this field they have led to discussion about what is called climate justice (see box).

Climate justice

Reduction of greenhouse gas emissions into the atmosphere, also known as mitigation, is a vital requirement for combating climate change and ensuring that global temperature does not rise to unacceptable levels (above 1.5° Celsius as compared to the preindustrial level). This has been reiterated in a large number of policy documents and (international) climate negotiations, perhaps most prominently in the Paris Agreement (signed in 2016) and the Conference of Parties (COP26) in Glasgow (signed in 2021).

A key question that has arisen in discussions about climate change is how to fairly distribute the responsibility for mitigation and how to fairly deal with losses and damages due to global warming. This has led to the articulation of what is called "climate justice." Climate justice addresses a range of questions, including the following ones:

1 How to justly distribute the responsibility for mitigating climate change over countries?
2 How to justly distribute the "right" to emit greenhouse gases, or in other words how to justly distribute the available "greenhouse gas budget?"
3 How to justly distribute, or compensate for, the damages and losses caused by climate change?

With respect to the first question, several principles of justice have been proposed (Caney, 2021). The most prominent one is probably the polluter pays principle (PPP), which is also discussed in the text. According to this principle, the countries that have emitted most greenhouse gases in the past (absolutely or per capita) have the greater responsibility for mitigating climate change. One objection that might be raised against the PPP is that past greenhouse gas emissions may not always have been morally blameworthy. First, in the far past, people might have been unaware of the greenhouse effect; second, some greenhouse gas emissions are necessary for subsistence and therefore would not be blameworthy.

An alternative justice principle that has therefore been proposed is the "beneficiary pays" principle. This principle says that those countries that have unjustly benefited from (large) greenhouse gas emissions should bear the costs of mitigation. The resulting distribution of responsibility over countries may not be very different from the PPP, but the underlying moral principle is different.

A third proposed principle is "the ability to pay" principle. According to this principle, those countries who can pay for mitigating climate change should do so. This principle may be attractive from a more utilitarian or pragmatic point of view, as it may be (more) effective in ensuring climate mitigation. However, it may be felt that it is unjust to ignore past contributions to greenhouse warming.

With respect to the second question, one proposed approach is to distinguish between subsistence emissions and so-called luxury emissions, where the first has priority over the latter (Shue, 1993). Another often proposed approach is to distribute the available greenhouse budget equally per capita (Caney, 2021). The disadvantage of such an approach is that it treats climate justice in isolation from other (in)justices. For people in developing countries, it might be much harder to attain a reasonable level of well-being with the same amount of emissions, as for people in developed countries; so from a broader justice perspective, there are reasons to give them a relatively larger share of the greenhouse gas budget.

The third question concerns how to distribute the damages and losses that are the result of greenhouse warming. Such damages and losses are already occurring and even if limiting greenhouse warming to 1.5 degree Celsius is feasible, more of them are to come. Oftentimes, the countries and people that have the least historical responsibility for emitting greenhouse gases are the ones who suffer the most from the consequences of global warming, for example, India and the Sahel countries. The PPP would seem to imply that the historically biggest emitters – rich and industrialized countries – should compensate the damages and losses of the countries heavily suffering from the consequences of climate change. However, politically this remains a very contentious issue.

8.4   Engineers and Sustainability

Although there is only limited agreement on what sustainable development is, and on how or whether it will be achieved, engineers can certainly make a contribution. Technical knowledge can be used in different ways to directly or indirectly solve or prevent environmental problems. This means engineers are responsible for making use of those opportunities. There are many ways in which engineers can live by their responsibility for the environment and for sustainability. We might, for example, expect them to do an environmental impact assessment for the technological projects they undertake, or at least to commission one if they themselves lack the expertise. Here we will focus on how engineers can contribute by designing for sustainability, and what challenges that brings.

8.4.1   Design for sustainability

Life phases   The phases through which a product goes during its "life": production phase, use phase, and removal phase.

Design for sustainability is aimed at reducing the environmental impact of a product during its **life phases,** that is, *extraction, refining, production, use,* and *disposal.*[5] Each phase raises its own environmental questions. The choices for the environmental impact of one phase can influence the environmental impact of another phase. As an example of how considerations of sustainability work in a practical design context, we shall discuss the design of a heat pump boiler (see box).

Case   Heat Pump Boiler

A heat pump boiler is a piece of equipment that provides the household requirements for hot water by means of a thermodynamic cycle process in which heat is extracted from a source to heat the water. It is a kind of reverse fridge.

One important characteristic of such a device is that heat can be pumped in a direction where it normally would not go. The heat pump boiler can extract heat from groundwater or from ventilation air and because this normally unused low-quality energy is utilized, the yield (defined as [high-quality energy out/ high quality energy in] * 100%) even rises above 100%. So this would seem to be a truly sustainable device.

During design a number of choices have to be made that have an impact on the sustainability of the device:

- Life span: what life span does the device require?
- Recyclability: this is a term that can be defined in many ways. In principle nearly everything is recyclable. The question is how much energy is involved in the recycling. Reuse is another option: a heat pump boiler could be designed in a modular fashion, so that components with shorter life spans can be replaced allowing the parts that still work to be used much longer.
- Energy source: should the heat pump boiler work on natural gas or electricity?
- Just like a refrigerator, a heat pump boiler contains a primary refrigerant – this is a substance that has thermodynamic properties suitable for a cycle process. Primary refrigerants often have an impact on the greenhouse effect and the ozone hole, but they differ greatly in the degree to which they display these effects (see Section 5.6).
- The source medium: this is where the heat has to come from.

Note that placing these design choices up front does not even begin to touch on the question of whether the heat pump boiler technology as such is sustainable. It is quite possible that some other technology offers a better combination of functions or that applying heat pump boilers leads to more environmental problems than the older technology. Sometimes the greatest sustainability benefits can be gained by scrapping an entire concept instead of holding onto all sorts of small design choices within a given concept. This plays a role in the discussion about the hydrogen cell car. This kind of car may be more sustainable than the present generation of petrol, diesel, and gas cars, but the question of whether sustainable transport should consist of car-like solutions is not answered this way. Sustainable technology development requires that the system boundaries that the design choices are framed in are not ignored unless there is reason to do so.

8.4.2   Specifying sustainability

Many people feel the Brundtland definition is vague. There are all sorts of concrete ways to fill it in. There has been an explosive growth in such ways and as a result, many believe that sustainability has turned into a rather hollow phrase. However, we should not give in to this pessimism. Even though the level of abstraction of the Brundtland definition is high, it does not mean that it cannot be made more concrete in a sensible fashion. We should take the definition for what it is: a foundation for further discussion about how we should take responsibility in time and space. In short, the Brundtland definition requires specification: it should be detailed into a number of concrete policy measures if we are talking about environmental policy or into concrete design guidelines when we are talking about concrete cases of sustainable design of technology. The *values hierarchy* (see Subsection 5.5.2) can be a useful tool for specifying the value sustainability into more concrete norms and design requirements that can guide the design of a new technology. As an example, we will apply the idea of a values hierarchy to biofuels (see box).

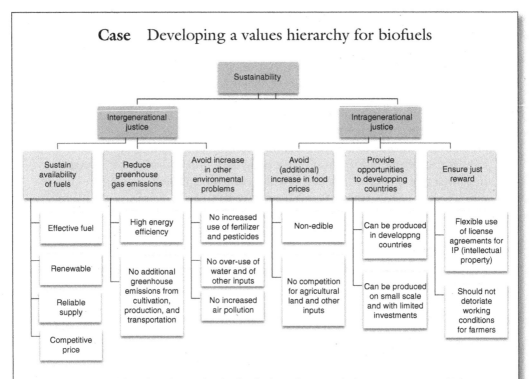

Case   Developing a values hierarchy for biofuels

Figure 8.2   Possible values hierarchy for biofuels. Values are dark grey, norms are light grey, and design requirements are white. Adapted from Van de Poel (2017a).

The UK Nuffield Council on Bioethics has in 2011 proposed an ethical framework for the development of biofuels that includes the following five ethical principles (Nuffield Council on Bioethics, 2011):

1 Biofuels development should not be at the expense of people's essential rights (including access to sufficient food and water, health rights, work rights, and land entitlements).
2 Biofuels should be environmentally sustainable.
3 Biofuels should contribute to a net reduction of total greenhouse gas and not exacerbate global climate change.

4 Biofuels should be developed in accordance with trade principles that are fair and recognize the rights of people to just reward (including labor rights and intellectual property rights).

5 Costs and benefits of biofuels should be distributed equitably.

These principles can be used to construct a values hierarchy for biofuels. To this end, these principles have been adopted for the context of the design and development of biofuels. Sustainability has been chosen as the overarching value and intergenerational justice and intragenerational justice as the values of which sustainability is compounded. Each of these compounding values is in Figure 8.2 translated into one or more general norms that are relevant in the case of biofuels and that help to attain the relevant value.

In the case of intragenerational justice, one general norm is the sustenance of the availability of fuels: biofuels offer an alternative to fossil fuels which will be exhausted at some point in the future. In addition, they may help decrease the emission of greenhouse gases that is caused by fossil fuels. (cf. the 3rd principle of the Nuffield Council). Third, they should not increase other environmental problems (cf. the 2nd principle of the Nuffield Council). The value of intragenerational justice relates to the 1st, 4th, and 5th principles of the Nuffield Council. In light of the 1st and 5th principles, it is undesirable that food prices increase due to the use of biofuels as this particularly affects the world's poor. In addition, one might want to require that biofuels help to increase intragenerational justice by offering developing countries new development opportunities and new means of income. The 4th principle of the Nuffield Council is here understood as the general norm: ensure just reward. Each of the general norms may, in turn, be translated into several more specific design requirements as illustrated in Figure 8.2.

Currently, there are no biofuels that meet all design requirements that are mentioned in Figure 8.2. So-called first-generation biofuels are edible. Second-generation biofuels are non-edible but they compete for land and other agricultural resources like fertilizer with food crops and, therefore, do not meet all requirements that derive from the norm to avoid increases in food prices. Third and fourth-generation biofuels are based on bacteria and (genetically modified) algae and are likely to meet the requirements with respect to the norm of not increasing food prices. However, they are still under development and expensive. They do therefore not (yet) meet the design requirement of competitive price.

From this values hierarchy for biofuels, it becomes clear that we face a value conflict (see Section 5.6) between intergenerational justice and intragenerational justice. While biofuels increase intergenerational justice by making fuels available to future generations and reducing greenhouse gas emissions, they reduce intragenerational justice, in particular by competing with food production and so having an upward effect on food prices. From the viewpoint of intergenerational justice, we should probably choose 1st or 2nd generation biofuels, and from the viewpoint of intragenerational justice we should likely prefer 3rd generation biofuels. The values of intergenerational and intragenerational justice are also not trumping each other because neither of them is overall more important than the other.

In designing for sustainability, life cycle analysis (LCA) is often proposed as a method to deal with such sustainability conflicts, as some LCA approaches propose an overall metrics for sustainability. We will therefore now first discuss LCA before we delve deeper into dealing with value conflicts in design for sustainability.

8.4.3 Life cycle analysis (LCA)

Life cycle analysis   An analysis that maps the environmental impact of a product across the entire cycle from raw material extraction (cradle) through disposal (grave).

LCA is a tool to assess the environmental impacts of a product across its life cycle. LCA can be very useful to understand design options, to compare them, and to see whether they meet design requirements. What is important is the integral nature of such analysis: fair comparisons between products, or better yet, of ways in which a series of functions can be carried out can only be achieved by looking at the whole life cycle.

The viewpoint is reflected in a number of instruments and design approaches – some are more quantitative and other more qualitative.

There are various software packages, like SimaPro and Gabi, suitable for a quantitative life cycle analysis. These software packages make use of a database in which various environmental consequences of production (obtaining raw materials, processing methods, and transport), use, and scrapping (reuse, recycling, and waste) are included. As a final result, all of the above is summarized in a score that expresses the total environmental impact of the product or design in question. An example of the result of a life cycle analysis is shown in Figure 8.3. To obtain better designs, engineers are given insight into which phases there is an impact and what is having an impact in the first place. Often conclusions can be drawn from this about the phase in which most can be done to improve environmental impact.

A disadvantage of the lifecycle analysis is that it focuses exclusively on the environmental impact. The question is whether the other aspects of sustainability – such as intragenerational and intergenerational justice – are sufficiently dealt with. Intergenerational justice is indirectly dealt with as the quality of the environment in the longer run is considered. The intragenerational justice also is indirectly addressed to the extent that environmental damage often is for the account of weaker parties in society, so a lower environmental impact helps that group in particular. However, this is quite something else from "giving priority to the essential needs" of the poorest people on the planet. That is not so strange considering if we remember that we are discussing the introduction of sustainability considerations in the design process. Outside that framework all sorts of steps must be taken to achieve true sustainability. However, lifecycle approaches have hardly any bearing on this.

A second disadvantage is related to the use of one-dimensional eco-indicators used to simplify the output generated by these kinds of packages, such as the Eco-indicator 99,[6] or the ReCiPe.[7] These indicators express the environmental impact on a one-dimensional scale (see Figure 8.4). Exhaustion of materials, pollution, and degradation are all aggregated in one measure by means of weighing factors. Aggregation supposes that it is always possible to compensate a loss in one environmental dimension by a gain in another. One might also wonder whether a gain in environmental impacts would compensate for an increase of hunger in the third world, or the other way around. As we already have seen in Section 5.6, some values are incommensurable, in the sense that they cannot be expressed on a common scale and that a loss in one value cannot be fully compensated by a gain in another value.

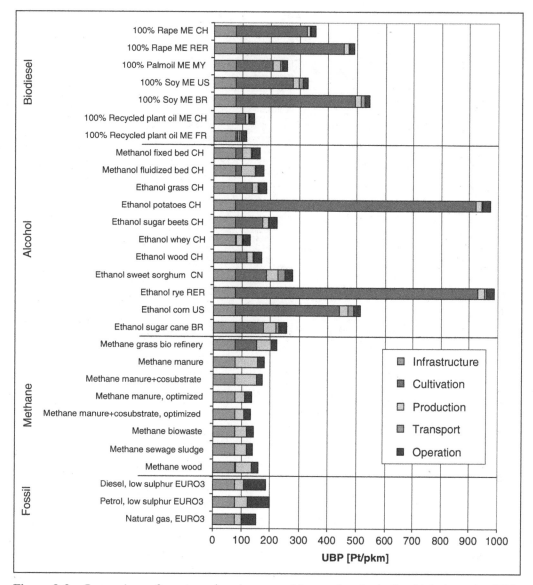

Figure 8.3 Comparison of aggregated environmental impact (method of ecological scarcity, UBP 06) of biofuels in comparison with fossil fuels (petrol, diesel, and natural gas). Zah et al. (2007)/ Reproduced with permission of EMPA.

A third disadvantage is that the normative judgments that are applied in the weighing factors often remain implicit. These weighing factors cannot be altered by the user of the software package. This gives a false sense of factuality and objectivity, while this is not justified. The weighing factors should be based on normative choices related to what should weigh most, which is of great importance for the final result.

Let's say that you want to determine when to replace a car on the basis of the environmental impact. If we look at the amount of energy used during the life cycle, we see that it is relatively small in the production phase compared to the energy used in the use phase. You could draw the conclusion that it would be more sustainable to buy a new car more often. The lower energy use of the new car easily compensates for the costs of its production.

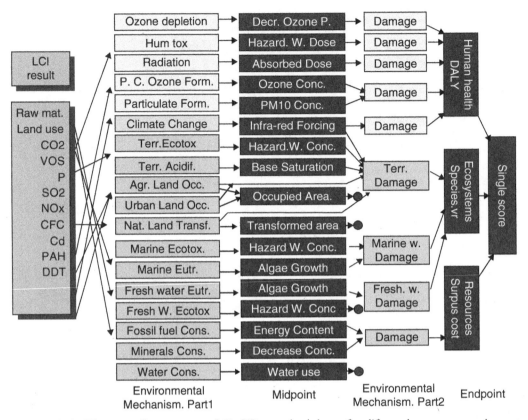

Figure 8.4 The overall structure of ReCiPe methodology for life cycle assessment impact assessment. Reproduced by permission of PRé Consultants B.V.

However, if we look at the output of damaging substances, this is relatively high during production compared to use. This would result in the opposite conclusion: do not replace your car too quickly. So the question is: should raw material exhaustion (use of fossil fuels) weigh heavier than air pollution? This is an important discussion with small variations for many products. However, the use of a quantitative lifecycle analysis with a one-dimensional output will not stimulate this discussion. The results, therefore, seem to be somewhat random, for they are based on the normative points of departure of the eco-indicator used. While there may be merits in concentrating this normative debate in the stage of constructing the eco-indicator, leading to clear-cut lifecycle comparisons between technological choices, we must nonetheless wonder whether the normative debate about weighing factors is best conducted at that level of abstraction, and whether it is a good idea to shield those who make concrete lifecycle assessments from this debate, often leaving them puzzled in the face of apparently randomly different results from different eco-indicators.

8.4.4   Value conflicts in design for sustainability

LCA is of limited use when it comes to dealing with value conflicts in design for sustainability. First, as we have seen in the previous section, when an LCA expresses environmental impact in one output metric, it is often based on assumptions that are untransparent and may be disputable. Second, and perhaps more importantly, in most cases, sustainability is not the only value that should be accounted for in design and engineering. We thus need approaches for dealing with conflicts between sustainability and other values as well, and not just LCA methods.

In fact, engineers are increasingly confronted with value conflicts that involve sustainability but also other values. One cause of this development is the fact that sustainability now is – or at least arguably should be – an important value in the design of any technology, whereas in the past the focus was often on specific technological sectors that cause severe environmental problems, like chemical engineering or transportation, or on technologies that can contribute to environmental protection or remediation.

This is not to say that sustainability is always sufficiently recognized as a value in all engineering fields. For example, sustainability does not yet seem a common concern when it comes to the development of digital technologies, like artificial intelligence. Some have therefore argued we should give this value much more prominence in the digital realm (van Wynsberghe, 2021). Blockchain – which is used by digital currencies like Bitcoin – has particularly raised concerns about energy consumption, but the issue is much broader. Belkhir and Elmeligi (2018) estimate that the carbon footprint of digital technologies, which includes not only energy use but also, for example, emissions from production, will increase from about 4% of the total worldwide carbon footprint in 2020 to 14% in 2040 (with a lower bound of 6 %). However, digital technologies like smart grids and smart meters may significantly contribute to a decrease in energy use in other sectors when properly designed.

The solution to some environmental problems, like greenhouse warming, might require more emphasis on sustainability as a value compared to other values. At the societal level, more emphasis on sustainability might come at the cost of the realization of other values like economic prosperity and comfort (at least in the short run). While dealing with such value conflicts may require political and societal choices, engineers are also increasingly confronted with similar value conflicts in researching, developing, and designing new technologies.

To give a few examples: smart energy meters may contribute to energy saving at home but may also violate the privacy of their users; cars might be made more sustainable by decreasing their weight (to decrease fuel consumption) but that might make them less safe because less (active and passive) safety systems can then be added; people might be stimulated to behave more sustainably through behavior-steering technology and nudging but this may diminish their freedom and (moral) autonomy; circular design and recycling will generally decrease environmental pollution and the waste of resources but may also occasionally introduce new risks.

The increasing emphasis on sustainability may also lead to the reconsideration of technological options to meet certain societal needs. For example, climate change and the need for an energy transition toward low-carbon energy technologies have led to a renewed interest in nuclear energy as a possibility for energy generation.

At the same time, the energy transition makes clear that there might be new values that have not yet been (sufficiently) considered in the choice for and design of energy technologies (see box). For example, the transition to renewable energy technologies may come with considerable costs and an increase in energy prices, which has led to concerns about energy poverty, that is, the situation that some social groups cannot afford enough energy (or have to spend a considerable amount of their income on energy). Some have consequently pleaded for more attention for values like energy justice and energy democracy in the energy transition, which may also have consequences for engineering and the design of energy technologies.

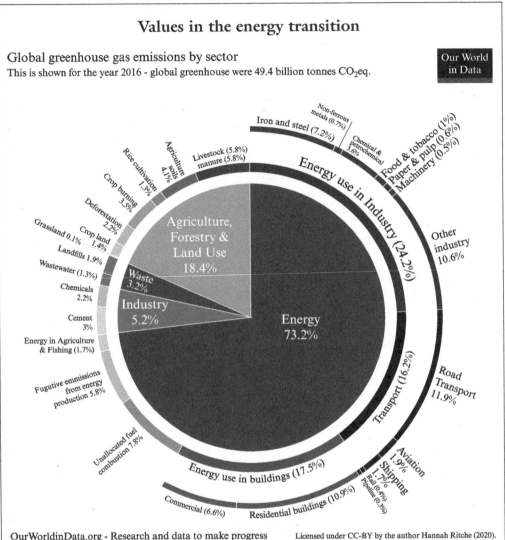

Values in the energy transition

Global greenhouse gas emissions by sector
This is shown for the year 2016 - global greenhouse were 49.4 billion tonnes CO₂eq.

Figure 8.5 Global greenhouse gas emissions by sector. From https://ourworldindata.org/emissions-by-sector#energy-electricity-heat-and-transport-73-2 / CC BY 4.0.

Mitigation of greenhouse warming will, among others, require changes in energy production technologies. As Figure 8.5 shows most of the global greenhouse gas emissions are somehow related to energy. Mitigation, therefore, requires a transition of current, strongly fossil-based energy systems toward more sustainable ones, with substantially fewer emissions. This may be achieved by adopting non-fossil technologies for energy generation, but it will also require a reduction in energy consumption and new energy carriers (like hydrogen and biofuels). In terms of energy generation, it will require energy from renewable sources such as the sun, water, wind, the earth, and biomass (e.g., solar energy, wind energy, hydro energy, tidal energy, geothermal energy, bioenergy), but it has been suggested by some that nuclear energy may also be required, at least in a transition phase to fully renewable energy sources.

The so-called energy transition has economic, technological, and institutional dimensions, but it relates to values as well. At the most fundamental level, it is aimed at achieving the value of sustainability, a value that has only emerged in the scientific and policy discourse on energy policy since the 1980s (De Wildt et al., 2022). Other values play a role in the energy transition as well, including security and reliability, social justice and fairness, autonomy and power, safety, privacy, and esthetics and landscape embedding (Demski et al., 2015). These values are not static but can change over time, for example, because a value may become more important or it may lose relevance. A value that seems to have gained importance, due to a transition to more decentralized energy generation and the advancement of community energy systems, is energy autarky (e.g., Müller et al., 2011), that is, the value of being able to producing locally (without external dependencies) enough energy. We also witness the emergence of new values in the energy transition. For example, energy justice and energy democracy have emerged as new values in the last decade in relation to the energy transition (Jenkins et al., 2016; Szulecki, 2018).

Source: Based on Van de Poel & Taebi (2022), Jenkins et al. (2016), and Szulecki (2018).

When it comes to dealing with value conflicts in design for sustainability, some of the approaches discussed in Chapter 5 may be helpful. Below, we briefly discuss the methods of *respecification* and *innovation* for making decisions in cases of value conflict in design (see Section 5.6), as well as the use of so-called value dams and value flows.

Respecification

With respecification, a new specification is made of the higher-order values. In the earlier discussed case of biofuels, this would mean that the values of intergenerational justice and intragenerational justice are kept, but that the way these values are specified in terms of design requirements is changed compared to Figure 8.2. As we have seen in Chapter 5, usually more than one specification of a value is tenable. This means that we can often respecify the values in a values hierarchy and still have a justifiable specification. The trick now is to aim for a specification so that the value conflict is resolved or at least diminished. One might, for example, argue that the value of intergenerational justice would also allow biofuels that are more expensive than current fuels because this is a reasonable sacrifice to ask from the current generation, so relaxing the requirement competitive price. Alternatively, one could argue that intragenerational justice does not require that there is no competition for agricultural land and other inputs between biofuels and food crops but only that this competition is kept to a minimum.

Respecification may, as these examples already suggest, be controversial. People may disagree whether a certain respecification is indeed tenable. Moreover, there is no guarantee that a respecification is available that solves the value conflict. Even if we allow somewhat more expensive biofuels, no actual biofuels may still meet all the new requirements. Nevertheless, respecification avoids some of the pitfalls of the aggregation methods like life cycle analysis (LCA), cost-benefit analysis (CBA) and multiple criteria analysis (MCA) (see also Chapter 5). First, if the initial values hierarchy contains all relevant values, it does not leave out crucial values. Second, it avoids the application of weighting factors and aggregation and does not treat values as directly commensurable. Third, it makes the normative value judgments that are made to solve the value conflict explicit and open for discussion.

This does, of course, not guarantee agreement but it makes the normative discussion possible, instead of proceeding on the basis of largely implicit normative assumptions.

Innovation

Another alternative is innovation, the development of new options. Such new options might meet all the design requirements and so solve the value conflict. In the earlier mentioned conflict between sustainability and privacy for smart meters, an innovative privacy-by-design approach might avoid privacy intrusions while maintaining the sustainability advantages of smart meters. Also in the case of biofuels, we can see how the value conflict has led to innovation. A motive behind the development of so-called 3rd generation biofuels is the fact that current biofuels do not meet all the design criteria mentioned in Figure 8.2. The case of biofuels, however, also shows that innovation does not always solve a value conflict. Currently, there are no biofuels that meet all the design requirements. Still, ongoing R&D efforts may eventually lead to innovations so that all the mentioned requirements can be met.

Innovation may also have disadvantages. It may, for example, lead to new products which have drawbacks. Fourth-generation biofuels, for example, based on genetic manipulation of algae, may introduce new hazards. In the case of sustainability, the rebound effect is often cited as a possible disadvantage. If a solution is available without major direct disadvantages against comparable or lower costs, it may lead to additional consumption of the product; so if a good biofuels alternative is available for traditional fuels, this might frustrate attempts to cut the fuel consumption of, for example, cars or even lead to higher consumption, so that the initial problem is not solved and sometimes even aggravated.

Value dams and value flows

Design for sustainability is also complicated because different stakeholders may not agree on how to best conceptualize and specify sustainability.

An interesting approach to deal with such cases has been proposed by Miller et al. (2007). They propose "value dams" and "value flows" to manage conflicts among stakeholders. A value dam occurs when certain design features are strongly opposed by one (or more) stakeholders, and a value flow occurs when a design feature is supported by a large number of stakeholders. Similarly, in design for sustainability, a value dam would occur with respect to those design features that are strongly opposed by at least one relevant conception of sustainability, and a value flow would occur with respect to design requirements that fit a large number of conceptions of sustainability.

8.5   Chapter Summary

Two arguments can be given why we should care for the environment. One is anthropocentrism, which states that the environment is only valuable as a means to human well-being. The second argument is that the environment is intrinsically valuable. In both positions, the environment needs to be protected, albeit for different reasons and probably to different extents.

A core notion nowadays in the discussion about the environment is "sustainable development." In 1987, the Brundtland committee gave the following definition: "Sustainable development is development that meets the needs of the present without compromising the ability of future generations to meet their own needs." This definition raises questions about how sustainable development is to be justified and how it is to be operationalized (specified), especially in engineering contexts.

With respect to justification, two basic values lay at the basis of sustainability: intragenerational justice and intergenerational justice. Intragenerational justice refers to the just distribution of resources within a generation, for example between the developed countries like the United States and Europe and developing countries in, for example, Africa. Intergenerational justice refers to the just distribution of resources between generations, for example, between us and our grandchildren. As we saw in the biofuel example at the beginning of the chapter both types of justice may conflict. Biofuels may contribute to intergenerational justice, by decreasing problems like oil shortages and the greenhouse effect for future generations, but at the same time they may very well decrease intragenerational justice, because they will likely result in higher food prices that especially damage the already poor and hungry in the Third World.

With respect to operationalization, the following more specific principles to attain sustainable development have been proposed: the standstill principle, the precautionary principle, and the polluter pays principle.

We have seen that technical choices have far-reaching consequences for nature and the environment. They can contribute to the realization of sustainable development, but also to "unsustainability." On this basis, engineers have a special responsibility. This responsibility can be expressed by the development of special environmentally friendly and sustainable techniques. However, engineers can also take environment and sustainability into account in standard design processes. This can be realized by sustainable design, which can be given shape by a life cycle analysis. While life cycle analysis is a useful instrument in design for sustainability, it is often not the right approach to solve a value conflict if a range of incommensurable values is underlying the value conflict. Strategies like respecification and innovation might be more appropriate in such circumstances.

The responsibility of engineers, however, reaches beyond the design of sustainable products. For a technology to be accepted it is sometimes necessary for changes to be made at a societal level. This responsibility lies not only with the engineers, but in part it does. A more sustainable society is not necessarily sustainable. In fact, there is no guarantee that the steps that are taken to make technology more environmentally friendly will indeed result in a truly more sustainable society: a society that meets a defensible notion of the Brundtland definition. In many cases, even more, radical steps are necessary, such as the complete cessation of a certain activity or technology. These more radical steps do not always lie within the sphere of influence of engineers. However, the fact that the responsibility for the realization of sustainable development is largely in hands of other actors in society is no reason to be aloof as an engineer. A meaningful societal discussion about what is wise from the perspective of sustainability and what is technically possible can only be held if engineers are willing to contribute their specialized knowledge.

Study Questions

1 What is meant by operationalizing "sustainable development?" Mention three ways in which the Brundtland definition needs to be operationalized?
2 Mention two disadvantages of quantitative life cycle analyses.

3 Why can social and environmental problems not be separated?

4 In Subsection 6.7.1, it was indicated that the precautionary principle has four dimensions. Indicate what these four dimensions are in the application of the principle to environmental problems that were discussed in this chapter.

5 Explain what the standstill principle implies.

6 Which three kinds of environmental problems can be distinguished? Give an example of each kind. For each problem also indicate how it may be impacted by technology.

7 What is meant by internalizing the costs of environmental damage in the prices of products? Explain how this may contribute to the development of more sustainable technologies.

8 Which justification of sustainable development do you consider most convincing? Try to indicate what this justification implies for the operationalization of the notion?

9 Should engineers be concerned about considerations of intragenerational and intergenerational justice in the case of biofuels discussed in the beginning of the chapter? Which kind of considerations should they give more weight, and why?

10 Explain why respecification and innovation might be more appropriate to deal with value conflicts than life cycle analysis.

Discussion Questions

1 Do you believe that animals or the environment can be bearers of rights? If not: does this mean that humans have no moral obligations to animals and the environment or should and can these obligations be justified otherwise? How then?

2 Is compensation ever due to those harmed by the effect of pollution? Who would have to pay, and how should the amount to be paid be calculated? What problems do you envisage?

3 Have engineers a duty to design sustainable technologies? If so, what is the extent of this duty, and on what is it based?

Notes

1 https://sdgs.un.org, accessed October 16, 2022.

2 https://www.un.org/en/academic-impact/sustainability, accessed October 16, 2022.

3 See Singer (2002, Chapter 2). He refers to Locke (1986 [1690]).

4 This is Locke's theory of property. It should be noted that Locke believed it to apply in the state of nature, not when government comes into existence (Locke, 1986 [1690]).

5 Often more phases are distinguished, but here this seems sufficient.

6 https://pre-sustainability.com/articles/eco-indicator-99-manuals, accessed October 16, 2022.

7 https://www.rivm.nl/en/life-cycle-assessment-lca/recipe, accessed October 16, 2022.

9

Responsible Innovation

Having read this chapter and completed its associated questions, readers should be able to:

- Explain why it might be hard to open up the innovation process to societal stakeholders and moral concerns;
- Provide principled as well as instrumental (consequentialist) arguments for responsible innovation;
- Explain the relation, and potential tension, between the three dimensions of responsible innovation (process, product, societal challenges);
- Apply the four process criteria (anticipation, inclusiveness, reflexivity, responsiveness) for responsible innovation to concrete examples of innovation processes;
- Distinguish solution and negotiation strategies when it comes to dealing with societal challenges (through innovation) and know their pros and cons;
- Identify opportunities and challenges for implementing responsible innovation in industry;
- Discuss the four ways in which innovations can be disruptive and their consequences for the responsible innovation approach;
- Discuss the responsibility of engineers for responsible innovation.

Contents

9.1 **Introduction** 267

9.2 **Opening up the Innovation Process** 269

    9.2.1 Science, technology, and society 269

    9.2.2 The strategic importance of innovation 270

    9.2.3 The need for responsible innovation 271

9.3 **What is Responsible Innovation?** 272

    9.3.1 Responsible innovation as process 273

    9.3.2 Responsible innovation as product 273

    9.3.3 Responsible innovation and societal challenges 274

9.4 **Process Criteria for Responsible Innovation** 274

Ethics, Technology, and Engineering: An Introduction, Second Edition. Ibo van de Poel and Lambèr Royakkers.
© 2023 John Wiley & Sons Ltd. Published 2023 by John Wiley & Sons Ltd.

9.4.1 Anticipation 274
9.4.2 Inclusiveness 276
9.4.3 Reflexivity 278
9.4.4 Responsiveness 279

9.5 Responsible Innovations and Societal Challenges 280

9.6 Responsible Innovation in Industry 283

9.7 Disruptive Innovation 285

9.7.1 Market disruption 285
9.7.2 Social disruption 286

9.7.3 Regulatory disruption 287
9.7.4 Conceptual and normative disruption 288

9.8 Responsible Innovation and the Responsibility of Engineers 289

9.9 Chapter Summary 290

Study Questions 291

Discussion Questions 292

9.1 Introduction

Case Fairphone

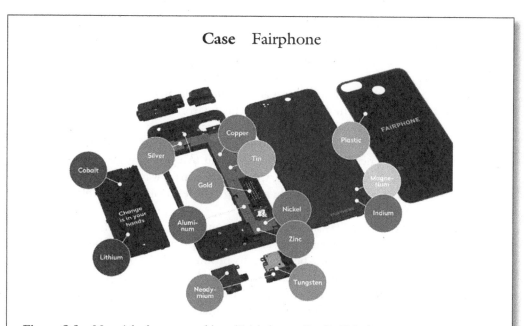

Figure 9.1   Materials that are used in a (Fair)phone. Credit: Fairphone.

Fairphone started as a social campaign against so-called conflict minerals. These are raw materials that are mined in conflict areas, such as the Democratic Republic of Congo, and of which the income is used to finance armed conflicts. Moreover, miners often work in very bad, slave-like, working conditions. Examples of conflict materials are tantalum, tin, and tungsten, which are used in electronic devices like mobile phones.

However, the initiative takers also wanted to show that it was actually possible to develop a more responsible phone, and therefore they developed a mobile phone using "conflict-free" minerals and a modular design so that parts can be repaired and replaced. In 2013, Fairphone became an independent company. Its aim is to have a "positive impact across the value chain in mining, design, manufacturing, and life cycle, while expanding the market for products that put ethical values first."[1]

The development of the Fairphone has focused on four areas of concern, namely fairer materials, better working conditions, longevity, and circularity. Since a smartphone contains tens of different materials, each with its own complicated supply chain, the company decided initially to focus on the ten most impactful materials, and to extend this number over time. Rather than turning away from countries where human rights and the environment are endangered, they decided to invest in them. For responsible mining, they focus on so-called artisanal and small-scale mining. These operate outside the formal mining sector and they "offer the most potential for directly improving the lives of miners and their families,"[2] according to the Fairphone website. For example, in 2020, the Fairphone company was one of the co-founders of the Fair Cobalt Alliance, which aims to improve the livelihoods and working conditions of artisanal cobalt miners and their communities in the Democratic Republic of Congo.

The Fairphone company also aims at improving working conditions across the entire supply chain, so not only for miners but also, for example, during assemblage. However, for cost reasons, the company decided to produce the Fairphone in China, where workers are reported to make still long working weeks (up to sixty hours).

When it comes to environmental impacts, a life cycle analysis commissioned by Fairphone suggested that the largest part of the environmental impacts for mobile phones comes from the production phase. This means that the best way to reduce environmental impact is to increase the longevity of the phone. Fairphone aims at producing mobile phones that will be used for about five years, rather than the average of two years for most traditional smartphones. The main way in which they try to do so is by making the phone repairable and upgradable through a modular design, which allows repairing or replacing parts.

The Fairphone company has won several environmental and business awards and obtained certificates for environmental and fair performance. Nevertheless, as the company itself admits, it has not been able to produce a 100% fair phone; but on most social and environmental criteria its scores clearly much better than traditional phones. The turnover is still small compared to large players in the mobile phone industry, but the company also hopes to trigger other phone producers to pay more attention to environmental and social values.

Source: Based on https://en.wikipedia.org/wiki/Fairphone, https://goodelectronics.org/company/fairphone https://www.fairphone.com/en; https://www.dw.com/en/the-fairphone-fair-deal-or-fairytale/a-18660137, and https://waag.org/en/project/fairphone, accessed August 22, 2022, and Sánchez et al. (2022).

The Fairphone case exemplifies three aspects that are key to current approaches in the ethics of technology. First, it is an example in which ethical issues are addressed proactively rather than reactively. Rather than addressing moral issues after technology has come into use, it is aimed at already addressing moral issues in the early phases of technological development. Second, moral concerns play a constructive role in this case; so ethics of technology here does not just criticize technology, but it also helps to improve technological development from a moral point of view. Third, ethical reflection in this case is not only aimed at preventing harm, but also at doing good, for example, by offering better opportunities to miners. The ultimate aim is to contribute to better technology in a better society.

In this chapter, we will discuss an approach to technological development and innovation that represents all three aspects: responsible innovation.[3]

Responsible innovation Approach that stimulates innovators to work with stakeholders during the research and innovation process, to better align the (expected) outcomes of innovation with the values, needs, and expectations of society.

Responsible innovation has come into vogue as a term in Europe, mainly due to its central role in the European Union's science funding program Horizon 2020 (where it is usually called Responsible Research and Innovation or RRI), but the term, and the underlying ideas, have now also become popular in, for example, the US and China (Von Schomberg and Hankins, 2019).

The core idea of responsible innovation is to better align technological innovation with the values, needs and expectations of society by opening up the innovation process to a broader range of stakeholders and moral concerns or values. Responsible innovation is aimed at making the innovation process more participative, democratic, and value-driven, so that it will result in better technologies that ultimately contribute to the solution of societal problems and a better society.

We start this chapter with explaining why it is difficult to open up the innovation process (Section 9.2). In Section 9.3, we will define the notion of responsible innovation which has three dimensions: a process dimension (Section 9.4); a product dimension (discussed in Chapter 5); and a dimension which aims at resolving grand societal challenges (Section 9.5). In Section 9.6 we will explain why it is useful for companies to have a responsible innovation strategy. New technologies, especially digital technologies are often socially disruptive, and can be challenging for responsible innovation, however can also offer opportunities. This will be dealt with in Section 9.7. In Section 9.8, we will discuss the responsibility of engineers for responsible innovation.

9.2 Opening up the Innovation Process

Responsible innovation is aimed at opening up the innovation process to societal stakeholders and moral considerations. There are two more fundamental reasons why this may be difficult. The first is that innovation is sometimes seen as a mere application of new scientific insights and therefore more or less value-neutral. The second is that innovation is often strategically key to companies and countries, which makes them reluctant for opening up the innovation process to outsiders and moral considerations. We discuss both issues in the following subsections. Next, we will discuss why it is nevertheless important to open up the innovation process.

9.2.1   Science, technology, and society

The idea that innovation is merely the application of new scientific insights is often based on certain assumptions about the relation between science, technology, and society. Roughly, the idea is that new insights are developed in science; these are then applied, for example, by industry, to develop technological innovations, and these innovations then have a certain societal impact. This idea is captured in the motto of the 1933 World's Fair in Chicago: "Science Finds, Industry Applies, Man Conforms" (Ganz, 2008, p. 57). As this motto testifies, the role of society is just to adapt to changes brought about by technological innovations, which in turn are based on new scientific findings. Responsible innovation aims at reversing this relation: society rather than science is supposed to be in the lead and determines what science we do and what innovations we develop.

Two more specific ideas are underlying the classical idea of science-driven innovation. One is the idea that "technology is applied science"; the other is what has been called "technological determinism." The idea of **technology as applied science** is that technology is nothing more than the application of scientific insights. This idea is mistaken for at least two reasons. One is that the objectives of science and technology are different. Science is primarily aimed at finding truth, technology at developing artifacts and systems that work. In technology, also all kind of practical considerations, like effectiveness, efficiency, user-friendliness, etc., play a role that has no obvious place in science. Social and

> **Technology as applied science**   The idea that technology is merely the application of scientific knowledge. This view overlooks that technology services practical ends and therefore requires not just theoretical scientific knowledge.

moral values therefore always have a place in technology. An implication is that also the kind of knowledge that is needed for technology, and that is developed in what is sometimes called technological science, is different from traditional science (Vincenti, 1990).

The idea that technology is a mere application of scientific insights is also mistaken for another reason. At least some technological innovations are not based on new scientific insights. One may think of all kinds of more practical innovations, but also for example a technology like the steam engine which was developed *before* the theory of thermodynamics was known (Kroes, 1992). Here, the relation was reversed: the technology of the steam engine spurred scientific investigations into thermodynamics. This is not to deny that there are also technologies that are primarily based on new scientific insights. Nuclear energy and the atomic bomb are clear examples, and attempts to develop a quantum computer are based on new scientific discoveries in quantum physics. One might even argue that nowadays science and technology have become very much intertwined and it is hard to say for research at for example technical universities where science stops and technology starts. Some, therefore, prefer to speak of **technoscience**.

> **Technoscience** Nowadays, science and technology are often so interwoven that they are hard to distinguish; the resulting complex is often called technoscience.

Technological determinism is the idea that technology develops autonomously and then has a determinant causal influence on society. It is based on two underlying ideas. The first is that the development of technology is autonomous and therefore cannot be influenced. This might be based on the idea that technology is just applied science, as discussed and criticized earlier. It can also be based on a view of technology as a kind of independent force that is not necessarily science-based. In either form, the idea is mistaken as technological development is in the end the result of human actions and choices, and therefore, at least in principle open to human choice, even if in a practical sense it may be hard to direct or modulate technological developments.

> **Technological determinism** The view that technology develops autonomously and then has a determinant causal influence on society.

The second underlying idea is that technology determines society. At first sight, this idea is not so strange. It is certainly true that throughout history technological developments have sometimes had a large impact on societal developments. New military technology may have decisive in wars between countries. Technological developments have made possible significant increases in hygiene and have made food production more efficient, and therefore food affordable for many. Obviously, some technological developments have contributed to human welfare. Still, it is too simplistic to say that technology *determines* society. First, societal development is a complex multi-causal process, in which technological development is just one of the many factors. Second, even if new technology has become available, there are still many choices to be made by humans about how it is used, and for what purposes, and these choices will affect the eventual societal impacts of technology.

9.2.2 The strategic importance of innovation

Innovation is often considered strategically key to companies and countries. Companies may gain a **competitive advantage** by being the first in developing a certain innovation. Think of companies like Google, Apple, and Meta (Facebook) that gained a dominant economic position by embarking early on certain digital technological innovations. Similarly, technological innovation may give countries a key advantage in military and economic respects. This

> **Competitive advantage** The relative advantage of companies or countries vis-a-vis their competitors, attained, for example, through (technological) innovation.

explains why Europe, the US, and China are heavily investing in artificial intelligence, and aim at a competitive lead in this field.

The strategic importance of innovation makes it more difficult to open up the innovation process and to make it more responsible for at least two reasons. First, the strategic reasons for which innovations are developed may conflict with moral considerations. They may be more aimed at profit maximization or competitive advantage than at moral concerns. Second, to maintain competitive advantage, innovators may choose to keep the knowledge that is required for a certain technology secret or to protect it through, for example, patents so that the innovation cannot be copied. Such secrecy is not only in tension with the idea that responsible innovation requires open science to make the results available to everyone (Von Schomberg, 2019), but also makes it harder to open up the innovation process to outsiders, as responsible innovation requires.

9.2.3   The need for responsible innovation

Despite the above-discussed difficulties, proponents of responsible innovation believe that it is desirable to open up the innovation process to societal stakeholders and moral considerations. Two types of arguments might be given for opening up the innovation process, namely more principled and more instrumental or consequentialist arguments. The first typically states that it is desirable to open up the innovation process for reasons of inclusiveness and democracy. The second refers to the conviction that opening up the innovation process will contribute to better innovations, in the sense of better products and technologies. We will discuss both types of arguments next.

We start with the principled arguments for responsible innovation. The underlying idea is that technological innovations nowadays affect individuals and society to such a large extent that people should have a say in their shaping. This means that it is desirable to include affected people somehow in the innovation process, and that innovation should be inclusive with respect to the values and needs of a wide range of stakeholders.

One might also formulate this principled argument in terms of democratic control over technological innovation. One might argue that the degree to which new technologies affect our society is comparable to the degree that laws and governmental policies affect society. The formulation of such laws and policies is, at least in liberal democratic states, subject to certain democratic procedures and safeguards. The idea is that consequently also the development of new technologies should be subject to democratic processes and safeguards. One might then argue that responsible innovation provides such a more democratic process of developing technology.

It should be noted that one might agree that democratic control of technology is needed without agreeing with the need for responsible innovation. Some might argue that responsible innovation is not needed for democratic control over technology, as there are other possibilities for democratic control, like general laws or governmental policies with respect to technology. Others might perhaps agree that responsible innovation is desirable but they may argue that it is not enough for democratic control, as it is not embedded in formal democratic procedures and institutions.

Apart from such more principled arguments, more instrumental or consequentialist arguments may be given for responsible innovation. Such arguments might start by pointing out the limited effectiveness of current democratic control over technological development through general laws and governmental policies (rather than opening up the innovation process itself). Two arguments may be given why such general laws and governmental policies are not very effective. First, laws and governmental policies tend to lag behind technological development; they are often only formulated after the social effects of new technologies have materialized and harm has been done. If we want to proactively include social and moral concerns in the development of new technology, we need to open up the innovation process itself and cannot wait for laws and governmental regulations. Second, governments that make laws and formulate policies are often not the central players when it comes to innovation. Rather, innovation is often developed in companies. Such companies might be in a better position to anticipate the social effects of new technologies and to proactively address moral concerns and values. Moreover, some companies operate worldwide and are therefore hard to regulate by national laws and policies.

The more consequentialist argument for responsible innovation basically states that responsible innovation will result in technologies and products that are both societally and morally "better." Better here has two aspects, namely the social acceptance and the moral acceptability of new innovations (Van de Poel, 2016; Taebi, 2017).

Social acceptance (of innovations) An innovation is socially acceptable if it is in line with the values and needs of relevant stakeholders.

An innovation is socially acceptable if it is accepted by relevant stakeholders. Social acceptance might be understood in terms of actual use of technology, but people may sometimes use technology to which they have objections because they have no alternative. Therefore, we will understand social acceptance in terms of congruence with the actual values of relevant stakeholders. This makes it easy to see that by including the values of relevant stakeholders, responsible innovation will likely increase the **social acceptance of new innovations**.

Moral acceptability (of innovations) An innovation is morally acceptable if it meets all relevant (minimal) moral standards.

An innovation may be called morally acceptable if it meets relevant moral standards. What these standards exactly are may be different for different innovations and open for discussion. What is important to be aware of is that **moral acceptability** is not the same as social acceptance, because the values of stakeholders might not coincide with moral standards. For example, doing justice to future generations is an important moral standard for nuclear waste technologies but future generations cannot be involved in the development of new technologies in this area (because they do not yet exist). If responsible innovation is to result in morally acceptable technology, it should therefore not just include values of relevant stakeholders but also, and foremost, moral considerations and standards.

9.3   What is Responsible Innovation?

The notion of responsible innovation expresses the idea that research and innovation should be better aligned with the values, needs, and expectations of society. Already during the early phases of technological research and development, innovators should anticipate potential uses and societal consequences, risks, and benefits of technologies and proactively aim to contribute with research and innovation to important societal challenges. Innovation processes should also be inclusive by addressing a range of potential societal and moral concerns and by including all relevant stakeholders. Moreover, innovation processes should become responsive to social needs and values.

There is a number of definitions of "responsible innovation." Some of these definitions emphasize both the innovation *process* as well as the *outcomes* of that process, that is, the innovations that are taken up by and embedded in society. For example, Von Schomberg (2012, p. 50) has defined responsible innovation as "a transparent, interactive process by which societal actors and innovators become mutually responsive to each other with a view on the (ethical) acceptability, sustainability and societal desirability of the innovation process and its marketable products."

So understood, responsible innovation has a process dimension (referring to the process of innovation) as well as a product dimension (referring to the outcomes of the innovation process). In addition to these two, we might distinguish a third dimension, which refers to the idea that responsible innovation has to contribute to the solution of societal challenges, rather than being driven by techno-scientific progress.

9.3.1 Responsible innovation as process

Four process criteria have been proposed for responsible innovation (Owen et al., 2013; Stilgoe et al., 2013):

- Anticipation: are the possible social consequences (risks and benefits) of innovations anticipated and fed back to the innovation process?
- Inclusion: are all relevant stakeholders included in the process of innovation?
- Reflexivity: do the innovators reflect on the social goals, values, expectations, and promises of their innovations and are they aware of their assumptions?
- Responsiveness: is the innovation process responsive to the needs, values, and expectations of society and to new insights that arise during innovations' development and implementation into society?

These four criteria are procedural in nature and can be used to assess innovation processes on how "responsible" they are, and to make innovation processes more responsible.

9.3.2 Responsible innovation as product

In addition to the process dimension, there is the product dimension of responsible innovation, which refers to products (services, systems) as the result of the innovation process. Eventually, these products diffuse into society and have certain social consequences. For example, Van den Hoven (2013, p. 82) defines responsible innovation as "an activity or process which may give rise to previously unknown designs ... which – when implemented – *expand the set of relevant feasible options regarding solving a set of moral problems*." The idea expressed here is that innovations often try to meet different, potentially conflicting values. Safer cars, for example, are often heavier, implying more fuel consumption and decreased sustainability (Van Gorp, 2005). Safety and sustainability are conflicting values in car design and create moral overload, that is, the impossibility to meet various moral requirements at the same time (Van den Hoven et al., 2012). According to Van den Hoven, responsible innovation means the overcoming of moral overload by innovation, that is, by developing new options that resolve value conflicts.

9.3.3   Responsible innovation and societal challenges

Technology push (innovation) Innovation based on new technical possibilities rather than market demand or societal needs.

Market pull (innovation) Innovation based on (new) market demands.

In the innovation literature, sometimes a distinction is made between **"technology push"** and **"market pull"** when it comes to innovation (Dosi, 1982). In case of "technology push," innovations are primarily based on new technoscientific insights, rather than on demands in the market. For "market pull," the reverse applies: innovations are primarily driven by new demands in the market rather than by new technological possibilities.

Responsible innovation is more in line with market pull, but it extends this pull beyond economic demand and the direct users of a technology, to a broader range of stakeholders and ultimately to society as a whole. The idea is that innovation should be driven by social problems, or so-called societal challenges. Societal challenges are hard to solve and urgent societal problems, such as the energy transition, or the abatement of hunger in the world. Several governments and other funders of technoscience have now formulated societal challenges as the intended drivers of new technological innovations.

The third important dimension of responsible innovation then is the idea that innovation should be driven by the desire to solve societal challenges, rather than being driven by progress in technoscience or developments in the market. While this has consequences for both the process and products of innovation, it adds also something to these dimensions. It requires not just a process of innovation that meets the four mentioned criteria (anticipation, reflexivity, inclusion, and responsiveness) but also one that is driven by societal challenges from the start. In terms of products, it does not just require products that embed societal and moral values, but something more, namely a contribution to the solution of societal challenges: it is aimed not only at preventing harm, but also at doing good.

9.4   Process Criteria for Responsible Innovation

We will now look in more detail into the three dimensions of responsible innovation. The second (product) dimension has already been discussed in the chapter on design for values and will therefore get less attention in this chapter. Consequently, in this and the next section, we respectively focus on the process dimension of responsible innovation and the relation between responsible innovation and societal challenges. In this section, we discuss the four proposed process criteria for responsible innovation: anticipation, inclusiveness, reflexivity, and responsiveness.

9.4.1   Anticipation

The anticipation criterion for responsible innovation requires that innovation processes are so organized that they contribute to anticipating the possible consequences of new innovations, and that the resulting insights are taken into account in the innovation process. To understand what this criterion practically means, it is important to distinguish between anticipation and prediction of the social impacts of new technology.

Since the second world war, attempts have been made to better predict the consequences of new technological developments for society. This was done in (new) disciplines like Impact Assessment and Technology Assessment. Initially, the ambition was to make objective predictions about the impacts of technology on society (Wynne, 1975). However, this has proven hard

> **Anticipation (as requirement for responsible innovation)** An innovation process is anticipatory if it is organized in such a way that possible consequences, including side-effects and risks, are anticipated and fed back into the innovation process.

if not impossible to do reliably. One reason is that the real consequences of new technologies often only become clear once these technologies are used on a substantive scale and have become well embedded in society. This is the Collingridge dilemma that we already discussed in Chapter 1.

To overcome the Collingridge dilemma, one may try to *anticipate* rather than predict the potential consequences of new technologies in society. The idea of anticipation is that while we may not be able to predict the future, we might be able to explore and prepare for different *possible* futures. In fact, innovations are often developed with the motivation or the expectation to realize a certain desirable future. However, what is sometimes forgotten in traditional innovation processes is that innovations usually also have all kinds of side effects and risks. Responsible innovation minimally requires that such side-effects and risks are also anticipated as good as possible and are taken into account in the innovation process. Taking into account can mean a number of things here. It can mean that the innovation is adjusted to avoid or minimize risks and side effects. It can also mean taking mitigating measures to better deal with the anticipated risks and side effects; or it can mean that an innovation is not pursued because of certain side effects or risks.

Anticipation can be pursued in several ways, for example through carrying out a risk assessment (see Chapter 6), or through doing, for example, an environmental life-cycle assessment of a product. Anticipation can also be done by creating various scenarios (see box). Such scenarios are not aimed at predicting the future, not even in probabilistic terms, but they sketch possible futures. These possible futures may reveal concerns that are not yet addressed in the current innovation process. They may also reveal that in order to achieve a certain desirable future, the innovation should be adjusted, or other accompanying measures are required.

Scenarios

Different kinds of scenarios or scenario techniques can be used for responsible innovation. For example, sociotechnical scenarios try to anticipate developments in a technological sector; they sketch relevant developments at different levels (micro, mesa, macro) and how these might interact resulting in different possible scenarios (Rip and te Kulve, 2008). Technomoral scenarios have the specific aim of descriptively anticipating changes in norms or moral views. Creating such scenarios typically involves three steps, namely (1) sketching the existing moral landscape; (2) anticipating new moral controversies that a technology may give rise to; and (3) anticipating possible closure of these controversies that might result in a change in norms and values (Boenink et al., 2010).

A third method is value scenarios, which are used in value-sensitive design (Nathan et al., 2007). Value scenarios can be used to discover new values, of which the realization may be affected by a technological application, and which should thus be taken into account in the design process. In developing value scenarios, one might ask questions like: What will happen if this technology is used on a large scale? Or what will happen if it is transferred to another cultural context? One might also sketch scenarios of how a technology will be used and how it will affect direct and indirect stakeholders. Czeskis et al. (2010) present examples of value scenarios for the (hypothetical) example of mobile phone apps that can be used to monitor the life of one's teenagers. One hypothetical app for which they develop a value scenario is PhoneTracker, which is "designed to help parents keep track of their teens. Once installed on a mobile phone, parents can use the application to surreptitiously turn on the phone's microphone or to read text messages on the teen's phone at any time" (Czeskis et al., 2010, p. 4). The scenario they develop features father Paul and his 14-year-old son Ben. Paul has been raising Ben since his mother passed away six years ago; they have a good relationship of trust. When Paul first hears of PhoneTracker, he is not enthusiastic about it, but his colleagues at work convince him that it is a useful app that a responsible parent should use. Paul starts doubting himself and eventually decides to secretly install PhoneTracker on Ben's phone. Over the next months, Paul observes Ben's behavior and observes nothing peculiar; he learns a lot though about Ben's social life, and with whom his son hangs out, particularly about Ben's best friend Jon. He starts wondering whether he is not spying on Jon too, and whether he should tell Jon's parents. (How would he feel if they were following Ben through Jon indirectly without telling?). One day, Ben finds out that Paul spies on him and is very upset. Their relationship is never quite the same after that. This scenario suggests values like privacy and trust, of which the realization may be affected by the use of Phonetracker, and which should be taken into account in its design.

9.4.2   Inclusiveness

Inclusiveness (as requirement for responsible innovation)   All relevant stakeholders are involved in the innovation process and have a say.

Responsible innovation processes need not only be anticipatory but also inclusive. Inclusive means that all relevant stakeholders are included and have a say in the process. Relevant stakeholders are not restricted to the developers and direct users of technology, but include everyone potentially affected.

Two issues are pertinent when it comes to further specifying when an innovation process is inclusive. First, how do we determine what stakeholders are relevant and should be included? One criterion for relevance is that stakeholders should be somehow affected by the innovation, but should they already be actually affected or only potentially affected? And who determines whether they are affected or not? Since relevance may be disputed or changed over time, one might also want to require that innovation processes are open, in the sense that new stakeholders can acquire a place at the table in the course of the process.

A second issue is how much influence different stakeholders should have on the innovation process and its outcomes to be inclusive. Since stakeholders have different and sometimes conflicting perspectives and stakes, there is no easy answer to this question. Some might argue that the main criterion for inclusiveness is that the concerns and needs of all stakeholders are somehow addressed in the innovation process rather than they themselves have a (decisive) say in the process.

Methods for public engagement

- *Referenda*: in a referendum people can vote for or against a particular option. All people have one vote and the outcome is usually binding.
- *Public opinion surveys*: through a written questionnaire or telephone survey, people can express their views and opinions on an issue.
- *Public hearing*: Presentation of plans by agencies in an open forum. People can express their opinions, but usually have no direct impact on the recommendations.
- *Focus groups*: planned discussions among a small group of stakeholders facilitated by a moderator.
- *Citizen juries*: lay panel to obtain a recommendation from informed citizens. Usually not open to a wider audience beyond the panel.
- *Consensus conferences*: public inquiries with citizens who assess a particular, often controversial, topic. Usually open to a wider audience.
- *Participative scenario analysis*: an interactive process of scenario development and exploration with stakeholders to identify key issues.
- *Policy exercises*: aimed at developing policy options or ideas with a heterogeneous group of stakeholders based on knowledge from various resources, for example, in a gaming environment.
- *Participatory modeling*: the active involvement of model users in the modeling of a phenomenon to gather a variety of viewpoints.
- *Participatory Value Evaluation (PVE)*: a method where participants can allocate a budget over different options and so express their preferences and values.

Source: Based on Rowe and Frewer (2000), Roeser and Pesch (2016), and Mouter et al. (2019).

Inclusiveness can be achieved in several ways. One way is through more participatory design approaches or through open innovation, in which external stakeholders become one of the actors in the innovation process. Oftentimes, however, stakeholders are involved through some form of stakeholder or public engagement (see also the box for several methods for public engagement). Rowe and Frewer (2000) have proposed five criteria that a good method for public engagement should meet:

- Representative: the involved participants should form a representative sample of the affected public;
- Independent: the engagement should be carried out in an independent, unbiased way;
- Early involvement: the public should be involved at an early stage, not when the main decisions already have been made;

- Impact: the output of the process should have a real impact on policy or technological development;
- Transparent: the process should be transparent to outsiders.

Guidelines for fruitful deliberation

- Instead of one-way communication, create a two-way dialogue in which all participants are treated equally. Make sure to create a symmetric setup for discussions, for example, not giving experts a more prominent place than laypeople.
- Talk about values and emotions: values and emotions are an important part of the deliberation, as they may point to important moral considerations. They should be articulated not suppressed.
- Convey respect and ask questions: if people react emotionally or say things that you strongly disagree with, try to ask questions to understand their view rather than to react dismissively.
- Have a dialogue among all people: make sure to include all relevant stakeholders in the deliberation; also make sure that during the deliberation all viewpoints can be expressed and are heard, particularly ones that are marginalized.
- Have a clear procedure: there should be a clear procedure so that all participants know what to expect and understand their role. It should also be clear whether the procedure will result in an official recommendation and what role that recommendation will have in the further process, otherwise, people might lose trust.

Source: Adapted from Roeser and Pesch (2016).

9.4.3 Reflexivity

Reflexivity (as requirement for responsible innovation) An innovation process is reflexive if it is organized in such a way that it stimulates (first- and second-order) reflection on relevant goals, values, and (underlying) assumptions.

An innovation process is reflexive if it stimulates the involved actors and stakeholders to reflect on the values, goals, and needs that are to be served by an innovation, as well as on underlying assumptions. **Reflexivity** requires not only the articulation of values, needs, goals, and assumptions, but also that an ongoing learning process with respect to them takes place.

A distinction can be made between first- and second-order learning (Schot and Rip, 1997; Van de Poel and Zwart, 2010). In **first-order learning**, actors learn how to better achieve established goals and values. It takes the innovation goals and values as given,

First-order learning Learning how to better achieve given goals and values.

and focuses on how to better achieve these goals and values. It, for example, involves technical and scientific learning to improve the innovation or learning about the contextual factors that affect the uptake of an innovation.

Second-order learning Learning about what goals and values to aim for.

Second-order learning is learning about goals and values themselves, which may result in new or changed goals and values. Second-order learning can, for example, occur through anticipation: by considering a certain possible scenario, innovators may become aware that certain values may be at stake and they may then include this as an additional value (or goal) for the innovation process.

Innovation processes are often primarily aimed at first-order learning; second-order learning is more rare. Second-order learning may, for example, be promoted by the inclusion of stakeholders with another view. Second-order learning may also occur if certain social groups object to or protest against an innovation (Rip, 1987). Innovators may then become aware, for example, of certain side effects and may reconsider the goals and values that are central to the innovation process. Second-order learning thus does not require consensus among the stakeholders and may in fact be sparked by controversy over a technology.

9.4.4 Responsiveness

Responsiveness requires that the innovation process is so organized that it (1) responds to the values and needs of society and (2) responds to new insights that are gained along the innovation pathway. Responsiveness thus requires innovation processes that are aimed at learning along the way and are adaptive to new developments, including new needs and values in society.

> **Responsiveness (as requirement for responsible innovation)** An innovation process is responsive if it is so organized that it (1) responds to the values and needs of society and (2) responds to new insights that are gained along the innovation pathway.

As you might remember from Chapter 1, the Collingridge dilemma is the dilemma that in the early phases of technological development, we know too little to direct technology in societally desirable directions, while at the later stages technology is already too embedded in society to change its course. We may try to overcome this dilemma by acquiring knowledge of possible consequences through anticipation as was discussed earlier. But, we may also try to postpone the embedding of a new innovation in society so that we have more time to learn. A responsive innovation process may follow the latter approach by stimulating learning along the way and avoiding making irreversible decisions too soon.

One way in which we may try to do this is by organizing the innovation process as an on-going experimental learning process. Such a process might scale up from experiments in a lab setting to experiments in the field, to small-scale experiments in a part of society, to monitored full-scale introduction in society. Such an experimental upscaling trajectory is not uncommon in the medical field (see box), but it is still much rarer in other fields, like digital technologies.

Experimental phases for medical drugs

New medical drugs or other health interventions need to undergo testing before they are allowed on the market. Usually, a distinction is made between four phases:

- Phase 1 tests out a drug or another medical intervention for its safety on a small group of healthy volunteers, typically 20 to 100 people. Before phase 1 can be entered, extensive preclinical trials, involving both in vitro (test tubes or cell cultures) and in vivo (in animals) experiments have to be done, and they need to show that a new drug is likely to be safe and effective.

- Phase 2 involves tests for efficacy and side effects on a larger group, of about 100 to 300 people, who usually have a specific disease.
- Phase 3 involves largely scale testing through randomized clinical trials and may involve 300 to 3000 participants. Such trials typically involve a group that gets the treatment and a control group that gets a placebo. The distribution of people over both groups is random and people do not know in which group they are. The idea is that in this way it can be reliably shown that the drug or treatment is effective and has no major side effects.
- Phase 4 was initially not seen as a separate experimental phase, but is now often recognized as such. It concerns postmarketing surveillance, that is, the reporting of (long-term) adverse effects (or lack of effectiveness) after a drug has been approved (after phase 3) and comes into use. Several drugs have been withdrawn or restricted in use due to effects that only became apparent in phase 4; an example is Rofecoxib, a painkiller better known under its brand name Vioxx.

Source: Based on https://en.wikipedia.org/wiki/Phases_of_clinical_research, https://www.fda.gov/patients/drug-development-process/step-3-clinical-research#Clinical_Research_Phase_Studies, and https://www.cancer.org/treatment/treatments-and-side-effects/clinical-trials/what-you-need-to-know/phases-of-clinical-trials.html, accessed August 11, 2022

If we organize innovation as an experimental learning process, we also need to think beforehand about how and what kinds of things need to be learned along the way. Previously, we have seen that learning requires reflexivity about the goals and values underlying an innovation. More specifically, we want to suggest that a responsive innovation process requires three types of learning along the way:

1 Impact learning, that is, learning about the possible and actual impacts of an innovation on society;
2 Normative learning, that is, learning about the moral and political issues that a new technological raises, and learning about possible new values, goals and principles to properly address the moral and political issues;
3 Institutional learning: learning about what institutions, organizations, laws, and regulatory frameworks are needed to properly embed an innovation in society.

9.5   Responsible Innovations and Societal Challenges

Responsible innovation can also be understood as innovation that is aimed at resolving grand societal challenges, such as the Sustainable Development Goals (SDGs) of the UN (see Subsection 2.2.2).

Innovation that is aimed at addressing grand societal challenges is not primarily motivated by technological possibilities or profit-making but primarily oriented to a societal mission. This does not mean that such innovation cannot require sophisticated technology or cannot help companies to make a profit, but these are not the main drivers of the innovation.

Societal challenges and SDGs are best understood as ill-structured problems (see Chapter 4). They are complex, uncertain and value-laden (Voegtlin et al., 2022). They are complex because their attainment requires a combination of social, institutional, technical, economic and political measures and takes place in sociotechnical systems that we only partially understand. They are uncertain, because of these complexities but also because there is often no agreement on how they can be best understood. This makes them also value-laden, in the sense that value judgments are required in both understanding and addressing them.

Ludwig et al. (2022) distinguish two kinds of strategies when it comes to addressing societal challenges: solution strategies and negotiation strategies. The first kind of strategies assumes that the societal challenge is well understood and such strategies look for a straightforward solution based on the understanding of the societal challenge by dominant actors. Following a **solution strategy** in responsible innovation for societal challenges is risky for three reasons: (1) the solution is likely to reflect the values and interests of the innovators rather than those suffering from the problem; (2) the focus on the solution might lead to an ignorance of risks and side effects of the innovation; and (3) consequently the innovation proposed may not be accepted by users or may not solve the underlying problem.

> **Solution strategy (for addressing societal challenges)** A solution strategy takes the understanding of a societal challenge, and of possible solutions, by a dominant actor as given without considering possible other problem formulation by other stakeholders.

An alternative is to follow a so-called **negotiation strategy**. Such a strategy does not assume beforehand that the societal challenge is understood but rather takes this to be the object of negotiation between the relevant stakeholders. This can make the innovation process more inclusive (it includes more stakeholders),

> **Negotiation strategy (for addressing societal challenges)** A negotiation strategy takes the understanding of societal challenges as object of negotiation between the relevant stakeholders.

Case Golden rice

Figure 9.2 Golden rice. Photo: Laila Austria.

Golden rice has been genetically modified to help abate vitamin A deficiency among children. According to estimates by the WHO, in 2012 about 250 million preschool children suffered from a vitamin A deficit, and 2.7 million children were dying unnecessarily from it. Golden rice contains amounts of beta carotene, which can be converted into vitamin A by the human body. Since many children who suffer from vitamin A deficiency eat rice on a daily basis, the idea was that golden rice could be a major help in addressing this societal challenge.

While golden rice was thus developed for noble reasons, it has met quite a bit of opposition from organizations like Greenpeace, Friends of the Earth, and MASIPAG, a network of organizations based in the Philippines. These opponents pointed out that golden rice is not necessary, and argued that there are more cost-effective means to abate vitamin A deficiency, like vitamin A supplements or eating green vegetables. For example, WHO malnutrition expert Francesco Branca has suggested that supplements, the fortification of existing foods with vitamin A, and growing carrots or leafy vegetables might be more promising to fight the problem. Opponents also argued that children would need to eat quite large amounts of rice to avoid vitamin A deficiency. While this was true for the initially developed variety of golden rice, it did not apply to a variety that was later developed by the biotechnology company Syngenta and which contains up to 23 times more beta carotene. Opponents also criticized the fact that golden rice ignored broader social, cultural, and economic issues, and ignored poverty as an underlying problem. It was thus based on a solution strategy that assumed that vitamin A deficiency could be solved (at least partly) by genetic manipulation of rice without a need to look at other causes of the underlying problem and without paying attention to local contextual factors. In addition, particularly organizations like Greenpeace saw golden rice as nothing more than a PR stunt for promoting genetically modified food, which had met with public resistance, particularly in Europe.

Against these arguments, proponents pointed out that several studies had shown clear human health benefits of golden rice. They also argued that golden rice was not developed from a profit motive but to help solve an important societal challenge. The technology was free to use for farmers and could be freely transferred to countries in which rice was cultivated. They also felt that some opponents, like Greenpeace, were motivated by what they considered to be groundless fears of genetically modified organisms (GMOs). In 2016, more than a hundred Nobel laureates signed a letter urging Greenpeace to stop their campaign against GMOs, and particularly against golden rice.

What is striking is that local stakeholders were largely absent from the debate about golden rice. Among the proponents, there were organizations like the International Rice Research Institute (IRRI) from the Philippines, while among the opponents there were organizations like MASIPAG, a farmer-led network in the Philippines, but still local concerns were not prominent in the debate and the development of golden rice, while they are likely to be crucial for the success (or failure) of golden rice. As argued by Stone and Glover (2017), the effectiveness of golden rice will in part depend on the varieties into which it is crossed locally. (Initial testing largely focused on varieties that were not used on a large scale in countries with a vitamin A deficiency.) Another question, as they point out, is whether the beta carotene in the rice would (fully) survive local habits of storage, milling, and cooking; storage might, for example, be quite long. Yet another uncertainty is how healthy children need to be in order to effectively convert beta carotene into vitamin A. It is known that this effectiveness will depend, among others, on the fat percentage (which is likely low for undernourished children), and most experiments to test health benefits have focused on healthy participants. Such questions might have gotten more priority if a negotiation strategy had been chosen from the start to address the societal challenge of abating vitamin A deficiency. And perhaps, following a negotiation strategy rather than a solution strategy, as proponents now did, and consulting with local stakeholders would also have led to less resistance to the innovation.

Source: Based on Mayer (2005), Enserink (2008), and Stone and Glover (2017).

more reflexive (goals and values are not taken for granted but need to be together learned), and also more anticipatory (side effects and risks are anticipated rather than ignored). It thus better fits the process criteria for responsible innovation discussed before.

9.6　Responsible Innovation in Industry

In Chapter 2, we discussed the responsibility of companies, and initiatives for Corporate Social Responsibility (CSR). Insofar as enterprises have a large impact on society through the innovative products they bring to the market, responsible innovation certainly fits under the broad umbrella of CSR.

A survey among UK nanotechnology companies suggests that in their CSR policies, most of these companies "focus on goals of regulatory compliance and supporting commercialization, rather than on wider issues regarding the need to seek from stakeholders some form of informed consent to bear risks and uncertainties, or regarding the need to shape innovation in accordance with agreed-upon societal priorities" (Groves et al., 2011, p. 547). While these companies thus practice CSR they are not yet adopting a responsible innovation approach. Doing so, would require these companies to move beyond such reactive, defensive, or accommodative policies to deal with social, environmental, and ethical issues, and adopt a more proactive stance. This would imply that they actively anticipate the issues that may affect their processes, products, or services, which groups of people they may have an impact on or what environmental footprint they will have, that they assume responsibility for this, and that they aim not only at doing no harm to society, but also at doing good.

Although several companies, mainly large ones, have assumed responsibility for many parts of their operations through CSR, they have only done so to a limited extent for their R&D and innovation processes. Figure 9.3 shows the various phases of the product development and life cycle. Current CSR efforts that relate to this product development and life cycle usually focus on later phases, for example, the manufacture, use, and disposal of products. Responsible innovation emphasizes the earlier phases of R&D, innovation, and design when addressing responsibility in product development and its life cycle phases. This is also important for industry because the outcome of the innovation, namely its limitations and its effects on users and wider society, depends on decisions made in these

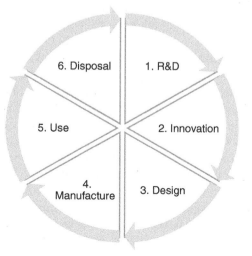

Figure 9.3　Product development and life cycle phases. Van de Poel (2017)/MDPI/CC BY 4.0.

earlier phases of the life cycle. Changing the product at later phases of the product development is likely to induce higher costs. It is therefore important to think about what societal challenges and what values need to be included during the early product development phases, which can lead to the development of products that better meet such challenges and values, and are, as a result, socially, environmentally, and financially more successful.

At the same time, the foreseeability of social and ethical impacts increases as the life cycle proceeds. Therefore, some benefits, values, risks, and concerns are hard to foresee in the early phases. What is needed to resolve this dilemma is a balance between foreseeability and the cost of change. This brings responsible innovation into the equation for companies. Through responsible innovation, companies can identify early in the product development and life cycle process of new innovations what potential social effects are associated with a potential innovation and how to accommodate these before the technology's embedding is irreversible or can only be undone at high costs and with delays in market launch.

Responsible innovation may have clear advantages for companies and contribute to competitive advantage (Lees and Lees, 2017; Scholten and van der Duin, 2015). It helps them to better know and understand their clients and stakeholders, and thus to identify their needs and concerns and translate these into the development of products that will be better accepted and more morally acceptable. This process creates more value for both users and society. Furthermore, because this process encourages timely consideration of potential issues that products may raise, it sidesteps or mitigates scenarios in which companies face public criticism. More generally, it may help to build a relation of trust with society (Nathan, 2015), which may make companies less sensitive to resistance or protest. In addition, it may help them to show governments that they take their responsibility seriously, and thus be allowed to make voluntary agreements rather than being regulated. However, this may also result in mere window-dressing if companies do not at the same time seriously take on board a proactive responsibility to develop more responsible technologies.

In the literature, it has been argued, that currently CSR is often disconnected from **corporate strategy** (Porter and Kramer, 2006). Some companies undertake CSR activities in areas that are not central to their core business activities. For instance, a chemical company that supports local social initiatives but does not invest in making its production process more sustainable. It has been suggested that instead CSR and responsible innovation should aim at creating shared value together with social actors (Porter and Kramer, 2006; van de Poel et al., 2017).

> **Corporate strategy** The strategy in which a company tries to distinguish itself from its competitors to sustain long-term profitability.

In terms of developing a responsible innovation strategy, this means that companies should establish in which areas they can have added value to society. These areas vary depending on the company, for example, how a company can add value depends on the technological area, the type of market, and the company's resources and capabilities. It also means that responsible innovation strategies should be closely attuned to the corporate strategy of the company. Companies should be selective in their responsible innovation strategy and look for those areas where they can add value to society while making a profit. Such strategies are typically focused not only on avoiding harm, but also on doing good.

For individual companies, responsible innovation may be difficult because it is costly and may lead to a competitive disadvantage if a company follows more strict moral guidelines than its competitors. These barriers may be partly overcome by initiatives at the branch level like a code of conduct that is to be followed by all companies in an industry branch (see box).

Responsible Care

Responsible Care is a voluntary self-regulatory code for the chemical industry that has been adopted by an increasing number of chemical companies since the 1980s. Responsible Care is aimed at improving the environmental and safety performance of the chemical industry, also in innovation; it can therefore be seen as an example of responsible innovation *avant la lettre*. One of the main reasons to develop Responsible Care was an awareness that the public image of individual chemical companies depends not only on the performance of that company, but also on the sector as a whole. Responsible Care may thus be seen as an attempt to create a level playing field that improves the safety and environmental performance of the chemical industry as well as its public image. At the same time, it has been criticized as an attempt to postpone government regulation. One of the lessons learned from Responsible Care is that it is important to have some form of external auditing to check whether companies indeed live up to voluntary self-regulatory codes.

Source: Based on King and Lenox (2000), Moffet et al. (2004), and Givel (2007).

9.7 Disruptive Innovation

Innovation is often incremental, that is, it is aimed at relatively small improvements of existing technological products, services, or systems, taking the current underlying technological principle and customer base as given. However, some innovations are more disruptive, which creates additional challenges and opportunities for responsible innovation. Disruptive innovation may be challenging for responsible innovation because existing regulatory or ethical frameworks may no longer suffice; sometimes even existing values and concepts may be disrupted by new technology. Disruptive innovation can also offer opportunities for responsible innovation, for example, when it comes to addressing grand societal challenges through responsible innovation. Disruptive innovation may be required to address some of these societal challenges, like reducing greenhouse warming or abating hunger in the world.

Innovations can be disruptive in various ways. Here we distinguish between four different kinds of disruptiveness: (1) disruption of (economic) markets; (2) disruption of social practices and institutions; (3) regulatory disruption; and (4) conceptual and normative disruption. We discuss each of these briefly below, and set out what the implications are for responsible innovation.

9.7.1 Market disruption

New technologies may disrupt existing economic markets because they may offer newcomers to such markets opportunities for getting a more dominant position or even taking over certain markets. The reason is that existing companies, incumbents, often focus on incrementally improving their products on the basis of existing technology and existing customer needs. Newcomers may introduce more radical innovation based on new technology and serving needs of customers that are not yet served or are only latent (Christensen, 2013).

An example is the introduction of smartphones. The traditional market for mobile phones was dominated by companies like Nokia. Of course, such companies introduced new models and improved their products incrementally, but the smartphone was based on new innovative technology, and it created a device with many more functionalities than traditional mobile phones. It also served customer needs that had not yet been clearly articulated in relation to mobile phones and were thus largely latent. It is no coincidence that the smartphone was introduced by companies like Apple that had not been active in the mobile phone industry before but rather had a background in the computer industry. These companies quickly gained a dominant position in the mobile phone market at the expense of companies like Nokia.

Market disruption creates challenges for responsible innovation, and for the companies developing and introducing innovations. From a business point of view, a main challenge for companies is when to choose for more disruptive innovations and how to make them successful; in terms of responsible innovation, more radical or disruptive innovations might well raise more concerns with respect to social acceptance and moral acceptability. This may make the responsible innovation approach more pertinent but also more difficult to follow (because ethical issues and requirements are harder to foresee) for disruptive innovations.

Market disruption may also offer new opportunities for responsible innovation. For example, it may offer the possibility to serve certain customer groups that are not served by current technology. In the light of responsible innovation, companies may for example choose to aim at products for what has been called the bottom of the pyramid (BOP); the BOP refers to the poorest, but largest socio-economic group worldwide, who live on less than 2.50 dollar a day. While this group has limited purchasing power, it may nevertheless be an interesting market for companies due to its size, particularly if they develop so-called frugal innovations, which are aimed at offering similar user value as traditional products, services and systems but against lower costs, particularly for markets in developing countries (Van Beers et al., 2020).

9.7.2   Social disruption

New technologies may also have disruptive effects on social practices and institutions. Such social practices and institutions are characterized by certain norms or rules that people (have to) follow when engaging in them and that are constitutive of the social practice or institution. For example, the social practice of driving a car on public roads is characterized by such rules as:

- You should have a driver's license to drive a car on public roads;
- Your car should meet certain requirements that are regularly checked;
- You should drive on the right side of the road (at least in most countries);
- Your driving should not unnecessarily endanger other road users.

Some of these rules are formal and may even be laid down in the law; others are more informal but nevertheless constitutive for the social practice or institution. With the introduction of self-driving cars, both types of rules may come under pressure. For example, in a system with only self-driving cars, drivers may no longer need a driver's license, or perhaps they may still need one, but the requirements for a license might be quite different from today. The rule that one should not endanger other road users may still be relevant, but with the introduction of self-driving cars, it might become quite unclear who is responsible for this. Is it still the driver, who should intervene if the car is about to make a dangerous maneuver? Or is the car to be programmed in a way that it prevents accidents? And who is to be held accountable in case of an accident?

The disruption of existing social practices and institutions may be harmful because such practices and institutions often help us to realize shared ends (O'Neill, 2022). For example, the social practices and institutions around driving cars on public roads help to coordinate individual behavior in a way that makes it possible for all to effectively and safely travel from A to B. If a social practice or institution is disrupted, this may no longer be possible, and new social practices and institutions may be needed to achieve the shared end.

Of course, not all social practices and institutions are worth preserving. Some may serve immoral ends (like certain criminal or mafia practices); others may be outdated and serve ends that are no longer relevant, or they may need changing due to the need to abate new problems, like climate change.

Responsible innovation as an approach should take into account the potentially disruptive effect of technology on social practices and institutions. As indicated, this disruptive effect may sometimes be undesirable, which could be a reason not to develop a technology, to adjust its design or to take additional measures during its introduction. In other cases, disruptions may be desirable and even be an explicit aim of responsible innovation.

9.7.3 Regulatory disruption

New technologies may also disrupt existing rules and regulations. For example, technical codes and standards for the safety of pressure vessels were traditionally based on the assumption that such vessels are made of metal. When it became possible to make pressure vessels with plastics and composites these standards were no longer applicable, and new ones had to be formulated in order to ensure the safety of pressure vessels (Van Gorp, 2005).

Also for some legal rights, it may unclear whether they apply to new technology. For example, when e-mail was introduced, it was not immediately clear whether the letter secret, which guarantees the privacy of correspondence, also legally applied to it, as it was typically formulated in terms of physical letters, and e-mails are not physical. In such cases, it might require new laws, or at least new legal interpretations by judges (jurisprudence) to ensure that legal rights also apply to new technology. In such cases, the new law or the jurisprudence might be based on an underlying right (or value), that is not or less technology-specific, in this case the privacy of correspondence.

New technologies may also be regulatory disruptive because they create new types of (moral) problems that as such did not exist before. For example, social media have led to new forms of fake news, cancelling of people, and sextortion. Such new issues may require new laws and regulations to adequately address them. Drafting and getting accepted (in parliament) such new laws would typically take several years, so that legal rules and regulations tend to lack behind technological development. Another difficulty might be that some technologies – like social media – span the whole world and therefore do not fall under one jurisdiction, which makes them hard to regulate for individual governments, while international rule-making tends to be slower and may be harder to effectuate than national laws.

When innovations are (potentially) regulatory disruptive it has a number of implications for responsible innovation. A main implication is that existing regulatory and legal frameworks may no longer be a sufficient or appropriate guide for responsible innovation. This is particularly true for more detailed provisions that are based on current technology. Often underlying legal and moral values and norms – like, for example, privacy or safety – may still be applicable. However, sometimes technology may be so disruptive that it requires new values and moral concepts for responsible innovation, a category of disruption to which we now turn.

9.7.4   Conceptual and normative disruption

New technologies may also disrupt the very concepts and values by which we understand and evaluate their impacts (Swierstra, 2013; Hopster, 2021). Conceptual disruption occurs if our intuitive and unreflective way of applying concepts no longer holds. For example, under normal circumstances, we classify entities that we encounter without further reflection as either a human person or not. However, when we watch an embryo or a fetus (e.g., on an echo during pregnancy), we might be uncertain whether we should classify it as a human person or not; similarly, when we encounter a humanoid robot, we may also be conceptually uncertain whether it classifies as a person or not (perhaps, it is a "non-human person")

Conceptual disruption may have normative implications. For example, if we are unsure whether a certain robot is (like) a person, we are likely also uncertain about how we should treat it. Should we treat it with respect as we do with fellow human persons and, for example, get upset or angry when someone kicks the robot? Or should we treat it as a lifeless object that can be kicked without remorse like a football?

Also values may be disrupted by new technological developments. For example, economic and technological developments have caused large-scale environmental problems, which have led to the emergence of the value of sustainability, which has gained further prominence since the 1980s (Van de Poel and Kudina, 2022). In response to concerns about autonomous weapon systems (and self-driving cars), a value like "meaningful human control" has been articulated and proposed as a guide for the design of autonomous systems, in order to ensure that humans retain responsibility for the functioning of such systems (Santoni de Sio and Van den Hoven, 2018). And in response to the advance of machine learning techniques in artificial intelligence, which allow learning in ways that are often opaque to humans, the value of "explainability" has been proposed as guiding value for the design of algorithms that advice or guide humans in morally relevant decisions (High Level Expert Group on AI, 2019). In all these cases, technology created new moral concerns, that – at least according to some – required new values that should be accounted for in the development and design of these technologies.

The fact that technology may trigger conceptual and normative disruption has several consequences for responsible innovation. A main question is to what extent it is appropriate to judge the moral acceptability of new innovations on the basis of current moral values and normative concepts. On the one hand, we know that values can change and in particular we should be open for new moral concerns that are raised by new technology, which may sometimes require the articulation of new values. On the other hand, we should not too easily give up existing values and concepts, as these might be the product of decades or even centuries of human experience and moral reflection. When it comes to the design of concrete technological products or systems, one possibility might be to design them in a flexible way if we are unsure about what future values might become important, so that they can be adapted later on the basis of new insights.

9.8 Responsible Innovation and the Responsibility of Engineers

After reading the previous sections, you might wonder what exactly the responsibility of engineers is for responsible innovation. In particular, you might worry if it is not too demanding to ask from engineers to make innovation fully responsible by some of the criteria for responsible innovation discussed earlier. After all, engineers are only one of the actors involved in innovation processes. Moreover, some of the normative criteria for responsible innovation that we discussed, like inclusiveness, are quite demanding and beyond the control of (individual) engineers.

Indeed when we look at the conditions for responsibility discussed in Chapters 1 and 7, several are typically not met in innovation processes (Swierstra and Jelsma, 2006; van de Poel and Sand, 2021):

- The *causal* condition for responsibility is rarely met in innovation as a multiplicity of actors is involved and there are long causal chains between the innovation process and the eventual social impacts. Outcomes are affected by actions of a variety of actors and it may be hard, if not impossible, to detect a causal connection between the actions of individual engineers and the overall consequences of innovation processes.
- Also, the *freedom of action* condition can be undermined in innovation in various ways. For one thing, engineers often work in companies or large research organizations, which constrain their freedom through corporate statutes and other measures of internal regulation. Moreover, if the freedom condition is understood in terms of the availability of options to the agent to avoid undesirable outcomes, it may be even harder to fulfill. This is because of the causal inefficiency of individual agents in such settings.
- The *foreseeability* condition is also rarely met in innovation. Innovations involve many uncertainties and unknowns. The social consequences of an innovation may not only be unforeseen but in principle be unforeseeable.

Does this mean that engineers have no responsibility for innovation? A first thing to note is that engineers have at least some *active* responsibilities in innovation, along the lines discussed in previous chapters. They should, for example, avoid conflict of interests and hold paramount the safety, health and welfare of the public in their decisions (Chapter 2). They should also pay attention to risks (Chapter 6) and to values in design (Chapter 5).

We believe, however, that responsible innovation adds some new and additional responsibilities for engineers in addition to the ones discussed in previous chapters. We are here primarily thinking of the active responsibility of engineers to contribute to innovation processes that meet the discussed criteria for responsible innovation (anticipation, inclusiveness, reflexivity, responsiveness) and to contribute to addressing societal challenges.

This does not mean that engineers are also morally accountable or blameworthy when an innovation has negative societal impacts. For the reasons discussed earlier, such attribution of passive responsibility for the negative consequences of innovations to engineers might be unjust and inappropriate. The kind of responsibility of engineers we are referring to is instead active, it is about the responsibility to help bring about more responsible innovation processes. It is moreover a co-responsibility, that is, a responsibility engineers bear together with others in the innovation process, including managers, users, relevant stakeholders, and the government.

You might still wonder whether an innovation process can ever be fully responsible by the criteria discussed previously. Can all possible consequences be anticipated? Can all relevant stakeholders be included? (What about future generations for example?) It seems indeed true that the criteria for responsible innovation can be quite demanding, but that does not mean that one cannot aim for *more* responsible innovation processes, and that one can, also as an engineer, assume active responsibility for contributing to such more responsible innovation processes.

Obviously, responsible innovation also requires certain actions on the part of managers, CEOs of companies, users, governments, and other stakeholders. One might even argue that responsible innovation to be successfully introduced in society requires more large-scale economic, political, and institutional changes. These are clearly beyond the responsibility of individual engineers as engineers, although they may still contribute to them collectively, for example, through professional organizations, or as citizens.

9.9   Chapter Summary

Responsible innovation is innovation that addresses the values, needs, and expectations of society by opening up the innovation process to a broader range of stakeholders and moral concerns. Such opening up of the innovation process may be difficult for at least two reasons. One is that technological innovation is sometimes seen as nothing more than the application of new scientific insights and therefore as value-neutral. The second reason is that innovation is of key strategic importance to companies and countries, which may therefore be reluctant to open up the innovation process.

Despite these barriers, there are good reasons to open up the innovation process. Some of these reasons are of a more principled nature: they state that since innovations have a major impact on people's lives, people have a democratic right to be involved in their shaping. Other arguments are more instrumental or consequentialist: opening up the innovation process may result in innovations (products, services, systems) that are better in the sense of being more socially accepted and more morally acceptable.

Responsible innovation has three complementary dimensions: process, product, and societal challenges. As product, responsible innovation refers to innovative products, services or systems that reflect social and moral values, as also discussed in Chapter 5 on design for values. When it comes to the process dimension of responsible innovation, four procedural criteria are important: anticipation, inclusiveness, reflexivity, and responsiveness. The third dimension relates to societal challenges that should be addressed in responsible innovation. We discussed two types of strategies that can be followed in addressing societal challenges: solution strategies and negotiation strategies. We argued that the latter is more in line with the procedural criteria for responsible innovation.

Responsible innovation is not yet practiced on a significant scale in industry. While companies may be reluctant to open up the innovation process or shy away because of unfamiliarity or expected costs, there are also potential benefits for companies. It may help to reduce the costs of innovation in terms of lack of acceptance or public resistance, and it may even provide companies with a competitive advantage. The latter requires companies to develop a specific responsible innovation strategy that is tied to their core business, rather than an add-on, and that offers value to society as well as to the company.

While innovation is often incremental, it may sometimes be disruptive which creates additional challenges and opportunities for responsible innovation because we may no longer be able to rely on existing regulatory or moral frameworks. We have discussed four ways in which innovations can be disruptive: market disruption, social disruption, regulatory disruption, and conceptual and normative disruption.

A final issue we discussed was the responsibility of engineers for responsible innovation. The influence that engineers have on the innovation process, and the eventual societal consequences of innovative products is limited. They can therefore usually not be blamed or held liable when innovations have negative consequences. Still, they have an active responsibility for making the innovation process more responsible and inclusive; this responsibility they share with others involved in the innovation process; it is a co-responsibility.

Study Questions

1 The introduction mentions three aspects that are key to current approaches in the ethics of technology:
    a. What are these three aspects?
    b. Do all three aspects also apply to the design for values approach discussed in Chapter 5?
2 What is meant by the idea that technology is a form of applied science? Give an example (other than the one mentioned in the text) that belies the idea that technology is applied science. Can you also give an example in which a technological innovation became possible due to new scientific insights (other than the ones mentioned)?
3 What is meant by "technological determinism?" Search the Internet or a newspaper for a view or quote that clearly represents the view of technological determinism, and explain why it does so. Do you agree with the view or quote?
4 Give an example in which a company clearly gained a competitive advantage vis-à-vis its competitors through a technological innovation. Do you think this company would have been reluctant to opening its innovation process to a wider range of stakeholders and moral concerns? Why? How might this reluctance be possibly overcome?
5 Does responsible innovation better align with ideas of direct democracy (in which citizens themselves decide about important issues) or indirect, representative democracy (in which citizens elect representatives that decide about important issues)? Explain why. What would be required to make innovation more democratic from the viewpoint of direct democracy? And what from the viewpoint of indirect democracy?
6 Explain why an innovation that is morally acceptable (according to the definition given in the text) is not necessarily the morally best innovation.
7 Can you think of an innovation process that meets the process criteria for responsible innovation but still results in products that are either not socially accepted or morally unacceptable?
8 In what respects is anticipation different from prediction? Do you think that anticipation might really help to overcome the Collingridge dilemma as suggested in the text? Argue your answer.
9 Is an innovation process that is inclusive with respect to a wide range of societal stakeholders also necessarily inclusive with respect to all relevant moral considerations? Explain why you think so. Can you think of moral considerations that are ignored despite a wide range of stakeholders being involved?
10 What is the difference between first-order and second-order learning? Why is second-order learning more difficult and rare? Can you think of an experience in which second-order learning took place?
11 It might be argued that for large companies, anticipation is a more feasible strategy, while for small companies, like start-ups, responsiveness is more feasible when it comes to responsible innovation. Why do you think this is the case? Do you think that nevertheless large companies should also aim to be responsive and smaller also to be anticipatory?
12 For your own area of technology, give an example of disruptive technology. In which of the four aspects mentioned in the text is this technology disruptive?

Discussion Questions

1   In the text, the example is given of the different phases of testing and experimentation (for risks and side effects) that new drugs (or other medical interventions) have to go through. These phases are (partly) based on strict legal requirements. Similar requirements do not seem to apply for, for example, social media.

   a. Why do you think that legal requirements for the introduction of new medical drugs into society are different from requirements for introducing, for example, new social media?

   b. What are potential risks and side effects of social media? Could these be tested for or experimented with before new social media are introduced into society on a large scale?

   c. Can you think of certain advantages and disadvantages of having more strict testing requirements on side-effects and risks in the case of social media?

   d. Do you think that there should be stricter legal requirements for introducing social media (or comparable technologies) into society?

2   Look for a company that states to contribute to the solution of the SDGs through their innovative products. Does this company in your view follow a solution or a negotiation strategy in contributing to the SDGs? Explain why.

3   As an engineer, do you feel a responsibility for making innovation processes more responsible? Why (not)?

4   According to Grinbaum and Groves, whether the actions of innovators will have good or bad consequences will often depend on luck, which undermines the attribution of responsibility. As they write: "[T]here is no guarantee that moral luck in the uncertain future will not mean that one's efforts to act responsibly will not turn out to have unintended consequences. Whatever choices are made, the final verdict on a distinction between responsible and irresponsible innovation is not in our capacity to make." (Grinbaum and Groves, 2013, p. 139).

   a. Why does luck undermine the attribution of responsibility?

   b. To which dimension of responsible innovation does the argument by Grinbaum and Groves refer (product, process, or societal challenges)?

   c. Does their argument also affect the possibility to judge responsible innovation as a process by criteria like anticipation, inclusiveness, reflexivity, and responsiveness?

   d. Do you agree with Grinbaum and Groves that is not "in our capacity" to distinguish irresponsible from responsible innovation?

5   Responsible innovation has been criticized for being based on a too narrow techno-economic view of innovation which assumes that technological innovation is always desirable rather than questioning innovation itself (Blok and Lemmens, 2015).

   a. What is meant by a techno-economic view of innovation? In what sense is such a view narrow?

   b. Do you agree that a responsible innovation approach is based on the assumption that innovation is always desirable? Argue your answer.

   c. How should the responsible innovation be adapted (if at all) if such criticism is to be taken seriously?

6   Some have suggested that the advance of digital technologies has led to "the end of privacy." One way to interpret this is to say that digital technologies have led to the disruption of the value of privacy, which should therefore be given less weight in the digital age. Another, radically different, interpretation is that digital technologies have made it much harder to realize the value of privacy but that from a moral point of view the value has not become less important, and perhaps even more, important. What may be needed according to this second line of reasoning are new normative understandings of privacy and new regulatory instruments to safeguard privacy. (An example of the latter might be the European GDPR: General Data Protection Regulation).

   a. Explain why both lines of reasoning recognize the disruptive character of digital technologies.

   b. What types of disruptions are central in both lines of reasoning?

   c. Which line of reasoning do you find most convincing? Or would you argue for yet another viewpoint?

Notes

1 https://www.fairphone.com/en/about/about-us/, accessed 2 March 2023.
2 https://www.fairphone.com/en/impact/fair-materials/, accessed 2 March 2023.
3 It should be noted that if we call an innovation process "responsible" as we will do now, we mean something different than when we call a person "responsible." In the latter case, we usually mean to say that a person acts in a certain (laudable) way or has a certain (praiseworthy) character. Innovations or innovation processes, however, do not act and do not have a (psychological) character, but they may have characteristics that make them morally good or desirable, like aligning with the values and needs of society.

Appendix I: Engineering Qualifications and Organizations in a Number of Countries

Table AI.1 Overview

Country	USA	UK	Australia	Netherlands	Europe
Titles/qualifications	PE	CEng IEng	NER IRAQ Chartered credential	Ir. Ing.	EUR ING
Regulatory or licensing body	State licensing boards	Engineering Council	Engineers Australia BPEQ	–	ENGINEERS EUROPE
Other important organizations	NSPE NAE ABET IEEE ASCE ASME AIChE NIEE OEC	RAEng IET ICE IMechE IChemE IED	Consult Australia Professionals Australia	KIVI	

United States

Titles and qualifications

Engineers in the United States can get licensure as Professional Engineer (PE). "To become licensed, engineers must complete a four-year college degree, work under a Professional Engineer for at least four years, pass two intensive competency exams and earn a license from their state's licensure board. Then, to retain their licenses, PEs must continually maintain and improve their skills throughout their careers." "Only a licensed

Ethics, Technology, and Engineering: An Introduction, Second Edition. Ibo van de Poel and Lambèr Royakkers.
© 2023 John Wiley & Sons Ltd. Published 2023 by John Wiley & Sons Ltd.

engineer may prepare, sign and seal, and submit engineering plans and drawings to a public authority for approval, or seal engineering work for public and private clients" (www. nspe.org/Licensure/WhatisaPE/index.html[1]).

For many engineering jobs, especially in industry, licensure is not required.

For more information, see www.nspe.org/Licensure/WhatisaPE/index.html

For an overview of state licensing boards, see https://www.nspe.org/resources/licensure/licensing-boards.

Important organizations

NSPE   (National Society of Professional Engineers): www.nspe.org
NAE   (National Academy of Engineering): www.nae.edu
ABET   (Accreditation Board for Engineering and Technology): www.abet.org
IEEE: www.ieee.org
ASCE   (American Society of Civil Engineers): www.asce.org
ASME   (American Society of Mechanical Engineers): www.asme.org
AIChE   (American Institute for Chemical Engineers): www.aiche.org
NIEE: (National Institute for Engineering Ethics): www.niee.org
OEC   (Online Ethics Center): onlineethics.org

Codes of conduct

ASCE: https://www.asce.org/career-growth/ethics/code-of-ethics
ASME: https://www.asme.org/getmedia/3e165b2b-f7e7-4106-a772-5f0586d2268e/p-15-7-ethics.pdf
IEEE: https://www.ieee.org/content/dam/ieee-org/ieee/web/org/about/corporate/ieee-code-of-ethics.pdf
AIChE: https://www.aiche.org/about/governance/policies/code-ethics
NSPE: see Appendix II

UK

Titles and qualifications

"The Engineering Council is the UK regulatory body for the engineering profession. We hold the national registers of over **229,000** Engineering Technicians (EngTech), Incorporated Engineers (IEng), Chartered Engineers (CEng) and Information and Communications Technology Technicians (ICTTech). In addition, the Engineering Council sets and maintains the internationally recognised standards of professional competence and ethics that govern the award and retention of these titles. This ensures that employers, government and wider society – both in the UK and overseas – can have confidence in the knowledge, experience and commitment of professionally registered engineers and technicians." (www.engc.org.uk)

CEng (Chartered Engineer)

"The CEng title is open to anyone who can demonstrate the required professional competences and commitment, as set out in the professional standard UK-SPEC. Individuals generally develop these through education and working experience.

The application process for CEng registration is more straightforward for those with exemplifying academic qualifications. For CEng this is one of the following:

- A Bachelors degree, with Honors, in engineering or technology, accredited for CEng, plus an appropriate and accredited Masters degree or Engineering Doctorate (EngD), or appropriate further learning to Masters level
- An accredited integrated MEng degree

Please note:

- If you gained academic qualifications which, at the time you gained them, were recognised as the exemplifying qualifications for CEng, these are still considered to be so.
- Applicants holding the exemplifying qualifications are automatically eligible for interim registration.
- To check whether your academic qualification is accredited for CEng please click here.

You can still become a Chartered Engineer without these academic qualifications. Further information about the individual, work-based, assessment process can be found in UK-SPEC and from your professional engineering institution." (https://www.engc.org.uk/ceng)

IEng (Incorporated Engineer)

"The IEng title is open to anyone who can demonstrate the required professional competences and commitment, as set out in the professional standard UK-SPEC. Individuals generally develop this through education and working experience.

The application process for IEng registration is more straightforward for those with exemplifying academic qualifications. For IEng this is one of the following:

- An accredited Bachelors or honors degree in engineering or technology
- An accredited HNC [Higher National Certificate] or HND [Higher National Diploma] in engineering or technology (for programs started before Sept 1999)
- An HNC or HND started after Sept 1999 (but before Sept 2010 in the case of the HNC) or a Foundation Degree in engineering or technology, plus appropriate further learning to degree level
- An NVQ4 [National Vocational Qualification 4] or SVQ4 [Scottish Vocational Qualification 4] that has been approved for the purpose by a licensed engineering institution, plus appropriate further learning to degree level

(…) You can still become an IEng without these academic qualifications. Further information about the individual, work-based, assessment process can be found in UK-SPEC and from your professional engineering institution."
(https://www.engc.org.uk/IEng)

Important organizations

ECUK	(Engineering Council UK):	www.engc.org.uk
RAEng	(Royal Academy of Engineering):	www.raeng.org.uk
ICE	(Institution of Civil Engineers):	www.ice.org.uk
IET	(Institution of Engineering and Technology):	www.theiet.org
IMechE	(Institution of Mechanical Engineers):	www.imeche.org

IChemE   (Institution of Chemical Engineers): www.icheme.org
IED   (The Institution of Engineering Designers): www.ied.org.uk

For other relevant institutions see: www.engc.org.uk/about-us/our-partners/professional-engineering-institutions.aspx

Codes of conduct

Royal Academy of Engineering: https://raeng.org.uk/media/cz5du0gl/engineering_ethics_in_practice_full.pdf
IET: https://www.theiet.org/about/governance/rules-of-conduct
IMechE: https://www.imeche.org/about-us/imeche-governance/governance-and-finance-reviews/code-of-conduct-explained
IchemE: https://www.icheme.org/about-us/governance/code-of-professional-conduct

Australia

Titles and qualifications

"Engineering is an innate part of everyday life. Without thinking, we trust that the cars, roads, buildings and electronic devices we use will work and won't harm us. Because of that trust, preserving the integrity of engineering practice is vital.

Registering engineers is important because it ensures engineering professionals meet benchmarked education, training, professional conduct and competency standards. It allows consumers to feel confident in the abilities of the engineers they hire and the products they use.

In Australia, the states and territories are responsible for the statutory registration of engineers. How this is managed is up to each state government. Some states don't require an engineer to be registered to practise and others have mandated registration. Visit the state registration page to learn more about how statutory registration works.

As Australia's peak body for engineers, Engineers Australia has created the National Engineering Register (NER). The NER works alongside statutory state registration and provides a mechanism for engineers to have their qualifications and experience recognised across the country. This mechanism also ensures those who are unsuitable to work as an engineer are excluded.

Being registered on the NER means that an engineer:

- has been assessed and has met the strict competencies for their occupational category
- is committed to their ongoing training and professional development
- is endorsed by Engineers Australia.

Once registered, engineers can use the Registered credential as part of their post-nominals. (https://www.engineersaustralia.org.au/credentials/registration)

To be eligible to apply for an NER credential, you need to meet the following criteria:

- have an Engineers Australia recognised engineering qualification or a successful Engineers Australia migration skills assessment or membership competency assessment
- have five or more years of relevant full-time equivalent engineering work experience in the last 10 years, with at least four years of full-time equivalent post graduate experience
- meet all the competencies for independent practice in your occupational category."
- (https://www.engineersaustralia.org.au/credentials/registration/national-engineering-register#accordion-15691)
- Another register for professional engineers is the Registered Professional Engineer Queensland (RPEQ) assessment of the Board of Professional Engineers Queensland (BPEQ). Since 2016 the NER is consistent with the RPEQ.

Besides the NER and RPEQ there is the Chartered credential. "The Chartered credential is the highest available technical credential for an engineering professional. It's nationally and internationally recognised as a measure of excellence and signifies a certain level of skill, talent and experience.

When you hold a Chartered credential, it puts a solid value on the years you've devoted to the profession. It gives prospective employers and clients immediate respect and confidence in your abilities.

A Chartered credential also opens the door to career progression, leadership development and opportunities to work overseas." (https://www.engineersaustralia.org.au/credentials/chartered)

Important Organizations

Engineers Australia:	www.engineersaustralia.org.au
NERB	(National Engineering Registration Board): www.nerb.org.au
BPEQ	(Board of Professional Engineers of Queensland): www.bpeq.qld.gov.au
Consult Australia	(formerly known as the Association of Consulting Engineers Australia (ACEA)): https://www.consultaustralia.com.au
Professionals Australia	(formerly known as the Association of Professional Engineers, Scientists, and Managers, Australia): https://www.professionalsaustralia.org.au

Code of conduct

Engineers Australia:	https://www.engineersaustralia.org.au/publications/code-ethics
Consult Australia:	https://www.consultaustralia.com.au/home/about-us/governance/codes-of-ethics

The Netherlands

Titles and qualifications

Ir. is the title for engineers holding a Master's degree from a university and *Ing.* for engineers holding a Bachelor's degree from a professional school. There is no system for licensing or registration of engineers in the Netherlands.

Organization

KIVI (the Royal Netherlands Society of Engineers) is the Dutch association for engineers and engineering students. With 25,000 members KIVI is the largest engineering association in the Netherlands. All engineering disciplines are organized within KIVI. This organization plays a serious role in steering the future of engineering, and serves as the sole organization granting the professional and international status of Chartered Engineer and Incorporated Engineer in The Netherlands. See www.kivi.nl

Code of Conduct

KIVI: https://www.kivi.nl/uploads/media/5a587110c2160/2018-01 Code of Ethics. pdf

Europe

Titles and qualifications

European engineers can qualify for the title EUR ING of ENGINEERS EUROPE, formerly FEANI (European Federation of National Engineering Associations) with the following requirements:

After a secondary education at a high level validated by one or more official certificates, normally awarded at the age of about 18 years, candidates:

- must have exemplifying formal qualifications (degrees, diplomas of Higher Education Institutions) in combination with some years of professional experience. Formal qualifications will have been reviewed by the relevant ENGINEERS EUROPE National Member against the agreed standards (inclusion in the ENGINEERS EUROPE European Engineering Education Database – EEED), OR
- must have no exemplifying formal qualifications, but will have engaged in professional Career Learning and peer review via the individual route, OR
- are renewing their EUR ING Certificate upon the submission of evidence of continuous professional development (CPD) in the five years following the first issue of their EUR ING Certificate. Renewal after five years is mandatory.

In addition to these formation requirements, EUR INGs are required to comply with a Code of Conduct respecting the provisions of the ENGINEERS EUROPE Position Paper on Code of Conduct: Ethics and Conduct of Professional Engineers." (https://www. engineerseurope.com/sites/default/files/Code_of_Conduct_ENGINEERS_EUROPE. pdf)

Organization

ENGINEERS EUROPE: https://www.engineerseurope.com/

Code of conduct

ENGINEERS EUROPE: see Appendix III

General Links

Titles and qualifications

https://en.wikipedia.org/wiki/Regulation_and_licensure_in_engineering

Codes of conduct

https://onlineethics.org/cases/ken-pimple-collection/why-all-these-rules
http://ethicscodescollection.org
https://online-learning.tudelft.nl/courses/ethical-dilemmas-in-professional-engineering

Note

1 All the websites in this appendix were last accessed on November 2, 2022.

Appendix II: NSPE Code of Ethics for Engineers

Preamble

Engineering is an important and learned profession. As members of this profession, engineers are expected to exhibit the highest standards of honesty and integrity. Engineering has a direct and vital impact on the quality of life for all people. Accordingly, the services provided by engineers require honesty, impartiality, fairness, and equity, and must be dedicated to the protection of the public health, safety, and welfare. Engineers must perform under a standard of professional behavior that requires adherence to the highest principles of ethical conduct.

I. Fundamental Canons

Engineers, in the fulfillment of their professional duties, shall:

1 Hold paramount the safety, health, and welfare of the public.
2 Perform services only in areas of their competence.
3 Issue public statements only in an objective and truthful manner.
4 Act for each employer or client as faithful agents or trustees.
5 Avoid deceptive acts.
6 Conduct themselves honorably, responsibly, ethically, and lawfully so as to enhance the honor, reputation, and usefulness of the profession.

II. Rules of Practice

1 Engineers shall hold paramount the safety, health, and welfare of the public.
   a. If engineers' judgment is overruled under circumstances that endanger life or property, they shall notify their employer or client and such other authority as may be appropriate.
   b. Engineers shall approve only those engineering documents that are in conformity with applicable standards.

Ethics, Technology, and Engineering: An Introduction, Second Edition. Ibo van de Poel and Lambèr Royakkers.
© 2023 John Wiley & Sons Ltd. Published 2023 by John Wiley & Sons Ltd.

   c. Engineers shall not reveal facts, data, or information without the prior consent of the client or employer except as authorized or required by law or this Code.

   d. Engineers shall not permit the use of their name or associate in business ventures with any person or firm that they believe is engaged in a fraudulent or dishonest enterprise.

   e. Engineers shall not aid or abet the unlawful practice of engineering by a person or firm.

   f. Engineers having knowledge of any alleged violation of this Code shall report thereon to appropriate professional bodies and, when relevant, also to public authorities, and cooperate with the proper authorities in furnishing such information or assistance as may be required.

2 Engineers shall perform services only in the areas of their competence.

   a. Engineers shall undertake assignments only when qualified by education or experience in the specific technical fields involved.

   b. Engineers shall not affix their signatures to any plans or documents dealing with subject matter in which they lack competence, nor to any plan or document not prepared under their direction and control.

   c. Engineers may accept assignments and assume responsibility for coordination of an entire project and sign and seal the engineering documents for the entire project, provided that each technical segment is signed and sealed only by the qualified engineers who prepared the segment.

3 Engineers shall issue public statements only in an objective and truthful manner.

   a. Engineers shall be objective and truthful in professional reports, statements, or testimony. They shall include all relevant and pertinent information in such reports, statements, or testimony, which should bear the date indicating when it was current.

   b. Engineers may express publicly technical opinions that are founded upon knowledge of the facts and competence in the subject matter.

   c. Engineers shall issue no statements, criticisms, or arguments on technical matters that are inspired or paid for by interested parties, unless they have prefaced their comments by explicitly identifying the interested parties on whose behalf they are speaking, and by revealing the existence of any interest the engineers may have in the matters.

4 Engineers shall act for each employer or client as faithful agents or trustees.

   a. Engineers shall disclose all known or potential conflicts of interest that could influence or appear to influence their judgment or the quality of their services.

   b. Engineers shall not accept compensation, financial or otherwise, from more than one party for services on the same project, or for services pertaining to the same project, unless the circumstances are fully disclosed and agreed to by all interested parties.

   c. Engineers shall not solicit or accept financial or other valuable consideration, directly or indirectly, from outside agents in connection with the work for which they are responsible.

   d. Engineers in public service as members, advisors, or employees of a governmental or quasi-governmental body or department shall not participate in decisions with respect to services solicited or provided by them or their organizations in private or public engineering practice.

   e. Engineers shall not solicit or accept a contract from a governmental body on which a principal or officer of their organization serves as a member.

5 Engineers shall avoid deceptive acts.

   a. Engineers shall not falsify their qualifications or permit misrepresentation of their or their associates' qualifications. They shall not misrepresent or exaggerate their responsibility in or for the subject matter of prior assignments. Brochures or other presentations incident to the solicitation of employment shall not misrepresent pertinent facts concerning employers, employees, associates, joint venturers, or past accomplishments.

b. Engineers shall not offer, give, solicit, or receive, either directly or indirectly, any contribution to influence the award of a contract by public authority, or which may be reasonably construed by the public as having the effect or intent of influencing the awarding of a contract. They shall not offer any gift or other valuable consideration in order to secure work. They shall not pay a commission, percentage, or brokerage fee in order to secure work, except to a bona fide employee or bona fide established commercial or marketing agencies retained by them.

III. Professional Obligations

1 Engineers shall be guided in all their relations by the highest standards of honesty and integrity.
   a. Engineers shall acknowledge their errors and shall not distort or alter the facts.
   b. Engineers shall advise their clients or employers when they believe a project will not be successful.
   c. Engineers shall not accept outside employment to the detriment of their regular work or interest. Before accepting any outside engineering employment, they will notify their employers.
   d. Engineers shall not attempt to attract an engineer from another employer by false or misleading pretenses.
   e. Engineers shall not promote their own interest at the expense of the dignity and integrity of the profession.
2 Engineers shall at all times strive to serve the public interest.
   a. Engineers are encouraged to participate in civic affairs; career guidance for youths; and work for the advancement of the safety, health, and well-being of their community.
   b. Engineers shall not complete, sign, or seal plans and/or specifications that are not in conformity with applicable engineering standards. If the client or employer insists on such unprofessional conduct, they shall notify the proper authorities and withdraw from further service on the project.
   c. Engineers are encouraged to extend public knowledge and appreciation of engineering and its achievements.
   d. Engineers are encouraged to adhere to the principles of sustainable development[1] in order to protect the environment for future generations.
3 Engineers shall avoid all conduct or practice that deceives the public.
   a. Engineers shall avoid the use of statements containing a material misrepresentation of fact or omitting a material fact.
   b. Consistent with the foregoing, engineers may advertise for recruitment of personnel.
   c. Consistent with the foregoing, engineers may prepare articles for the lay or technical press, but such articles shall not imply credit to the author for work performed by others.
4 Engineers shall not disclose, without consent, confidential information concerning the business affairs or technical processes of any present or former client or employer, or public body on which they serve.
   a. Engineers shall not, without the consent of all interested parties, promote or arrange for new employment or practice in connection with a specific project for which the engineer has gained particular and specialized knowledge.
   b. Engineers shall not, without the consent of all interested parties, participate in or represent an adversary interest in connection with a specific project or proceeding in

which the engineer has gained particular specialized knowledge on behalf of a former client or employer.

5 Engineers shall not be influenced in their professional duties by conflicting interests.

   a. Engineers shall not accept financial or other considerations, including free engineering designs, from material or equipment suppliers for specifying their product.

   b. Engineers shall not accept commissions or allowances, directly or indirectly, from contractors or other parties dealing with clients or employers of the engineer in connection with work for which the engineer is responsible.

6 Engineers shall not attempt to obtain employment or advancement or professional engagements by untruthfully criticizing other engineers, or by other improper or questionable methods.

   a. Engineers shall not request, propose, or accept a commission on a contingent basis under circumstances in which their judgment may be compromised.

   b. Engineers in salaried positions shall accept part-time engineering work only to the extent consistent with policies of the employer and in accordance with ethical considerations.

   c. Engineers shall not, without consent, use equipment, supplies, laboratory, or office facilities of an employer to carry on outside private practice.

7 Engineers shall not attempt to injure, maliciously or falsely, directly or indirectly, the professional reputation, prospects, practice, or employment of other engineers. Engineers who believe others are guilty of unethical or illegal practice shall present such information to the proper authority for action.

   a. Engineers in private practice shall not review the work of another engineer for the same client, except with the knowledge of such engineer, or unless the connection of such engineer with the work has been terminated.

   b. Engineers in governmental, industrial, or educational employ are entitled to review and evaluate the work of other engineers when so required by their employment duties.

   c. Engineers in sales or industrial employ are entitled to make engineering comparisons of represented products with products of other suppliers.

8 Engineers shall accept personal responsibility for their professional activities, provided, however, that engineers may seek indemnification for services arising out of their practice for other than gross negligence, where the engineers' interests cannot otherwise be protected.

   a. Engineers shall conform with state registration laws in the practice of engineering.

   b. Engineers shall not use association with a nonengineer, a corporation, or partnership as a "cloak" for unethical acts.

9 Engineers shall give credit for engineering work to those to whom credit is due, and will recognize the proprietary interests of others.

   a. Engineers shall, whenever possible, name the person or persons who may be individually responsible for designs, inventions, writings, or other accomplishments.

   b. Engineers using designs supplied by a client recognize that the designs remain the property of the client and may not be duplicated by the engineer for others without express permission.

   c. Engineers, before undertaking work for others in connection with which the engineer may make improvements, plans, designs, inventions, or other records that may justify copyrights or patents, should enter into a positive agreement regarding ownership.

   d. Engineers' designs, data, records, and notes referring exclusively to an employer's work are the employer's property. The employer should indemnify the engineer for use of the information for any purpose other than the original purpose.

Note 1: "Sustainable development" is the challenge of meeting human needs for natural resources, industrial products, energy, food, transportation, shelter, and effective waste management while conserving and protecting environmental quality and the natural resource base essential for future development.

As Revised July 2019

By order of the United States District Court for the District of Columbia, former Section 11(c) of the NSPE Code of Ethics prohibiting competitive bidding, and all policy statements, opinions, rulings, or other guidelines interpreting its scope, have been rescinded as unlawfully interfering with the legal right of engineers, protected under the antitrust laws, to provide price information to prospective clients; accordingly, nothing contained in the NSPE Code of Ethics, policy statements, opinions, rulings, or other guidelines prohibits the submission of price quotations or competitive bids for engineering services at any time or in any amount

Statement by NSPE Executive Committee

In order to correct misunderstandings which have been indicated in some instances since the issuance of the Supreme Court decision and the entry of the Final Judgment, it is noted that in its decision of April 25, 1978, the Supreme Court of the United States declared: "The Sherman Act does not require competitive bidding."

It is further noted that as made clear in the Supreme Court decision:

1 Engineers and firms may individually refuse to bid for engineering services.
2 Clients are not required to seek bids for engineering services.
3 Federal, state, and local laws governing procedures to procure engineering services are not affected, and remain in full force and effect.
4 State societies and local chapters are free to actively and aggressively seek legislation for professional selection and negotiation procedures by public agencies.
5 State registration board rules of professional conduct, including rules prohibiting competitive bidding for engineering services, are not affected and remain in full force and effect. State registration boards with authority to adopt rules of professional conduct may adopt rules governing procedures to obtain engineering services.

As noted by the Supreme Court, "nothing in the judgment prevents NSPE and its members from attempting to influence governmental action …"

Note: In regard to the question of application of the Code to corporations vis-à-vis real persons, business form or type should not negate nor influence conformance of individuals to the Code. The Code deals with professional services, which services must be performed by real persons. Real persons in turn establish and implement policies within business structures. The Code is clearly written to apply to the Engineer, and it is incumbent on members of NSPE to endeavor to live up to its provisions. This applies to all pertinent sections of the Code.

Source: Downloaded from www.nspe.org/Ethics/CodeofEthics/index.html (accessed March 15, 2022). Reprinted by Permission of the National Society of Professional Engineers (NSPE) www.nspe.org.

Appendix III: ENGINEERS EUROPE Position Paper on Code of Conduct: Ethics and Conduct of Professional Engineers

Approved by the FEANI/ENGINEERS EUROPE General Assembly on 07/10/2022.

Ethical Principle

The decisions and actions of engineers have a large impact on the environment and on the wellbeing of society. This has an enhanced and growing importance as the twenty-first century progresses. The engineering profession thus has an obligation to ensure that it works in the public interest with particular regard for equality, diversity, health, safety, and sustainability.

In doing so they are required to maintain and promote high ethical standards and challenge unethical behaviour. There are four fundamental principles for ethical behaviour and decision-making. These are honesty and integrity; respect for life, law, the environment and public good; accuracy and rigour; and leadership and communication.

Framework Statement

National associations of engineers, and FEANI with regard to EURING registrants, have codes of conduct which have much in common and which have the intent of implementing the above ethical principle. As a result of this convergence the European engineering profession as a whole can make a universal statement regarding the conduct of professional engineers. Individual engineers have a personal obligation to act with integrity, in the public interest, and to exercise all reasonable skill and care in carrying out their work.

In so doing engineers:

- Shall maintain their relevant competences at the necessary level and only undertake tasks for which they are competent
- Shall not misrepresent their educational qualifications or professional titles

Ethics, Technology, and Engineering: An Introduction, Second Edition. Ibo van de Poel and Lambèr Royakkers.
© 2023 John Wiley & Sons Ltd. Published 2023 by John Wiley & Sons Ltd.

- Shall provide impartial analysis and judgment to employer or clients, avoid conflicts of interest, and observe proper duties of confidentiality
- Shall carry out their tasks so as to prevent avoidable danger to health and safety, and prevent avoidable adverse impact on the environment
- Shall accept appropriate responsibility for their work and that carried out under their supervision
- Shall respect the personal rights of people with whom they work and the legal and cultural values of the societies in which they carry out assignments
- Shall be prepared to contribute to public debate on matters of technical understanding in fields in which they are competent to comment
- Shall challenge unethical behaviour when observed or experienced and enforce best ethical practices in their work assignments.

Codes of Conduct

The pan-European statement on engineering ethics and conduct presented earlier is best implemented through the codes issued by national engineering associations. These codes can, and in general already do, incorporate the listed objectives in a form which reflects national circumstances and allow additional objectives to be added as required by national practice.

Source: Downloaded from https://www.engineerseurope.com/sites/default/files/Code_of_Conduct_ENGINEERS_EUROPE.pdf (accessed February 24, 2023). Reprinted by Permission of ENGINEERS EUROPE.

Appendix IV: Examples of Corporate Codes of Conduct

Pfizer

Pfizer, one of the world's premier biopharmaceutical companies, headquartered in Manhattan, New York City, has an informative and extensive code of conduct covering many aspects of working at the company (https://cdn.pfizer.com/pfizercom/investors/corporate/Pfizer_2020BlueBook_English_08.2021.pdf[1]). It starts with the introduction of Pfizer's purpose the four core values: Courage, Excellence, Equity, and Joy, and their definitions:

> At Pfizer, we do the right thing because patients' lives depend on us. We act with integrity in everything we do, and our Values guide us in making the right decisions ethically, thoughtfully, and responsibly so that our business can appropriately meet patient and societal needs. Ethical decisions promote trust and accountability for doing the right thing, both internally and externally.

> To fully realize Pfizer's purpose – breakthroughs that change patients' lives – we have established clear expectations regarding what we need to achieve for patients and how we will achieve those goals. The "how" is represented by our four powerful Values – Courage, Excellence, Equity, and Joy – that define our Company and our culture.

The four core values are defined as follows:

> **Courage**. Breakthroughs start by challenging convention, especially in the face of uncertainty or adversity. This happens when we think big, speak up, and are decisive.

> **Excellence**. We can only change patients' lives when we perform at our best together. This happens when we focus on what matters, agree who does what, and measure outcomes.

> **Equity**. We believe that every person deserves to be seen, heard, and cared for. This happens when we are inclusive, act with integrity, and reduce healthcare disparities.

> **Joy**. We give ourselves to our work, and it also gives to us. We find Joy when we take pride, recognize one another, and have fun.

Ethics, Technology, and Engineering: An Introduction, Second Edition. Ibo van de Poel and Lambèr Royakkers.
© 2023 John Wiley & Sons Ltd. Published 2023 by John Wiley & Sons Ltd.

From there, the code is broken down into four sections. Each section discusses one of the four core values containing guidelines for employees on how to act in specific situations. For example, with respect to the core value "Excellence," the following norms are included toward their patients:

- We consider how our interactions with customers may appear and do not engage in illegal or unfair activities, such as false or misleading advertising, bribery, or corruption, or making unfair comments about competitors' products.
- We are committed to acting with integrity in all marketing practices, including labeling, promotional programs, product samples, and communications with stakeholders.
- We provide timely and honest product information to patients, consumers, healthcare professionals, and regulators worldwide, providing appropriate uses for our products and the efficacy and safety data relating to those uses.
- We recognize our interactions with healthcare professionals can cause apparent or actual conflicts of interest; therefore, we support the disclosure of financial and other interests and relationships with healthcare professionals in research, education, or clinical practice.

Visa

Visa is an American multinational financial services corporation headquartered in San Francisco, California. Visa has developed a detailed and wide-reaching code of business conduct and ethics with the slogan: "Integrity. Everyday. Everywhere."

> Our Code of Business Conduct and Ethics reflects our commitment to the highest ethical standards and is core to the Visa brand. Our Code applies to everyone working with or on behalf of Visa, including employees, contingent staff and the Board of Directors.

The code is centered around the following six principles:

1 We honor the code
2 We speak up
3 We foster a culture of integrity
4 We safeguard our assets and information
5 We uphold the law
6 We connect the world

For example, the principle "We speak up" includes:

- Respecting each other as professionals
- Prohibiting discrimination, harassment, retaliation, or intimidation of any kind. Under no circumstances may an employee have a close personal relationship (dating, romantic, or sexual relationship) with someone they supervise, or whose conditions of employment they may influence.
- In all other cases of a close personal relationship with another person at work, disclosure to HR is required.

Each section, discussing one of the above-mentioned principles, offers essential norms and rules, examples of applications of these rules and norms, and guidelines about what you should do when facing a compliance concern or an ethical dilemma. For example, with regard to the principle "We uphold the law," the following norms are included:

When doing business with government, always:

- Learn the rules around the procurement process if you are dealing with government contracts
- Consider legal and local government requirements that may apply to your work, and contact Government Engagement or the Legal Department if you have questions
- Submit accurate, timely, and complete documents
- Follow applicable policies, including those related to gifts and entertainment and anti-bribery and anti-corruption
- In the United States, be mindful of state and local "pay-to-play laws" that can impact your personal political contributions.

An example of the application of a norm related to the accuracy in recordkeeping is:

> A coworker asked you to change some information on an invoice, but you never received any documentation to support the change. Should you make the changes he requested?

> No. You are required to record transactions ethically and honestly. You should ask him to provide the supporting documentation, and if he does not provide it, contact your manager, the Corporate Controller or our Hotline for help with the situation.

Noticeable is the comprehensive list of resources that employees can use to find answers to questions and raise concerns if employees see or suspect any activity that violates the code. The list consists of the issues and concerns together with the corresponding authoritative source and its contact information. For example, for any questions related to data privacy, employees must contact Global Privacy Office, and for questions related to speaking on Visa's behalf, social media or media inquiries, employees must contact Corporate Communications.

Source: https://usa.visa.com/dam/VCOM/global/about-visa/documents/code-of-business-conduct-and-ethics.pdf

Chevron

Chevron Corporation is an American multinational energy corporation headquartered in San Ramon (California). The Business Code and Ethics Code starts with Chevron's vision: "our vision is to be the global energy company most admired for its people, partnership and performance," purpose: "our purpose is to develop the affordable, reliable, ever-cleaner energy that enables human progress," and the five core values: "diversity and inclusion"; "leading performance"; "trust and integrity"; "partnership"; and "people and the environment".

The five core values form the basis of their code of conduct. From there, the code is separated into twelve areas:

1 *About the business conduct and ethics code.* The code helps us understand how Chevron's values are put into practice every day;

2 *Our role and responsibility.* Each of us has a responsibility to speak up;

3 *Our employees.* We value the uniqueness of individuals and the various perspectives and talents they provide;

4 *Human rights.* Chevron's support for universal human rights is a core value in the Chevron way;

5 *Company records and internal controls.* Fair and accurate books and records are essential for managing Chevron's business;

6 *Avoiding conflicts of interest.* We expect each other to act in the best interests of the company;

7 *Antibribery, international trade and antiboycott laws.* Wherever Chevron operates, we respect and comply with the local laws and regulations;

8 *Government affairs and political involvement.* Chevron's participation in the political arena is conducted in accordance with ethical standards;

9 *Operational excellence.* Workforce safety and health, process safety, reliability, and integrity, environment, efficiency, security, and stakeholders;

10 *Antitrust/competition laws.* We always operate not only according to the letter but also the spirit of all applicable laws;

11 *Data privacy.* All employees must exercise care and discretion in handling personal data;

12 *Protection of information and intellectual property.* We all have a responsibility to understand the risks when our information assets are compromised.

Each area details the company's expectations of its employees. It also has a recurring feature known as "Questions & Answers" that answers questions regarding potential situations where the code should guide the employees to the desired outcome and to whom they should report concerns regarding breaks in the law or company policies.

Example of "Questions & Answers" of the area "Operational excellence."

Question:
My supervisor asked me to perform a task that I believe violates environmental regulations. What should I do?

Answer:
Never guess about environmental regulations. If you are uncertain, check with your supervisor to be sure you have understood the request. If you still feel the request violates environmental regulations, report the concern to local management or the Chevron Hotline.

Question:
I have a work order that specifically outlines a task to be performed. As I began to do the task, I discovered that conditions are different from those expected when the job was planned. I have a feeling that continuing the job as outlined in the work order will be unsafe. What should I do?

Answer:
Employees have the responsibility and authority to stop or not begin work that they believe may be unsafe. You should communicate your concerns to your supervisor. Your supervisor has the responsibility to investigate, understand and resolve the issue.

Source: https://www.chevron.com/-/media/shared-media/documents/chevronbusinessconductethicscode.pdf

Boeing

Boeing is a leading global aerospace company headquartered in Chicago. The Boeing Code of Conduct is part of Boeing's "our principles." These principles are "Our Values"; "Sustainabilty"; "Environment"; "Compliance & Ethics"; "Human Rights"; "Diversity & Inclusion"; "Employee Safety"; "Education"; "Military & Veteran Engagement"; "Community Engagement". The Code of Conduct is introduced in "Compliance & Ethics":

Compliance and Ethics

Code of Conduct – Our Shared Commitment

Across our global enterprise, Boeing employees are united by a shared commitment to our values – safety, quality, integrity, and transparency – above all else. We believe that compliance and ethical behavior are everyone's responsibility. This means we must hold ourselves – and one another – accountable to our values and to creating an open and inclusive workplace. Boeing leadership encourages employees to proactively seek out issues, speak up and report concerns, and engage with transparency.

Boeing's Global Compliance organization drives compliant company performance across all geographic locations, encourages integrity and transparency, and demonstrates our commitment to compliant and ethical business practices. Boeing's Chief Compliance Officer works closely with the Board of Directors, senior company leaders, our employees, and external stakeholders to advance Boeing's compliance and ethics culture throughout the company.

Every year, Boeing employees reaffirm their commitment to do their work in a compliant and ethical manner, and respect one another, by reading and signing the Boeing Code of Conduct. Because our Code of Conduct guides the way we do our work every day, we provide the code in 18 languages to reach employees in their native language.

In addition, every employee participates in a company-wide training called Recommitment, which features real-life examples of compliance issues and consequences, and highlights how adherence to our values and doing business with integrity is critical to the company's success.
    Click here to learn more about our values or view our Code of Conduct.

The Code of Conduct reads as follows:

Boeing Code of Conduct

At The Boeing Company, our first commitment is to the people and customers who rely on our products and services to protect, connect, and explore our world and beyond. We are each personally responsible for honoring that commitment and for serving as stewards of our company's legacy of aerospace excellence and innovation. We do that by committing to our values, and by holding ourselves to the highest standards of conduct in how we do our work, and how we treat one another.

We understand that observing the highest ethical business standards is not only the right thing to do, but is critical to our long-term success as a company.

I commit that:

- I will comply with all applicable laws, rules, and regulations. If I do not understand them, I will seek guidance.
- I will prioritize safety, quality, and integrity above profit, schedule, or competitive edge. If I see something that raises a safety concern, I will speak up immediately.
- I will engage all regulators – including employees who act under delegated authority – and customers with candor, transparency, and respect at all times.
- I will treat my colleagues with respect and understand that harassment will not be tolerated.
- I will work to support Boeing's mission to build an inclusive culture in which diverse experiences and voices are heard, respected, and incorporated into the most important issues that we face as a company.
- I will not engage in any activity that creates a conflict of interest for me or for the company.
- I will not act in a way that calls into doubt Boeing's honesty, impartiality, or reputation, or that otherwise causes embarrassment for the company.
- I understand that I am entrusted with sensitive information and I need to honor that trust. I will protect Boeing information and ensure that non-Boeing proprietary information is handled appropriately.
- I will not take advantage of or abuse my Boeing position to seek personal gain, including through the inappropriate use of Boeing or non-public information.
- I will promptly report any illegal, improper, or unethical conduct to my management or through other appropriate channels.
- I will never retaliate against or punish anyone who speaks up to report a concern.

Please note that the Code of Conduct is not intended to prohibit or infringe on an employee's rights to discuss wages, hours, working conditions, or other terms and conditions of employment or to otherwise engage in protected concerted activity under Section 7 of the National Labor Relations Act. F70284 REV (07 MAR 2022)

Source: https://www.boeing.com/resources/boeingdotcom/principles/ethics_and_compliance/pdf/english.pdf

General information about corporate code of conduct:

https://www.valamis.com/hub/code-of-conduct
https://study.com/academy/lesson/code-of-business-conduct-ethics-standards-examples.html
https://www.britannica.com/topic/corporate-code-of-conduct

Note

1   All the websites in this appendix were last accessed on November 2, 2022.

Appendix V: DSM Alert Royal DSM Whistleblower Policy

This document explains our whistleblower policy and procedure to support our employees and third parties in expressing their concerns about suspected serious misbehavior at or related to activities of DSM (hereinafter the "Policy").

V.1 General

Today's society expects that persons, who have serious concerns about what is happening at their work, should feel free to report such concerns. In order to protect those persons when reporting their concerns, guidelines, and legislation (e.g. in the Netherlands the Dutch Corporate Governance Code and specific legislation for Whistleblowers), have been established.

Within DSM, we wish to maintain the highest standards of business conduct and ethical behavior. We have embedded the opportunity to file a complaint in our Code of Business Conduct and in the DSM Corporate Requirements (together: the DSM Regulations).

We encourage our employees to report any Breach (as defined hereunder) without being worried of any retaliation, punishment or unfair treatment. This Policy describes what a person should do when he suspects or observes a Breach. Before reporting a Breach, a person always has a possibility to consult an advisor. Also third parties can raise issues under this procedure, using the provided links on our corporate website www.dsm.com.

DSM's Managing Board appointed an officer responsible for handling reported Breaches (hereinafter the "Alert Officer"). The Alert Officer (Senior Vice President Corporate Operational Audit) is supported by the DSM Alert Committee (hereinafter the "DAC"). The DAC consists of the Alert Officer, the VP Corporate Affairs, the SVP Group Legal Affairs and the EVP People & Organization/member EC. The Alert Officer chairs the DAC. Other designated DSM employees can support the DAC.

Ethics, Technology, and Engineering: An Introduction, Second Edition. Ibo van de Poel and Lambèr Royakkers.
© 2023 John Wiley & Sons Ltd. Published 2023 by John Wiley & Sons Ltd.

V.2 Breaches

A "Breach" is a violation or the suspicion of a violation on reasonable grounds of any legislation and/or DSM Regulations by any DSM employee, contractor, agent, or distributor operating on behalf of DSM or commissioned by DSM. Breaches are not limited to fraud, theft, corruption, discrimination, or harassment, but can regard any other ethical or behavioral complaint as well.

For us it is important to know what is happening within our organization. Therefore, we encourage persons to engage in a discussion with colleagues who show behavior considered a Breach. Everyone is also encouraged to report any Breach to his/her management or the Alert Officer.

V.3 The Essentials of Whistle Blowing

V.3.1 Non-retaliation

We protect any person, who reports a situation or occurrence, which he/she reasonably believes is a Breach. We shall, in no way, harass him/her because of a report. Retaliation against a person for reporting in accordance with this Policy is a serious violation of the Policy itself. If this occurs, the violator will be subject to appropriate disciplinary sanctions. Any such retaliation must be reported to the Alert Officer at once.

V.3.2 Confidentiality

We recognize that individuals who observe a Breach, and wish to report it, will do so under assurance of confidentiality. We will handle all reports confidentially and we equally expect persons reporting a Breach to keep this confidential too.

We do however acknowledge that in some situations the investigation process may reach a point where the person who reported the Breach needs to make a statement or provide further evidence. Under such circumstances, when maintaining someone's privacy hinders finding the truth, we may not be able to guarantee full confidentiality to the reporting person.

If a person feels there is no other way than filing an anonymous report and applicable local law allows for it, we will always take appropriate protective action.

V.3.3 Abuse of the policy

We encourage persons to report Breaches and assume this is done in good faith. If after an investigation Breaches can't be confirmed or can't be substantiated, no action is taken against persons raising such Breaches. We will, however, take appropriate action if the person reporting knows, or could have known that a reported alleged Breach is false. Malicious reports without any factual foundation, will lead to disciplinary action.

V.4   Reporting Levels

V.4.1   General

There are three reporting levels:

Level 1: Line Management
Level 2: Alert Officer
Level 3: Chairman of the Managing Board/Chairman of the Supervisory Board

At each level, all reported Breaches are handled carefully, confidentially and promptly. If a Breach isn't reported at the appropriate level, the person receiving the report will forward it to the appropriate level and inform the reporting person.

V.4.2   Level 1: Line Management

As a general rule, persons should report Breaches first within their own working environment or organization. Open discussion with line management or the local HR business partner is the starting point. If reporting to line management is not possible, because it would be inappropriate or unfeasible, the report can be made at Level 2.

If a specific local complaint procedure is in place, a Breach can also be reported through this procedure and to the persons mentioned in that specific procedure.

The decision by line management is not open for appeal at the next level, being the Alert Officer. In case the handling of the complaint by line management, or the decision taken is in itself considered a Breach, the person can report such case as a new case to the Alert Officer.

V.4.3   Level 2: Alert Officer

Notwithstanding the procedures mentioned in the previous section, a person should report:
   Breaches directly to the Alert Officer if:

- the Breach relates to any of the following subjects that may harm DSM (not limitative):
   o criminal acts in relation to DSM or our assets, such as fraud or theft;
   o a (potential) danger to the health, safety, and security of persons or to the environment;
   o corrupt and dishonest behavior;
   o harassment;
   o discrimination;
   o inappropriate accounting practices, or lack of internal accounting controls;
   o abuse of authority by management; and
   o any other behavior that could have a detrimental effect on our reputation and/or financial position.
- reporting to line management is not possible (because it would be inappropriate or unfeasible).

The role and tasks of the Alert Officer and the DAC are described below in the paragraph "Procedures." If an alleged Breach comes to the attention of the DAC, other than through a report by a person, the DAC has the authority to treat the matter in accordance with this Policy.

A decision taken by the DAC is not open for appeal at the next level, being the Chairman of the Managing Board or the Chairman of the Supervisory Board. In case the handling of a complaint by the DAC, or the decision taken is in itself considered a Breach, the person can report such case as a new case to the Chairman of the Managing Board or to the Chairman of the Supervisory Board.

V.4.4 Level 3: Chairman of the managing board/chairman of the supervisory board

If a Breach is reported concerning the Alert Officer and/or one or more other members of the DAC or a member of the Managing Board, not being the Chairman, the reporting person should report directly to the Chairman of the Managing Board.

If a Breach is reported concerning the Chairman of the Managing Board, then the reporting person should report directly to the Chairman of the Supervisory Board.

V.5 How to Report

Level 2 Breaches are reported to the Alert Officer via the Alert website, by telephone or by e-mail. Level 3 Breaches are directly reported to the Chairman of the Managing Board or the Chairman of the Supervisory Board in writing, depending upon the nature of the Breach.

The person reporting a Breach provides the background, history and reasons for his/her concern, together with names, dates, places, and as much other relevant information as possible. We will always try to support persons reporting a Breach to report in their native language, if preferred.

It is not necessary that a person reporting a Breach immediately proves all facts leading to a Breach, but he/she should be able to provide sufficient evidence to substantiate the assumption of a Breach. Individuals are encouraged to report Breaches at the earliest possible stage, in order to take timely action.

V.6 Procedures

The periods mentioned in this paragraph start on the day following the date on which the report is received at the appropriate reporting level, unless otherwise indicated.

V.6.1 A report at level 2

A level 2 report of a Breach is handled by the Alert Officer. Designated DSM persons, within the region where the Breach occurred, may support the Alert Officer. The Alert Officer will take the following actions:

- Confirm the receipt of the report to the reporting person within 7 days.
- If relevant, arrange an interview with the reporting person to get more details of the complaint.
- Inform the DAC as soon as possible after receipt of a report of a Breach. If the reporting person requests so, his/her name is kept confidential.

The role of the DAC

The DAC decides within ten (10) business days after receipt of a report, whether that report is admissible. A report by the reporting person is inadmissible if:

- The report clearly does not relate to a Breach; or
- The report is not sufficiently substantiated.

If the report is admissible, the DAC investigates the case. The internal investigation may be done by the Alert Officer, a representative of the DAC or another person appointed by the DAC, at its discretion. The person(s) performing the investigation may need to speak to the reporting person to clarify the information provided or may seek additional information from other persons.

When the investigation is finished, the DAC decides whether a Breach has occurred or not. In case of a Breach, the DAC will take a decision and/or provide a solution, or, at its discretion, ask line management to take a decision and/or to provide a solution. The DAC informs the reporting person in writing about the decision of the DAC within ten (10) business days after taking its decision.

If the investigation by the DAC takes more than three (3) months, the Alert Officer informs the reporting person and indicates how long it may take to provide a final feedback.

Any person involved in an investigation should cooperate with the assigned DAC investigator(s). Withholding relevant information will be regarded as serious misbehavior.

V.6.2   A report at level 3

If the Chairman of the Managing Board receives a report, he shall review and discuss this report with another member of the Managing Board who is not subject of that report. The Chairman of the Managing Board will decide within ten (10) business days whether the report is admissible. The criteria for inadmissibility of a report at Level 2 apply equally to reports at Level 3. The Chairman of the Managing Board may involve the Alert Officer, the DAC and other DSM persons, as well as external advisors or institutions in the investigation as required and as far as they are not subject of the report themselves.

The decision whether a Breach has occurred or not is taken within two (2) months after the Chairman of the Managing Board has received the report or – if two months is not reasonable – within an appropriate period. In case of a Breach, the Chairman of the Managing Board will ask the DAC to take a decision or the Chairman of the Managing Board will take a decision himself, together with another member of the Managing Board. The Chairman of the Managing Board informs the reporting person in writing about his decision within ten (10) business days after taking his decision.

If the Chairman of the Supervisory Board receives a report, he shall review and discuss it with the Audit Committee of the Supervisory Board. The Chairman of the Supervisory Board will decide within ten (10) business days whether the report is admissible. The criteria for inadmissibility of a report at Level 2 apply equally to reports at Level 3. The Chairman of the Supervisory Board may involve other members of the Supervisory Board who are not involved in the Breach, the Alert Officer, the DAC and other DSM persons, as well as external advisors or institutions in the investigation as required and as far as they are not subject of the report themselves.

The decision whether a Breach has occurred or not is taken within two (2) months after the Chairman of the Supervisory Board has received the report or – if two months is not reasonable – within an appropriate period. In case of a breach, the Chairman of the Supervisory Board will take a decision and/or provide for a solution himself, together with another member of the Supervisory Board. The Chairman of the Supervisory Board informs the reporting person in writing about his decision within ten (10) business days after taking his decision.

V.7   Subject of a Report

The Alert Officer may inform the person about whom a report is filed of such a report. In cases where there is a substantial risk that such notification would jeopardize the ability to effectively investigate the reported facts or to gather the necessary evidence, notification to the person about whom a report is filed can be delayed as long as such risks exist.

The information given to the subject of the report will contain the facts of the Breach as reported. He/she will be given the opportunity to provide an explanation, without the name of the person who reported the Breach being disclosed to him/her. The subject of the report may request access to his/her personal data held by the company via the Alert Officer. He/she has the right to have incorrect, incomplete and outdated data corrected or removed.

As soon as the investigation has been concluded, the subject of the report will be informed of any action to be taken as a result of the report. If the person about whom a report was filed is informed that no action will be taken, any suspension or temporary measure that has been imposed on him/her will automatically terminate.

V.8   Reporting by the Alert Officer

The Alert Officer will provide, at least bi-annually, an overview of the Alert cases to the members of the Managing Board and the Audit Committee of the Supervisory Board.

V.9   Confidentiality

The reporting person, the Alert Officer, and the DAC shall keep the final report confidential. Information relating to this report shall only be given to other persons within our company if they need this to execute their tasks under this Policy and/or to implement the conclusions of the investigation. The name of the reporting person will not be disclosed, unless this is necessary for the investigation and/or judicial procedures and only after informing the reporting person.

V.10   Privacy Issues and Retention Periods

Personal data relating to a report judged being "inadmissible" or "admissible but not valid" will be removed immediately. "Removed" means that the personal data are completely deleted or adapted in such a way that identification of the person involved is no longer possible. The Alert Officer will take the necessary technical and organizational measures to adequately safeguard personal data against loss or unauthorized access.

Personal data relating to reports that are "admissible and valid" will be kept for two (2) years, unless disciplinary action is taken or court proceedings are filed against a person. In these events, the data will be removed within two (2) years after the disciplinary action or the court proceedings have been finalized.

V.11   Final Provisions

This Alert Policy replaces all previous versions and will be effective as of December 18, 2017.

This English version of the Alert Policy will prevail over any other version.

Source: https://www.dsm.com/content/dam/dsm/corporate/en_US/documents/whistleblowing-policy-and-procedure.pdf, accessed September 29, 2022. Reprinted by Permission of DSM.

Appendix VI: Cases

Case 1 Herald of Free Enterprise

Figure VI.1 *Herald of Free Enterprise* capsized. Photo: PA Images / Alamy Stock Photo.

On March 6, 1987 the roll-on/roll-off passenger and freight ferry the *Herald of Free Enterprise* capsized just outside the Zeebrugge harbor (see Figure VI.1). Water rapidly filled the ship. Of the almost 600 people on board, 383 were eventually saved, 189 bodies were recovered, and four people were registered missing. The main cause of the disaster was the fact that the inner and outer bow doors were open when the ship set sail. The doors were sometimes deliberately left open for a while to allow all the car exhaust fumes to escape. That was done to prevent the passengers from feeling ill and getting headaches.

It was the job of the assistant boat-swain to close the doors, but he had fallen asleep. The first officer was to check whether the doors had indeed been closed and to report to the captain. However, the first officer was also expected to assist the captain on the bridge when setting sail. The

Ethics, Technology, and Engineering: An Introduction, Second Edition. Ibo van de Poel and Lambèr Royakkers.
© 2023 John Wiley & Sons Ltd. Published 2023 by John Wiley & Sons Ltd.

absence of warning lights made it impossible to see from the bridge whether the bow doors were closed. On at least two previous occasions, similar negligence had led to ships setting sail with their bow doors open but without disastrous results.[1] In the case of the *Herald*, as is often the case, it was human error that preceded the disaster, but it was the design that contributed to the occurrence of the disaster in the first place. Roll-on/roll-off ships are inherently unstable when water enters a deck.

It is likely that the maritime engineers who designed roll-on/roll-off ferries like the *Herald* were aware of the inherent instability of such ships. Moreover, there were, and are, simple technical solutions if one wants to prevent rapid capsizing when water enters a deck. Bulkheads created on the decks could easily impede the water and prevent rapid capsizing. However, such bulkheads increase the loading time of roll-on/roll-off ferries and this increase in loading time, in turn, implies an increase in transportation costs. Moreover, bulkheads decrease the efficient use of space, so reducing the transportation capacity of the ship. In Northwest Europe, shipping companies are in sharp competition with trains and planes, therefore they do not want to face increasing costs or longer loading times.

The official investigation into the sinking of the *Herald of Free Enterprise* commenced on April 27, 1987. The investigation was carried out by the Admiralty High Court, an investigation council of the British Supreme Court. The Admiralty High Court does not have the power to legally prosecute but it can recommend legal prosecution to ordinary courts of law. During the process much attention was given to the role of the assistant boatswain, Stanley, to the first navigating officer, Sabel, to captain Lewry and to Kirby, the shore captain. Kirby was directly responsible for drawing up the instructions to be followed on board, including the safety instructions. The role of the shipowner, to be precise of several of its directors, was also examined. The outcome of the investigation was that the ship's captain, Lewry, was suspended for a year and the first navigating officer, Sabel, for two years. The assistant boatswain, Stanley, got off free and the shipowner was officially reprimanded. The Admiralty High Court's written assessment of the shipowner was devastating. It identified a "disease of sloppiness" and negligence at every level of the corporation's hierarchy.

Despite all these resolute claims those responsible for the disaster were not immediately legally prosecuted. It was only after quite some time and pressure on the part of the Belgian authorities and the families of the victims that proceedings began in England. In 1989 it became clear that the shipping company was going to be accused of "corporate manslaughter," in other words, of deaths caused by a company rather than by an individual or individuals. By that time Townsend Thoresen had been taken over by the shipping company P&O. In September 1990 the case was taken to the Central Criminal Court at the Old Bailey in London. Eight people were accused. In addition to the shipping company, the assistant boatswain Stanley, the first officer Sabel and the captains Lewry and Kirby there were three directors who were accused: Develin, Ayres, and Alcindor.

When the case began the judge ordered the jury to forget everything they knew about the case. Even the Admiralty High Court's report, with its devastating criticism of the shipping company, should not be used. The prosecutor tried to prove that a number of mistakes had been made by the shipping company and the ship's crew alike and that it was "obvious" that sailing out of port with the ship's bow doors open would capsize the vessel. Most of the witnesses – experienced seamen included – did not find that this was so "obvious." The judge also maintained that there was insufficient evidence to support a verdict possibly to be voiced by the jury to the effect that there was an "obvious" connection between the open doors (cause) and the capsizing of the *Herald* (effect). On October 19, 1989, partly under pressure of the judge, the jury decided that there was insufficient evidence and so all the defendants were acquitted.

Source: This case description is mainly based on Van Gorp and Van de Poel (2001) and Baeyens (1992).

Possible questions related to Chapters 1 and 7:

1 Explain why this case can be considered as a problem of many hands.
2 Looking back, which model for allocating responsibility in organizations could be best applied in this case for safety?
3 Do you consider the assistant boatswain passively responsible for the Herald of Free Enterprise disaster? In answering this question, refer to the criteria for passive responsibility and use the information available. And what about the first officer and the maritime engineers?

Case 2 Gilbane Gold

Figure VI.2 Sewage treatment plant. Photo: © nonameman/Fotolia.com.

The city of Gilbane has been processing its waste water into manure for agriculture for 75 years. This yields a tax benefit of $300 per year per household. Given this tax advantage, the manure produced is called "Gilbane Gold." For the past 15 years, the city also has a company producing computer parts: Z-Corp. The city attracted the company to come by offering tax benefits. The company is important for the city because it creates opportunities for employment. However, the production process produces lead and arsenic, which are emitted through the waste water of the factory.

Lead and arsenic are heavy metals that built up in organisms and may cause negative health effects. If concentrations of arsenic and lead would cumulate in Gilbane Gold, this might have negative long-term effects. Therefore, the restrictions that the city has set on the concentration of arsenic and lead in the waste water are about ten times as strict as Federal Regulations.

Independent environmental consultant Tom Richards, who has been hired by Z-Corp, has found out that the conventional method for measuring arsenic and lead in the waste water used by Z-Corp measures lower concentrations than a new, more reliable, method. However, the old method is the one prescribed by City Regulations and city officials, after being informed on the issue, have not objected to the continued use of that method. Moreover, Z-Corp can easily stay within the limits of the City Regulations even with the new measurement method by diluting the waste water because the regulations only refer to concentrations and not to absolute amounts. Some consider this, however, "a major loophole in the law." When Richards further pursues the matter, Z-Corp decides not to renew his contract.

A young engineer, David Jackson, now becomes responsible for Z-Corp's emissions of arsenic and lead into the waste water. In the meantime, Z-Corp signs a contract with a Japanese firm which will result in a 500 percent increase in production. Jackson, who is really concerned now, raises the issue with management but is told that there is no money available to solve the problem: the factory is hardly profitable. Moreover, manager Diane Collins argues, as long as Z-Corp meets the law, it has no broader responsibility. Jackson nevertheless fears that the waste water treatment plant may not be able to deal with the larger amounts of arsenic and lead. Since he has a duty as professional engineer "to hold paramount the safety, health and welfare of the public," it might be appropriate to speak out in public. Indeed, Jackson is approached by Channel 13 a local television station about the issue.

Source: This is a fictional case based on a video produced by the National Society of Professional Engineers and the National Institute for Engineering Ethics.

Possible questions related to the application of the ethical cycle (see Chapter 4).

1 What exactly is Jackson's problem here? Explain why it is a moral problem.
2 Which moral values or principles are at stake here?
3 Mention three things Jackson can do to deal with his problem.

4 What would an analysis of this problem according to (Bentham's) classical utilitarianism look like? What would the utilitarian advice to Jackson be?

5 Would John Stuart Mill's modified form of utilitarianism lead to a different advice? Why or why not?

6 What would a Kantian ethicist recommend to Jackson? Motivate your answer.

7 What would a virtue ethicist recommend to Jackson? Motivate your answer.

8 What would a Confucian recommend to Jackson? Motivate your answer.

9 What would the NSPE code of conduct recommend to Jackson? Motivate your answer.

10 What do you think is the right thing to do for Jackson? Motivate your answer and also explain why you do not accept (some of) the advice from 4,5,6,7, 8, and 9.

Case 3 High Speed Train Disaster in Germany

Figure VI.3 Intercity Express train crash. Photo: Dpa picture alliance / Alamy Stock Photo.

On June 3, 1998, the German high speed Intercity Express (ICE) train 884 "Wilhelm Conrad Röntgen" derailed at a speed of about 200 kilometers an hour and ran into a bridge that fell down on the train (see Figure VI.3). The subsequent cars of the train jackknifed into the crashed cars in a zig-zag pattern. Overall, 101 people were killed and 88 severely injured. Investigations after the accident showed that the disaster was to a large extent due to a change in the wheel design of the train. Originally the train had been equipped with monobloc wheels, which are single casted. Such wheels decrease the rolling friction between the wheels and the rails, so increasing energy efficiency and lowering the costs of electricity consumption. However, such wheels may also decrease the comfort and may result in material stresses in the wheel. In some high speed trains, the comfort problem is solved by using a bogie (the framework carrying the wheels) with air suspension. The German ICE however was based on conventional bogies with steel springs. Once in service, it turned out that the wheel system caused severe resonance and vibration. This was considered a problem, especially in the dinner car where dinnerware "walked" over the tables. Engineers of the German railways then proposed to solve the problem with a new wheel design that had a rubber ring between the steel tire and the wheel body. This wheel design had been used in trams, but at significantly lower speeds. The new wheel design was not tested at high speed before it came into service, but was based on existing experience and materials theory. Nevertheless, after introduction the wheel solved the vibration problems and did not show any major problems until the fatal accident.

The new wheel design consisted of a steel wheel body surrounded by a 20 mm thick rubber damper and then a relatively thin steel tire. It was this tire that eventually failed and led to the disaster. This could have been caused by two mechanisms. One is metal fatigue due to the fact that the metal tire around the wheel is deformed a bit every rotation. The other mechanism is that the rubber damper extends due to heating up. Since rubber extends more than steel due to heat and since the rubber tire is locked in between the wheel body and the metal tire, this will result in a high pressure on the steel tire and, as a result, cracks may form from the inside in the metal tire. The first of these mechanisms was known at the moment that it was decided to change the wheel design, the second one not.

The wheels of the train were routinely checked every day. First, it was measured whether the wheels were still round enough or had become ellipsoids. The wheel out-of-roundness should not be larger than 0.6 mm. In practice, higher values were measured and allowed, probably because an out-of-round wheel was not seen as a safety issue but rather as an issue of comfort and wear. This might have been true for monobloc wheels, but it was certainly not true for the new wheel design. Second, the total wheel diameter should be at least 854 mm. A new wheel had a diameter of 920 mm. At the last check before the accident the wheels had a diameter of 862 mm. Later investigations of the Fraunhofer Institute in Darmstadt suggested that a norm of 890 mm would have been more appropriate given the new wheel design. Again, the norm was probably based on monobloc wheels, neglecting the peculiarities of the new wheel design. Third, the wheels were also inspected for cracks. Initially, advanced testing machines were used to inspect the wheels. These produced, however, a large number of false positives, that is, indications of a defect in the wheel while there actually was no problem. As result of these false positives, the use of the advanced inspection apparatus was discontinued and the wheels were only inspected visually with a flash light and audibly by sledging a hammer against the wheels. The latter method, which is often used for monobloc wheels, was probably inadequate for the new wheel design because the rubber damper will absorb the hammer stroke.

Three engineers, two from the German railways (Deutsche Bahn) and one from a supplier, who all had been involved in the certification of the wheels were tried and charged with manslaughter. After 53 days, the judge concluded that the three had not been grossly negligent. Three reasons were given for this. First, among experts there was not a principled objection against the new type of wheels. Second, the metal cracks that could cause an accident like this can develop rather quickly, so that adequate inspection procedures might not have prevented an accident like this. Third, it was not considered proven that the engineers had made gross mistakes in calculating the load on the wheels. Therefore, the case was dismissed on the condition that each defendant paid a fee of 10,000 Euros (a possibility that is specific to German criminal law).

Source: Case description is based on Brumsen (2006) and http://en.wikipedia.org/wiki/Eschede_train_disaster (accessed August 18, 2009).

Possible questions related to Chapters 5 and 6.

1 What values are at stake in the wheel design?
2 Describe the trade-off in this case.
3 Which method discussed in Section 5.6 did the engineers use to deal with the trade-off?
4 According to Brumsen (2006), the probability of the accident or consequences of it – even if the exact risk was unknown – could have been decreased if designers had taken into account some of the following four safety issues:
   a) Compounded wheels may be said to be inherently more dangerous than monobloc wheels because they introduce new potential failure mechanisms.
   b) It turned out that the minimally required diameter of the wheels was probably set too low.
   c) Some trains, for example, have sensors that monitor wheel breaks and that can also be used to stop the train automatically. Most trains also have an emergency brake that can be used by passengers. The emergency brake of the ICE could only be handled by the conductor, who refused to employ it despite warnings from passengers who had heard the wheel breaking. After the wheel had broken, the train drove for another two minutes without derailing.
   d) The rail track could have especially been designed for high-speed trains as is the case for the French TGV. Such tracks do not have switches and have curves that are adjusted to high speeds. For example, a French TGV that derailed on December 21, 1993 at a speed of 300 kilometers an hour, drove on just outside the rails for several kilometers causing only two minor injuries. In the case of the German ICE disaster, the train changed tracks only two hundred meter before the bridge into which it eventually ran, which was probably a major factor in the severity of the accident.

Show that each of these issues corresponds to one of the four strategies for safe design mentioned in Section 6.3.

Case 4   The Herbicide 2,4,5-T

Figure VI.4   Pesticide spraying. Photo: © Tyler Olson/Fotolia.com.

In the 1970s there was some unrest among agricultural workers in England concerning the safety of using herbicide 2,4,5-T on a large scale. The scientific Pesticides' Advisory Committee maintained that 2,4,5-T was not a health risk. They based themselves on laboratory studies on the toxicity of the substance. Complaints issued by the National Union of Agricultural and Allied Workers were set aside as imaginings. In its assessment, the scientific committee assumed that the laboratory situation was representative of actual practice: "pure 2,4,5-T offers neither hazards to users nor the general public ... provided that the product is used as directed" (cited in Wynne, 1989, p. 36).

More specifically, the following assumptions were the basis for believing that 2,4,5-T was harmless:

1 The production process is such that dioxin and other toxic substances never contaminate the main product.
2 The containers of 2,4,5-T always reach their destination with full and clear instructions for users (farmers and agricultural workers).
3 Farmers and agricultural workers always use the right solvents, pressure valves, spray nozzles, and safety clothing despite the inconvenience.

All three of these assumptions were problematic, but the third one in particular proved unfeasible. A scientist who investigated the working conditions of the farmers in question, described the scientists from the pesticide committee as "living in cloud-cuckoo land behind the laboratory bench" (cited in Wynne, 1989, p. 37). In contrast to what the pesticide committee stated, there were good reasons to believe that

2,4,5-T was a hazard to the health of agricultural workers – at least under the conditions in which they normally had to work.

Source: Wynne (1989).

Possible questions related to Chapter 5.

1 Explain how this case might be understood in terms of the difference between the embedded and the realized value of a design.
2 Who is, in your view, responsible for the dangers that the use of 2,4,5-T had in practice?
3 Should the designers have designed a product that could be used only in a safe way?
4 How can this kind of problem be prevented in the future?

Case 5   Hyatt Regency Hotel Walkway Collapse

Figure VI.5   Hyatt Regency walkway collapse. Photo: Dr. Lee Lowery, Jr. / Wikimedia Commons / Public domain.

On July 17, 1981, two walkways in the Hyatt Regency Hotel in Kansas City in the USA collapsed (see Figure VI.5) causing 114 fatalities. One reason for the collapse was that, as turned out during the investigation following the disaster, the construction did not meet the requirements of the Kansas City Building Code. As a result, two engineers lost their engineering registration.

However, the investigation showed also another major cause of the disaster that was due to lack of communication between designers and constructors. Figure VI.6 shows the original design of one of the walkways and the design as it was eventually implemented. In the implemented design the load on the bolts of the upper walkway are about twice as high as in the original design. This means that the walkway would probably not have collapsed had the design not been changed. The building contractors changed the design because the original design was difficult to build. They changed the drawings, but these changes were not noticed by the structural engineers who approved unwittingly the changes.

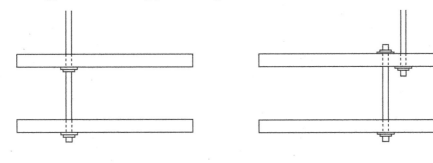

ORIGINAL DESIGN   FINAL IMPLEMENTATION

Figure VI.6   Original design and final implementation of Hyatt Regency Walkway.

Possible questions related to Chapters 1 and 7:

1 Can this case be considered as a problem of many hands? Why or why not?
2 Do you consider the building contractors passively responsible for the Hyatt Regency walkway collapse? In answering this question, refer to the criteria for passive responsibility.
3 Which concrete organizational measures can avoid these kinds of construction "mistakes" in the future?

Case 6   RFID Chips

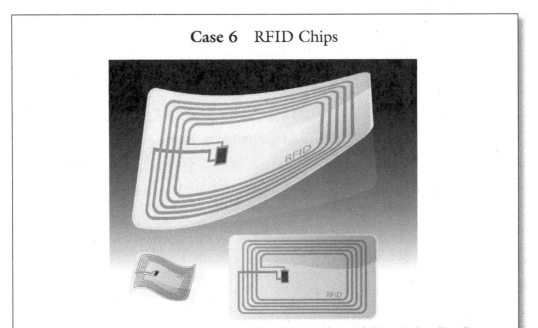

Figure VI.7   A Radio Frequency Identity Chip (RFID). Photo: © HuseyinBas/Fotolia.com. iStock/Getty Images.

The Radio Frequency Identity Chip (RFID) is a chip or tag consisting of a small integrated circuit and a very small radio antenna. Like bar codes, RFID chips have their own unique identification number. RFID chips are used, among other things, for tracking and tracing of objects, boxes and vehicles in logistic chains. Due to developments in nano-electronics, RFID chips will likely become smaller and may become invisible. Not only objects but also people may be tagged with RFID chips, as is now already the case for some small-scale applications. Eventually, people and objects may so become entangled in an "Internet of things," in which tiny devices exchange information without people noticing.

The philosophers Van den Hoven and Vermaas have argued that RFID chips raise privacy issues that are different from the way privacy is traditionally understood to be endangered by, for example, information technologies. An important metaphor in the traditional debate on privacy is the panopticon (see also Chapter 3 on Jeremy Bentham): a hemispherical prison in which the imprisoned are continuously watched from an authoritative point of view (the dome of the panopticon). While nano-electronics enables continuous surveillance, data storage is not necessarily central (although it can be central). In fact, information may be stored in individual tags, but by locally combining these bits of information, privacy-sensitive information may be revealed.

Many of the current attempts to protect citizen's privacy are focused on restraining the storage and processing of information in central databases, or in constraining the retrieval of information from such databases. With the advance nano-electronics this focus may be too limited. Attention should also be paid to the design of the hardware. Relevant issues for example include the reach of the antenna, the accessibility of the data on the chip for other devices, whether the chips are writable or not, and, if so, by whom.

Source: Van den Hoven and Vermaas (2007).

Possible questions related to Chapter 5.

1 Describe the value trade-off in this case.
2 Could the method "Respecification: reasoning about values" (see subsection 5.6.4) solve this value conflict?
3 Develop a values hierarchy (see Section 5.5.2) for the RFID tag five using the five guiding principles for RFID system creation and deployment by Simson Garfinkel (2002):
   a) The right to know if a product contains an RFID tag.
   b) The right to have embedded RFID tags removed, de-activated, or destroyed when a product is purchased.
   c) The right to first-class RFID alternatives. Consumers should not lose other rights (such as the right to return a product or travel on a particular road) if they decide to opt-out of RFID or exercise an RFID tag's kill feature.
   d) The right to know what information is stored inside their RFID tags. If this information is incorrect, there must be a means to correct or amend it.
   e) The right to know when, where, and why an RFID tag is being read.

Case 7   Obstetric Ultrasound

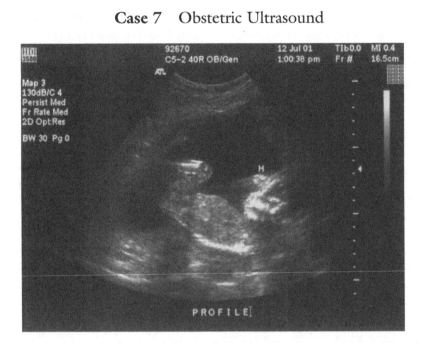

Figure VI.8   Obstetric ultrasound. Photo: © Melking/Fotolia.com. Melissa King / Adobe Stock.

Ultrasound is not simply a functional means to make visible an unborn child in the womb. It actively helps to shape the way the unborn child is given human experience, and in doing so it informs the choices his or her expecting parents make. Because of the ways in which ultrasound mediates the relations between the fetus and the future parents, it constitutes both fetus and parents in specific ways.

Ultrasound brings about a number of "translations" of the relations between expecting parents and the fetus, while mediating their visual contact. First of all, ultrasound isolates the fetus from the female body. In doing so, it creates a new ontological status of the fetus, as a separate living being rather than forming a unity with his or her mother. This creates the space to make decisions about the fetus apart from the pregnant woman in whose body it is growing.

Second, ultrasound places the fetus in a context of medical norms. It makes visible defects of the neural tube, and makes it possible to measure the thickness of the fetal neck fold, which forms an indication of the risk that the child will suffer from Down's Syndrome. In doing so, ultrasound translates pregnancy into a medical process; the fetus into a possible patient; and congenital defects into preventable suffering. As a result, pregnancy becomes a process of choices: the choice to have tests like neck fold measurements done at all, and the choice what to do if anything is "wrong." Moreover, parents are constituted as decision-makers regarding the life of their unborn child. To be sure, the role of ultrasound is ambivalent here: on the one hand it may encourage abortion, making it possible to prevent suffering; on the other hand, it may discourage abortion, enhancing emotional bonds between parents and the unborn child by visualizing "fetal personhood."

Source: Verbeek (2008).

Possible question related to Chapter 5, Section 5.7.

1 Explain why obstetric ultrasound mediates human perception.
2 Do you agree that obstetric ultrasound changes the relation between parents and fetus?
3 Do you think that it changes the decisions parents make with respect to, for example, abortion?
4 Can you think of other forms of "representing the unborn child" that would have other effects?
5 Explain why engineers should occupy themselves with taking responsibility for the future mediating roles of technologies in design.

Case 8   Intelligent speed adaption system

Intelligent speed adaption system is a system that constantly monitors vehicle speed and the local speed limit on a road and implements an action when the vehicle is detected to be exceeding the speed limit. This can be done through an advisory system, where the driver is warned (intelligent speed assistant) often via built-in car navigation systems, or through an intervention system where the driving systems of the vehicle are controlled automatically to reduce the vehicle's speed (intelligent speed authority) so that drivers cannot violate the speed limit. The greatest benefit of these systems, sought by the European Commission in particular, is in traffic safety. About **20,000** people were killed on EU roads in **2021**. The Commission aims to halve the total number of fatalities and serious injuries on European road by **2030**, as compared to 2020, as a milestone on the way to "Vision Zero" – zero fatalities and serious injuries by 2050.[2] This is a very ambitious goal, which can only be achieved by rigorous measures such as mandating the intelligent speed authority, since speed is a basic risk factor in traffic and has a major impact on the number of traffic accidents. According to the Royal Society for the Prevention of Accidents of the Netherlands, the number of fatalities would decrease by between 10 and 26 per cent, and the number of serious injury accidents by between 6 and 21 per cent, if all drivers adhered to speed limits. This could easily be addressed by mandating the intelligent speed authority. It is not expected that this variant will be implemented easily, because public acceptance of such an intrusive system is quite low. The car is considered in our culture as a symbol of freedom, and an intrusive system restricts the freedom of the driver, since the driver would be forced to have his or her car fitted with technology that keeps them within the speed limit. Policymakers should be aware that if they would like to introduce an intervention driver assistance system that controls automatically, the penetration level should be sufficient from the start to convince others to adopt such a system. For example, in the Ghent ISA trial in 2002, it was noted that most of the drivers were convinced of the effectiveness and were highly in favor of the intelligent speed authority but they stated that they would only use this system further when more or certain groups of drivers would also be forced to use the system.

Source: Royakkers and Van Est (2016).

Possible questions related to Chapters 5 and 6.

1   What values are at stake in the design of an intelligent speed system?
2   Provide a conceptualization and a specification for these values.
3   Are these values conflicting? If so, how and why?
4   What would in your view be the best way to deal with this value conflict?
5   Would you choose for the intelligent speed advisor or for the intelligent speed authority?
6   Are traffic risks personal or collective risks?
7   It seems that many people accept traffic risks. What could be the arguments for this acceptance? Do you agree with these arguments?

Case 9   Onshore Carbon Dioxide storage in Barendrecht (the Netherlands)

Since 2007, there have been plans to establish an onshore CCS (carbon capture and storage) demonstration project in the Dutch town of Barendrecht. This project, initiated by Shell, aims to store CO_2 from its nearby oil refinery in Pernis (in the Rotterdam harbor area) in two depleted gasfields largely located under Barendrecht. The plans caused debate between proponents and opponents, which delayed implementation of the project. In February 2011, the Minister announced that CCS onshore is cancelled until further notice, offshore CCS is optional.

Global demand for energy is increasing, and the challenge is to meet this demand while reducing our carbon footprint. Part of Shell's solution involves carbon capture and storage. The liquid CO_2 is injected more than two kilometers underground into a layer of rock filled with interconnected pores. The CO_2 becomes trapped within the pores, and locked in under many layers of solid water type rocks. The idea was to capture and store over one million tons of CO_2 every year under Barendrecht. That's equivalent to the emissions from about 250,000 cars.

Locals were worried because the government failed to reassure them that they were safe from leaks or potential explosions. Due to the protests of the Barendrecht residents, the CCS onshore was cancelled.

Source: Feenstra et al. (2010), and TheVJMovement, "Co2-Storage Offers a Controversial Solution to Global Warming," https://www.youtube.com/watch?v=cEQX3eDJRNw.

Possible questions related to Chapter 6

1   In an interview with residents of Barendrecht, one of them expresses her frustration of not being able to have a say in the plans, referring to a "dictatorship." To which principle does her reaction relate?
2   A topic for additional research was the impact of the project on human health, especially the psychosomatic effects (such as fear) on residents. The Dutch

National Institute for Public Health and Environment (RIVM) stated in its research report that no exact predictions could be made because no initial measurement of the psychosomatic situation had been performed nor were results available from comparable situations nationally or internationally. Explain why this lack of knowledge can be called ignorance or uncertainty.

3 This case has shown that "how not-in-my-backyard activism can trump efforts to stop global warming, even in countries with powerful green movements like the Netherlands" (Feenstra et al., 2010). The "not in my backyard" argument seems an argument to reject an activity if it involves a risk, however small, such as in this project. To which argument – mentioned in Section 6.5 why equally risks are not necessarily equally ethically acceptable – corresponds this "not in my backyard" argument?

4 An article by a national newspaper was published with the headline: "Unrest grows in Barendrecht about CO_2 storage." This piece featured an interview with a resident of Barendrecht, who said: "From an investigation, it seems that if the gas escapes everyone within the range of 16 to 23 km will be dead. That could be 10,000, more like 100,000 people, how can it [the project] be safe?" To which argument – mentioned in Section 6.5 why equally risks are not necessarily equally ethically acceptable – corresponds this statement (which cannot be substantiated).

5 A lot of lessons about risk communication could be learned, especially for CCS project developers. One lesson is that project developers should take into account not only the actual risks, but – maybe even more important – the perceived risks by the public. A large number of factors plays a role in how people perceive risks (see Section 6.6). Mention three of these factors which have played a role in how the residents of Barendrecht perceived the risks, and explain why these factors played a role.

Notes

1 MV Herald of Free Enterprise Report of Court No. 8074 Formal Investigation. London: Crown, 1987.
2 https://transport.ec.europa.eu/news/road-safety-european-commission-rewards-effective-initiatives-and-publishes-2021-figures-road-2022–10-17_en.

References

Abazi, V. (2020). The European Union Whistleblower Directive: A 'game changer' for whistleblowing protection? *Industrial Law Journal*, 49 (4), 640–656.

Abdullah, B., Faua'ad Syed Muhammad, S.A., Shokravi, Z. et al. (2019). Fourth generation biofuel: A review on risks and mitigation strategies. *Renewable and Sustainable Energy Reviews*, 107, 37–50.

Achterberg, W. (1994). *Samenleving, natuur en duurzaamheid. Een inleiding in de milieufilosofie* [Society, nature and sustainability. An introduction to environmental philosophy], Van Gorcum, Assen.

Achterhuis, H.J. (1995). De moralisering van de apparaten [The moralization of apparatuses]. *Socialisme en Democratie*, 52 (1), 3–12.

Adorno, R. (2004). The precautionary principle: A new legal standard for a technological age. *Journal of International Biotechnology Law*, 1 (1), 11–19.

Anderson, R.M., Otten, J. and Schendel, D.E. (1983). The Bay Area Rapid Transit (BART) Incident, in *Engineering Professionalism and Ethics* (eds J.H. Schaub, K. Pavlovic and M.D. Morris), John Wiley & Sons, New York, pp. 373–380.

Anderson, R.M., Perucci, R., Schendel, D.E. and Trachtman, L.E. (1980). *Divided Loyalties – Whistle-Blowing at BART*, Purdue Research Foundation, West Lafayette, IN.

Angwin, J., Larson, J., Mattu, S. and Kirchner, L. (2016). Machine bias: There's software used across the country to predict future criminals. And it's biased against blacks. ProPublica, May 23, 2016, https://www.propublica.org/article/machine-bias-risk-assessments-in-criminal-sentencing.

Arendt, H. (1965). *Eichmann in Jerusalem: A Report on the Banality of Evil*, Viking Compass, New York.

Aristotle (1980) [350 BC]. *Nicomachean Ethics* (trans W.D. Ross), Clarendon Press, Oxford.

Arkin, R.C. (2007). *Governing lethal behavior: Embedding ethics in a hybrid deliberative/reactive robot architecture* (Technical report GIT-GVU-07–11), Georgia Institute of Technology, Atlanta.

Baeyens, E.F. (1992). *Herald of Free Enterprise. Hét verslag van de ramp* [Herald of Free Enterprise. The Story of the Disaster], Van Geyt Productions, Hulst.

Balcerzak, A.P. and MacGregor Pelikánová, R. (2020). Projection of SDGs in codes of ethics – Case study about lost in translation. *Administrative Sciences*, 10 (4), 95, https://doi.org/10.3390/admsci10040095.

Baron, J. and Spranca, M. (1997). Protected values. *Organizational Behavior and Human Decision Processes*, 70 (1), 1–16.

Baxter, W.F. (1974). *People or Penguins. The Case for Optimal Pollution*, Columbia University Press, New York.

Beauchamp, T.L. (1984). On eliminating the distinction between applied ethics and ethical theory. *Monist*, 67, 514–531.

Beder, S. (1993). Engineers, ethics and etiquette. *New Scientist*, 25 (September), 36–41.

Begley, T.H., White, K., Honigfort, P., Twaroski, M.L., Neches, R. and Walker, R.A. (2005). Perfluorochemicals: Potential sources of and migration from food packaging. *Food Additives and Contaminants*, 22 (10), 1023–1031.

Belkhir, L. and Elmeligi, A. (2018). Assessing ICT global emissions footprint: Trends to 2040 & recommendations. *Journal of Cleaner Production*, 177, 448–463.

Bentham, J. (1948) [1789]. *An Introduction to the Principles of Morals and Legislation*, Hafner Press, Oxford.

Bielefeldt, A.R. (2018). Professional social responsibility in engineering (Chapter 3), in *Social Responsibility* (ed I. Muenstermann), InTech Publishing, London, https://www.intechopen.com/books/social-responsibility (accessed March 11, 2020).

Birch, D. and Fielder, J.H. (eds) (1994). *The Ford Pinto Case: A Study in Applied Ethics, Business and Technology*, State University of New York Press, Albany.

Blok, V. and Lemmens, P. (2015). The emerging concept of responsible innovation. Three reasons why it is questionable and calls for a radical transformation of the concept of innovation, in *Responsible Innovation 2: Concepts, Approaches, and Applications* (eds B.-J. Koops, I. Oosterlaken, H. Romijn, T. Swierstra and J. van den Hoven), Springer International Publishing, Cham, pp. 19–35.

Boele, R., Fabig, H. and Wheeler, D. (2001). Shell, Nigeria and the Ogoni. A study in unsustainable development: I. The story of Shell, Nigeria and the Ogoni people – Environment, economy, relationships: Conflict and prospects for resolution. *Sustainable Development*, 9, 74–86.

Boenink, M., Swierstra, T. and Stemerding, D. (2010). Anticipating the interaction between technology and morality: A scenario study of experimenting with humans in bionanotechnology. *Studies in Ethics, Law, and Technology*, 4 (2), 1–38.

Boers, C. (1981). *Wetenschap, techniek en samenleving. Bouwstenen voor een kritische wetenschapstheorie* [Science, technology and society. Building stones for a critical theory of science], Boom, Meppel.

Bovens, M. (1998). *The Quest for Responsibility. Accountability and Citizenship in Complex Organisations*, Cambridge University Press, Cambridge.

BP (2010). *Deepwater Horizon. Accident investigation report*, https://www.bp.com/content/dam/bp/business-sites/en/global/corporate/pdfs/sustainability/issue-briefings/deepwater-horizon-accident-investigation-report.pdf.

Brady, F.N. (1990). *Ethical Managing. Rules and Results*, Macmillan Publishing Company, New York.

Brätland, J. (2006). Toward a calculational theory and policy of intergenerational sustainability. *The Quarterly Journal of Austrian Economics*, 9, 13–45.

Brown, A.E. (2018). *Ridehail revolution: ridehail travel and equity in Los Angeles* (PhD Thesis), University of California, Los Angeles, https://escholarship.org/uc/item/4r22m57k.

Brumsen, M. (2006). Het ongeluk met de ICE-hogesnelheidstrein bij Eschede [The accident with the ICE high speed train at Eschede], in *Bedrijfsgevallen. Morele beslissingen van ondernemingen* (eds W. Dubbink and H. Van Luijk), Van Gorcum, Assen, pp. 45–55.

Buolamwini, J. and Gebru, T. (2018). Gender shades: Intersectional accuracy disparities in commercial gender classification. *Proceedings of Machine Learning Research*, 81, https://proceedings.mlr.press/v81/buolamwini18a/buolamwini18a.pdf.

Caney, S. (2021). Climate justice, in *The Stanford Encyclopedia of Philosophy*, Winter 2021 edn (ed E.N. Zalta), https://plato-stanford-edu.tudelft.idm.oclc.org/archives/win2021/entries/justice-climate.

Cave, S. and Dihal, K. (2020). The whiteness of AI. *Philosophy & Technology*, 33 (4), 685–703.

Ceva, E. and Bocchiola, M. (2019). *Is Whistleblowing a Duty?* Polity, Cambridge.

Chang, R. (ed) (1997). *Incommensurability, Incomparability, and Practical Reasoning*, Harvard University Press, Cambridge, MA.

Christensen, C.M. (2013). *The Innovator's Dilemma: When New Technologies Cause Great Firms to Fail* (The Management of Innovation and Change Series), Harvard Business Review Press, Boston, MA.

Coeckelbergh, M. (2022). The Ubuntu robot: Towards a relational conceptual framework for intercultural robotics. *Science and Engineering Ethics*, 28, article number 16, https://doi.org/10.1007/s11948-022-00370-9.

Collingridge, D. (1980). *The Social Control of Technology*, Frances Pinter, London.

Congressional testimony of schrage: The internet in China. (2006). http://googleblog.blogspot.com/2006/02/testimony-internet-in-china.html (accessed November 13, 2010).

Covello, V.T. and Merkhofer, M.W. (1993). *Risk Assessment Methods. Approaches for Assessing Health and Environmental Risks*, Plenum, New York.

Cranor, C.F. (1990). Some moral issues in risk assessment. *Ethics*, 101, 123–143.

Criado-Perez, C. (2019). *Invisible Women: Data Bias in a World Designed for Men*, Abrams Press, New York.

Crisp, R. (2021). Well-Being, in *The Stanford Encyclopedia of Philosophy* (ed N.E. Zalta), https://plato.stanford.edu/entries/well-being.

Cross, N. (1989). *Engineering Design Methods*, John Wiley & Sons, Chichester.

Cumming, M.L. (2006). Automation and accountability in decision support system interface design. *Journal of Technology Studies*, 32 (1), 23–31.

Czeskis, A., Dermendjieva, I., Yapit, H., Borning, A., Friedman, B., Gill, B. and Kohno, T. (2010). Parenting from the pocket: Value tensions and technical directions for secure and private parent-teen mobile safety. *Proceedings of the Sixth Symposium on Usable Privacy and Security*, Redmond, Washington.

Dancy, J. (1993). *Moral Reasons*, Blackwell Publishers, Oxford.

Daniels, N. (1979). Wide reflective equilibrium and theory acceptance in ethics. *Journal of Philosophy*, 76, 256–282.

Daniels, N. (1996). *Justice and Justification. Reflective Equilibrium in Theory and Practice*, Cambridge University Press, Cambridge.

Dann, G.E. and Haddow, N. (2008). Just doing business or doing just business: Google, Microsoft, Yahoo! and the business of censoring China's internet. *Journal of Business Ethics*, 79, 219–234.

Davis, M. (1998). *Thinking Like an Engineer. Studies in the Ethics of a Profession*, Oxford University Press, New York and Oxford.

Davis, M. (1999). Professional responsibility: Just following the rules? *Business & Professional Ethics Journal*, 18 (1), 65–87.

Davis, M. (2006). Engineering ethics, individuals, and organizations. *Science and Engineering Ethics*, 12, 223–231.

Davis, M. (2019). A whistle not blown: VW, diesels, and engineers, in *Next-Generation Ethics: Engineering a Better Society* (ed A. Abbas), Cambridge University Press, Cambridge, pp. 217–229.

De Boer, E. (1994). Zestig jaar Deltawerken: Dordrecht als opening en sluitpost [Sixty years Delta Works. Dordrecht as start and finish], in *Wonderen der techniek: Nederlandse ingenieurs en hun kunstwerken 200 jaar civiele techniek* (eds M.L. ten Horn-van Nispen, H. Lintsen and A.J. Veenendaal), Zutphen, Walburg, pp. 197–210.

De George, R.T. (1990). *Business Ethics*, Macmillan, New York.

Demski, C., Butler, C., Parkhill, K.A., Spence, A. and Pidgeon, N.F. (2015). Public values for energy system change. *Global Environmental Change*, 34, 59–69.

Derby, S.L. and Keeney, R.L. (1990). Risk analysis. Understanding "how safe is safe enough?" in *Readings in Risk* (eds T.S. Glickman and M. Gough), Resources for the Future, Washington, pp. 43–49.

De Reuver, M., Van Wynsberghe, A., Janssen, M. and Van de Poel, I. (2020). Digital platforms and responsible innovation: Expanding value sensitive design to overcome ontological uncertainty. *Ethics and Information Technology*, 22, 257–267.

Devon, R. (1999). Towards a social ethics of engineering: The norms of engagement. *Journal of Engineering Education*, 88 (1), 87–92.

Devon, R. (2004). Towards a social ethics of technology: A research prospect. *Techne*, 8 (1), http://scholar.lib.vt.edu/ejournals/SPT/v8n1/devon.html (accessed November 13, 2010).

Devon, R. and Van de Poel, I. (2004). Design ethics: The social ethics paradigm. *International Journal of Engineering Education*, 20 (3), 461–469.

De Wildt, T.E., Van de Poel, I. and Chappin, E.J.L. (2022). Tracing long-term value change in (energy) technologies: Opportunities of probabilistic topic models using large data sets. *Science, Technology, & Human Values*, 47 (3), 429–258.

Dieterich, W., Mendoza, C. and Brennan, T. (2016). *COMPAS risk scales: Demonstrating accuracy equity and predictive parity* (Technical Report, Northpointe Inc.), https://go.volarisgroup.com/rs/430-MBX-989/images/ProPublica_Commentary_Final_070616.pdf.

Dorst, C.H.M. and Royakkers, L.M.M. (2006). The design analogy: A model for moral problem solving. *Design Studies*, 27, 633–656.

Dosi, G. (1982). Technological paradigms and technological trajectories. A suggested interpretation of the determinants and directions of technical change. *Research Policy*, 11, 147–162.

Drèze, J. and Sen, A. (2002). Democratic practice and social inequality in India. *Journal of Asian and African Studies*, 37 (2), https://doi.org/10.1177/002190960203700202.

Dringenberg, E. and Purzer, Ş. (2018). Experiences of first-year engineering students working on ill-structured problems in teams. *Journal of Engineering Education*, 107 (3), 442–467.

Dworkin, R. (1977). *Taking Rights Seriously*, Harvard University Press, Cambridge.

Eilperin, J. (2006). Harmful PTFE chemical to be eliminated by 2015. Washington Post, January 26, 2006, http://www.washingtonpost.com/wp-dyn/content/article/2006/01/25/AR2006012502041.html (accessed November 13, 2010).

ElAlfy, A., Palaschuk, N., El-Bassiouny, D., Wilson, J. and Weber, O. (2020). Scoping the evolution of Corporate Social Responsibility (CSR) research in the Sustainable Development Goals (SDGs) era. *Sustainability*, 12, 5544, https://doi.org/10.3390/su12145544.

Englehardt, E., Werhane, P.H. and Newton, L.H. (2021). Leadership, engineering and ethical clashes at Boeing. *Science and Engineering Ethics*, 27 (1), https://doi.org/10.1007/s11948-021-00285-x.

Enserink, M. (2008). Tough lessons from golden rice. *Science*, 320 (5875), 468–471, https://doi.org/10.1126/science.320.5875.468.

Ernesto, D. and Xu, G. (2020). Google's dragonfly: The ethics of providing a censored search engine in China, in *SAGE Business Cases*, SAGE Publications, Ltd., Thousand Oaks, https://doi.org/10.4135/9781529781465 (accessed March 15, 2022).

Ewing, J. (2017). *Faster, Higher, Farther: The Inside Story of the Volkswagen Scandal*, Bantam Press, London.

Ewuoso, C. (2021). An African relational approach to healthcare and big data challenges. *Science and Engineering Ethics*, 27, article number 34, https://doi.org/10.1007/s11948-021-00313-w.

Fahlquist, J.N. (2006a). Responsibility ascriptions and public health problems. Who is responsible for obesity and lung cancer. *Journal of Public Health*, 14 (1), 15–19.

Fahlquist, J.N. (2006b). Responsibility ascriptions and vision zero. *Accident Analysis and Prevention*, 38, 1113–1118.

Fahlquist, J.N., Doorn, N. and Van de Poel, I. (2015). Design for the value of responsibility, in *Handbook of Ethics, Values, and Technological Design: Sources, Theory, Values and Application Domains* (eds J. van den Hoven, P.E. Vermaas and I. van de Poel), Springer, Netherlands, Dordrecht, pp. 473–490.

Fallah Shayan, N., Mohabbati-Kalejahi, N., Alavi, S. and Zahed, M.A. (2022). Sustainable Development Goals (SDGs) as a framework for Corporate Social Responsibility (CSR). *Sustainability*, 14 (3), 1222, https://doi.org/10.3390/su14031222.

FCC (2009). *V-chip: Viewing television responsibly* [Internet]. Federal Communications Commission, http://www.fcc.gov/cgb/consumerfacts/vchip.html (accessed March 25, 2009).

Feenstra, C.F.J., Mikunda, T. and Brunsting, S. (2010). *What happened in Barendrecht?! Case study on the planned onshore carbon dioxide storage in Barendrecht, the Netherlands* (Report ECN-E--10-057). ECN, Amsterdam.

Felt, U., Wynne, B., Callon, M., Gonçalves, M.E., Jasanoff, S., Jepsen, M., Joly, P.-B., Konopasek, Z., May, S., Neubauer, C., Rip, A., Siune, K., Stirling, A. and Tallacchini, M. (2007). *Taking European knowledge society seriously*. Report of the expert group on science and governance to the science, economy and society directorate, Directorate-General for Research, European Commission, Directorate-General for Research, Science, Economy and Society, Brussels.

Feng, F. (2020). The modern examination of Confucian filial ethics. *Open Journal of Social Sciences*, 8, 286–294.

Fischhoff, B., Lichtenstein, S. and Slovic, P. (1981). *Acceptable Risk*, Cambridge University Press, Cambridge.

Florman, S.C. (1976). *The Existential Pleasures of Engineering*, St. Martin's Press, New York.

Frankel, M.S. (1989). Professional codes: Why, how, and with what impact? *Journal of Business Ethics*, 8 (2), 109–115.

Frankena, W.K. (1973). *Ethics*, Prentice-Hall, Englewood Cliffs.

Friedman, B. and Hendry, D.G. (2019). *Value Sensitive Design: Shaping Technology with Moral Imagination*, The MIT Press, Cambridge.

Friedman, B., Kahn, P.H., Jr. and Borning, A. (2006). Value sensitive design and information systems, in *Human-Computer Interaction in Management Information Systems: Foundations* (eds P. Zhang and D. Galletta), M.E. Sharpe, Armonk, pp. 348–372.

Friedman, C. (2022). Ethical concerns with replacing human relations with humanoid robots: An ubuntu perspective. AI and Ethics, https://doi.org/10.1007/s43681-022-00186-0.

Friedman, M. (1962). *Capitalism and Freedom*, University of Chicago Press, Chicago.

Ganz, C. (2008). *The 1933 Chicago World's Fair: Century of Progress*, University of Illinois Press, Urbana.

Garfinkel, S. (2002). An RFID bill of rights. MIT Technological Review, October, https://www.technologyreview.com/2002/10/01/101947/an-rfid-bill-of-rights.

Ge, Y., Knittel, C.R., MacKenzie, D. and Zoepf, S. (2016). *Racial and gender discrimination in transportation network companies*. NBER Working Paper No. 22776, https://www.nber.org/papers/w22776.

Gert, B. (1984). Moral theory and applied ethics. *Monist*, 67, 532–548.

Gilligan, C. (1982). *In a Different Voice: Psychological Theory and Women's Development*, Harvard University Press, Cambridge, MA.

Githitho-Muriithi, A. (2010). Education for all and child labour in Kenya: A conflict of capabilities? *Procedia Social and Behavioral Sciences*, 2, 4613–4621.

Givel, M. (2007). Motivation of chemical industry social responsibility through Responsible Care. *Health Policy*, 81 (1), 85–92.

Goldberg, S. (1987). The space shuttle tragedy and the ethics of engineering. *Jurimetrics Journal*, 27 (Winter), 155–159.

Grauls, M. (1993). *Uitvinders van het dagelijks leven 2* [Inventors of daily life 2], CODA, Antwerp.

Griffin, J. (1986). *Well-being: Its Meaning, Measurement, and Moral Importance*, Clarendon Press, Oxford.

Grinbaum, A. and Groves, C. (2013). What is "responsible" about responsible innovation? Understanding the ethical issues, in *Responsible Innovation* (eds R. Owen, J. Bessant and M. Heintz), Wiley, Chistester, pp. 119–142.

Grother, P., Ngan, M. and Hanaoka, K. (2019). *Face Recognition Vendor Test (FRVT). Part 3: Demographic Effects*, National Institute of Standards and Technology, Gaithersburg, US.

Groves, C., Frater, L., Lee, R. and Stokes, E. (2011). Is there room at the bottom for CSR? Corporate social responsibility and nanotechnology in the UK. *Journal of Business Ethics*, 101 (4), 525–552.

Guo, P., Tang, C.S., Tang, Y. and Wang, Y. (2019). Gender-based operational issues arising from on-demand ride-hailing platforms: Safety concerns and system configuration. Available at SSRN: https://ssrn.com/abstract=3260427.

Habermas, J. (1981). *Theorie des kommunikativen Handelns* [Theory of communicative action], Suhrkamp Verlag, Frankfurt am Main.

Hanrahan, B., Ning, M. and Chien Wen, Y. (2017). The roots of bias on Uber. *Proceedings of 15th European Conference on Computer-Supported Cooperative Work – Exploratory Papers, Reports of the European Society for Socially Embedded Technologies*, https://doi.org/10.48550/arXiv.1803.08579.

Hansson, S.O. (1998). Should we avoid moral dilemmas? *The Journal of Value Inquiry*, 32 (3), 407–416.

Hansson, S.O. (2003). Ethical criteria of risk acceptance. *Erkenntnis*, 59, 291–309.

Hansson, S.O. (2004). Weighing risks and benefits. *Topoi*, 23, 145–152.

Hansson, S.O. (2007a). Safe design. *Techne*, 10 (1), 43–49.

Hansson, S.O. (2007b). Philosophical problems in cost-benefit analysis. *Economics and Philosophy*, 23, 163–183.

Hansson, S.O. (2009). Risk and safety in technology, in *Handbook of the Philosophy of Science. Vol. 9: Philosophy of Technology and Engineering Sciences* (ed A. Meijers), Elsevier, Amsterdam, pp. 1069–1102.

Hansson, S.O. (2020). How extreme is the precautionary principle? *Nanoethics*, 14, 245–257.

Hare, R.M. (1982). Ethical theory and utilitarianism, in *Utilitarianism and Beyond* (eds A. Sen and B. Williams), Cambridge University Press, Cambridge, pp. 23–38.

Hare, R.M. (1988). Why do applied ethics? in *Applied Ethics and Ethical Theory* (eds D.M. Rosenthal and F. Shehadi), University of Utah Press, Salt Lake City, pp. 71–83.

Harris, C.E., Pritchard, M.S. and Rabins, M.J. (2005). *Engineering Ethics. Concepts and Cases*, 3rd edn, Wadsworth, Belmont.

Harrison, R. (1993). Case study: Farm animals, in *Environmental Dilemmas: Ethics and Decisions* (ed R.J. Berry), Chapman & Hall, London, pp. 118–135.

Hassink, H., de Vries, M. and Bollen, L. (2007). A content analysis of whistleblowing policies of leading European companies. *Journal of Business Ethics*, 75 (1), 25–44.

Health Council of the Netherlands (2006). *Health Significance of Nanotechnologies*, Health Council of the Netherlands, The Hague.

Hellweg, S., Hofstetter, T.B. and Hungerbuhler, K. (2003). Discounting and the environment should current impacts be weighted differently than impacts harming future generations? *The International Journal of Life Cycle Assessment*, 8 (1), 8–18.

Herkert, J., Borenstein, J. and Miller, K. (2020). The Boeing 737 MAX: Lessons for engineering ethics. *Science and Engineering Ethics*, 26 (6), 2957–2974.

Herkert, J.R. (1999). ABET's engineering criteria 2000 and engineering ethics: Where do we go from here? Presented at the *OEC International Conference on Ethics in Engineering and Computer Science*, March 1999.

Herkert, J.R. (2001). Future directions in engineering ethics research: Microethics, macroethics and the role of professional societies. *Science and Engineering Ethics*, 7 (3), 403–414.

Herr, R.S. (2003). Is Confucianism compatible with care ethics? A critique. *Philosophy East and West*, 53 (4), 471–489.

High-Level Expert Group on AI (2019). *Ethics Guidelines for Trustworthy AI*, European Commission, Brussels, https://digital-strategy.ec.europa.eu/en/library/ethics-guidelines-trustworthy-ai.

Hillerbrand, R., Milchram, C. and Schippl, J. (2021). Using the Capability Approach as a normative perspective on energy justice: Insights from two case studies on digitalisation in the energy sector. *Journal of Human Development and Capabilities*, 22 (2), 336–359.

Hopster, J. (2021). What are socially disruptive technologies? *Technology in Society*, 67, 101750, https://doi.org/10.1016/j.techsoc.2021.101750.

Houkes, W., Vermaas, P.E., Dorst, K. and De Vries, M.J. (2002). Design and use as plans. An action-theoretical account. *Design Studies*, 23, 303–320.

House Committee on Transportation and infrastructure (2020). *Final Committee Report. The Design, development & Certification of the Boeing 737 MAX.* https://commons.wikimedia.org/wiki/File:Final_Committee_Report_on_the_Design,_Development_and_Certification_of_the_B737_MAX.pdf.

Hummels, H. and Karssing, E. (2007). Organising ethics, in *Ethics and Business* (ed R.J.M. Jeurissen), Van Gorcum, Assen, pp. 249–275.

Hung, L., Liu, C., Woldum, E. et al. (2019). The benefits of and barriers to using a social robot PARO in care settings: A scoping review. *BMC Geriatrics*, 19, article number 232, https://doi.org/10.1186/s12877-019-1244-6.

Hunter, T.A. (1997). Designing to codes and standards, in *ASM Handbook. Vol. 20. Materials Selection and Design* (eds G.E. Dieter, S.D. Henry and S.R. Lampman), ASM, Materials Park, pp. 66–71.

IEEE Global Initiative on Ethics of Autonomous and Intelligent Systems (2019). *Ethically aligned design: A vision for prioritizing human well-being with autonomous and intelligent systems.* IEEE, https://standards.ieee.org/wp-content/uploads/import/documents/other/ead_v2.pdf.

Ihde, D. (1990). *Technology and the Lifeworld*, Indiana University Press, Bloomington.

Inderwildi, O.R. and King, D.A. (2009). Quo vadis biofuels? *Energy & Environmental Science*, 2 (4), 343–346.

International Labour Organization and UNICEF (2021). *Child Labour: Global Estimates 2020, Trends and the Road Forward*, ILO and UNICEF, New York.

International Crisis Group (2008). *Nigeria: Ogoni land after shell.* Africa Briefing no. 54, September 18, 2008, https://www.files.ethz.ch/isn/92038/b54_nigeria___ogoni_land_after_shell.pdf.

Jenkins, K., McCauley, D., Heffron, R., Stephan, H. and Rehner, R. (2016). Energy justice: A conceptual review. *Energy Research & Social Science*, 11, 174–182.

Jeurissen, R.J.M. and Van de Ven, B. (2007). Values and norms in organisations, in *Ethics and Business* (ed R.J.M. Jeurissen), Van Gorcum, Assen, pp. 54–92.

Jing, S. and Doorn, N. (2020). Engineers' moral responsibility: A Confucian perspective. *Science and Engineering Ethics*, 26, 233–253.

Johnson, B.B. (1999). Ethical issues in risk communication. Continuing the discussion. *Risk Analysis*, 19 (3), 335–348.

Johnson, D.G. (2001). *Computer Ethics*, Prentice Hall, Upper Saddle River.

Johnson, D.G. (2020). *Engineering Ethics. Contemporary and Enduring Debates*, Yale University Press, New Haven and London.

Jungermann, H. (1996). Ethical dilemmas in risk communication, in *Codes of Conduct. Behavioral Research into Business Ethics* (eds M. Messick and A.E. Tenbrunsel), Russell Sage Foundation, New York, pp. 300–317.

Jungk, R. (1958). *Brighter Than a Thousand Suns*, Penquin Books, Middlesex.

Kant, I. (2006) [1784]. Beantwortung der Frage: Was ist Aufklärung? translated as "An Answer to the Question: What is Enlightenment?", in *Toward Perpetual Peace and Other Writings on Politics, Peace, and History* (ed P. Kleingeld et al.), Yale University Press, London, pp. 17–23.

Kant, I. (2002) [1785]. *Grundlegung zur Metaphysik der Sitten*, translated as *Groundwork for the Metaphisics of Morals*, Oxford University Press, Oxford.

Kenny, K. (2019). *Whistleblowing: Towards a New Theory*, Harvard University Press, Cambridge.

King, A.A. and Lenox, M.J. (2000). Industry self-regulation without sanctions: The chemical industry's responsible care program. *The Academy of Management Journal*, 43 (4), 698–716.

Kleinberg, J., Mullainathan, S. and Raghavan, M. (2017). Inherent trade-offs in the fair determination of risk scores. *Proceedings of the 8th Conference on Innovations in Theoretical Computer Science* (ITCS), https://arxiv.org/abs/1609.05807.

Kneese, A.V., Ben-David, S. and Schulze, W.D. (1983). The ethical foundations of benefit-cost analysis, in *Energy and the Future* (eds D. MacLean and P.G. Brown), Rowman and Littlefield, Totowa, pp. 59–74.

Koops, B.-J., Bryce Clayton, N., Timan, T., Škorvánek, I., Chokrevski, T. and Galič, M. (2017). A typology of privacy. *University of Pennsylvania Journal of International Law*, 38 (2), 483–575.

Kraakman, R., Davies, P., Hansmann, H., Hertig, G., Hopt, K., Kanda, H. and Rock, E. (eds) (2004). *The Anatomy of Corporate Law: A Comparative and Functional Approach*, Oxford University Press, New York.

Kraemer, F., Van Overveld, K. and Peterson, M. (2011). Is there an ethics of algorithms? *Ethics and Information Technology*, 13, 251–260.

Kremer, E. (2002). (Re)examining the Citicorp case: Ethical paragon or chimera. *Practice*, 6 (3), 269–276.

Kroes, P. (1992). On the role of design in engineering theories; Pambour's theory on the steam engine, in *Technological Development and Science in the Industrial Age* (eds P. Kroes and M. Bakker), Kluwer, Dordrecht, pp. 69–98.

Kroes, P. and Van de Poel, I. (2015). Design for values and the definition, specification, and operationalization of values, in *Handbook of Ethics, Values, and Technological Design: Sources, Theory, Values and Application Domains* (eds J. van den Hoven, P.E. Vermaas and I. van de Poel), Springer Netherlands, Dordrecht, pp. 151–178.

Krohn, W. and Weyer, J. (1994). Society as a laboratory: The social risks of experimental research. *Science and Public Policy*, 21 (3), 173–183.

Ladd, J. (1991). The quest for a code of professional ethics. An intellectual and moral confusion, in *Ethical Issues in Engineering* (ed D.G. Johnson), Prentice Hall, Englewood Cliffs, pp. 130–136.

Langat, S.K., Mwakio, P.M. and Ayuku, D. (2020). How Africa should engage Ubuntu ethics and artificial intelligence. *Journal of Public Health International*, 2 (4), 20–25.

Latour, B. (1992). Where are the missing masses? The sociology of a few mundane artifacts, in *Shaping Technology/Building Society* (eds W.E. Bijker and J. Law), MIT Press, Cambridge, pp. 225–258.

Lave, L.B. (1984). Eight frameworks for regulation, in *Technological Risk Assessment* (eds P.F. Ricci, L.A. Sagan and C.G. Whiplle), Martinus Nijhoff, The Hague, pp. 169–190.

Layton, E.T. (1971). *The Revolt of the Engineers. Social Responsibility and the American Engineering Profession*, The Press of the Case Western Reserve University, Cleveland and London.

Lee, G., Pittroff, E. and Turner, M.J. (2020). Is a uniform approach to whistle-blowing regulation effective? Evidence from the United States and Germany. *Journal of Business Ethics*, 163, 553–576.

Lee, M.S.A., Floridi, L. and Singh, J. (2021). Formalising trade-offs beyond algorithmic fairness: Lessons from ethical philosophy and welfare economics. *AI and Ethics*, 1 (4), 529–544.

Lees, N.J. and Lees, I. (2017). Competitive advantage through responsible innovation in the New Zealand sheep dairy industry. *International Food and Agribusiness Management Review*, 21 (4), 1–20.

Legge, J. (1879). *The Sacred Books of China. The Texts of Confucianism*, The Clarendon Press, Oxford.

Lenk, H. and Ropohl, G. (eds) (1987). *Technik und Ethik*, Reclam, Stuttgart.

Li, C. (2008). The philosophy of harmony in classical Confucianism. *Philosophy Compass*, 3 (3), 423–435.

Liu, J. (2017). Confucian robotic ethics. *International Conference on the Relevance of the Classics under the Conditions of Modernity: Humanity and Science*, Hong Kong, https://www.researchgate.net/publication/319391008_Confucian_Robotic_Ethics.

Locke, J. (1986) [1690]. *Second Treatise of Government*, Prometheus Books, New York.

Low, K.C.P. and Ang, S.L. (2013). Confucian ethics, governance and corporate social responsibility. *International Journal of Business and Management*, 8 (4), 30–43.

Ludwig, D., Blok, V., Garnier, M., Macnaghten, P. and Pols, A. (2022). What's wrong with global challenges? *Journal of Responsible Innovation*, 9 (1), https://doi.org/10.1080/23299460.2021.2000130.

Luegenbiehl, H.C. and Clancy, R.F. (2017). *Global Engineering Ethics*, Butterworth-Heinemann, Oxford.

MacIntyre, A. (1984a). *After Virtue*, Notre Dame University Press, Notre Dame.

MacIntyre, A. (1984b). Does applied ethics rest on a mistake? *The Monist*, 67, 498–513.

Martin, D.A., Conlon, E. and Bowe, B. (2021). A multi-level review of engineering ethics education: Towards a socio-technical orientation of engineering education for ethics. *Science and Engineering Ethics*, 27, Article number 60, https://doi.org/10.1007/s11948-021-00333-6.

Martin, K.E. (2008). *Google, Inc., in China* (Case BRI-1004), Business Roundtable, Institute for Corporate Ethics, Washington.

Martin, M.W. and Schinzinger, R. (1996). *Ethics in Engineering*, 3rd edn, McGraw-Hill, New York.

Mathiesen, K. (2013). The Internet, children, and privacy: The case against parental monitoring. *Ethics of Information Technology*, 15, 263–274.

Mayer, J.E. (2005). The golden rice controversy: Useless science or unfounded criticism? *BioScience*, 55 (9), 726–727.

McLinden, M.O. and Didion, D.A. (1987). Quest for alternatives. *ASHRAE Journal*, 29, 32–34.

Mhlambi, S. (2020). From rationality to relationality: Ubuntu as an ethical and human rights framework for artificial intelligence governance, discussion paper series 220-009. Car Center for Human Rights Policy, Cambridge, US, https://carrcenter.hks.harvard.edu/publications/rationality-relationality-ubuntu-ethical-and-human-rights-framework-artificial.

Mill, J.S. (1859). *On Liberty*, John W. Parker and Son, London.

Mill, J.S. (1979) [1863]. *Utilitarianism*, Collins, London.

Miller, H. and Engemann, K. (2019). The precautionary principle and unintended consequences. *Kybernetes: The International Journal of Systems & Cybernetics*, 48 (2), 265–286, https://doi.org/10.1108/K-01-2018-0050.

Miller, J.K., Friedman, B. and Jancke, G. (2007). Value tensions in design: The value sensitive design, development, and appropriation of a corporation's groupware system. *Proceedings of the 2007 International ACM Conference on Supporting Group Work*, ACM, Sanibel Island, Florida, pp. 281–290.

Mishan, E.J. (1975). *Cost-Benefit Analysis. An Informal Introduction*, Allen & Unwin, London.

Moffet, J., Bregha, F. and Middelkoop, M.J. (2004). Responsible care: A case study of a voluntary environmental initiative, in *Voluntary Codes: Private Governance, the Public Interest and Innovation* (ed K.R. Webb), Carleton Research Unit for Innovation, Science and Environment, Carleton University, Ottawa, ON, pp. 177–208.

Morgan, M.G. and Lave, L. (1990). Ethical considerations in risk communication practice and research. *Risk Analysis*, 10 (3), 355–358.

Morgenstern, J. (1995). The fifty-nine-story crisis. The New Yorker, May 29, 45–53.

Mostert, P. (1982). Reactorveiligheid [Reactor safety], in *Kernenergie in beweging. Handboek bij vraagstukken over kernenergie* [Nuclear energy on the move. Handbook for nuclear energy issues] (eds C.D. Andriesse and A. Heertje), Keesing Boeken, Amsterdam, pp. 61–73.

Mouter, N., Koster, P. and Dekker, T. (2019). *An introduction to participatory value evaluation*. Tinbergen Institute Discussion Paper 2019-024/V, https://ssrn-com.tudelft.idm.oclc.org/abstract=3358814.

Mugumbate, J. and Nyanguru, A. (2013). Exploring African philosophy: The value of Ubuntu in social work. *African Journal of Social Work*, 3 (1), 82–100.

Müller, M.O., Stämpfli, A., Dold, U. and Hammer, T. (2011). Energy autarky: A conceptual framework for sustainable regional development. *Energy Policy*, 39 (10), 5800–5810.

Myers, A. (2010). Camp Delta, Google Earth and the ethics of remote sensing in archaeology. *World Archaeology*, 42 (3), 455–467.

Nair, I. and Bulleit, W.M. (2020). Pragmatism and care in engineering ethics. *Science and Engineering Ethics*, 26, 65–87.

Nass, C.I. and Brave, S. (2005). *Wired for Speech: How Voice Activates and Advances the Human-Computer Relationship*, MIT Press, Cambridge, Mass.

Nathan, G. (2015). Innovation process and ethics in technology: An approach to ethical (responsible) innovation governance. *Journal on Chain and Network Science*, 15 (2), 119–134.

Nathan, L.P., Klasnja, P.V. and Friedman, B. (2007). Value scenarios: A technique for envisioning systemic effects of new technologies. *Extended Abstracts Proceedings of the 2007 Conference on Human Factors in Computing* Systems, *CHI 2007*, San Jose, CA, USA, https://doi.org/10.1145/1240866.1241046.

National Commission on the BP Deepwater Horizon Oil Spill and Offshore Drilling (2011a). Macondo: The Gulf oil disaster (Chief Counsel's report). Washington, DC: National Commission on the BP Deepwater Horizon Oil Spill and Offshore Drilling, https://www.loc.gov/item/2011505289 (accessed May 31, 2022).

National Commission on the BP Deepwater Horizon Oil Spill and Offshore Drilling (2011b). Deepwater. The Gulf oil disaster and future of offshore drilling. Washington, DC: National Commission on the BP Deepwater Horizon Oil Spill and Offshore Drilling, http://www.gpo.gov/fdsys/pkg/GPO-OILCOMMISSION/content-detail.html (accessed May 31, 2022).

Naylor, R.L., Liska, A.J., Burke, M.B., Falcon, W.P., Gaskell, J.C., Rozelle S.D. and Cassman, K.G. (2007). The ripple effect: Biofuels, food security, and the environment. *Environment*, 49 (9), 30–43.

Nelson, D. (1980). *Frederick W. Taylor and the Rise of Scientific Management*, University of Wisconsin Press, London.

Neufeld, M.J. (1995). *The Rocket and the Reich. Peenemunde and the Coming of the Ballistic Missile Era*, Harvard University Press, Cambridge, MA.

Ngwakwe, C.C. (2021). Niger Delta oil spill legal victory against the Shell Company: The future of corporate environmental responsibility. *Juridica*, 17 (2), 27–39.

Nissenbaum, H.F. (2010). *Privacy in Context: Technology, Policy, and the Integrity of Social Life*, Stanford Law Books, Stanford, CA.

Nuffield Council on Bioethics (2011). *Biofuels: Ethical Issues*, Nuffield Council on Bioethics, London.

Nussbaum, M. (2000). *Women and Human Development: The Capabilities Approach*, Cambridge University Press, Cambridge.

Nussbaum, M. (2011). *Creating Capabilities*, Harvard University Press, Cambridge.

Oberdöster, G., Oberdöster, E. and Oberdöster, J. (2005). Nanotoxicology: An emerging discipline evolving from studies of ultrafine particles. *Environmental Health Perspectives*, 113 (7), 823–829.

O'Neill, E. (2022). Contextual integrity as a general conceptual tool for evaluating technological change. *Philosophy & Technology*, 35, Article number 79, https://doi.org/10.1007/s13347-022-00574-8.

Oosterlaken, E.T. (2013). *Taking a capability approach to technology and its design: A philosophical exploration* (PhD-thesis). Delft University, Delft.

Otway, H.J. and von Winterfeldt, D. (1982). Beyond acceptable risk. On the social acceptability of technologies. *Policy Sciences*, 14, 247–256.

Owen, R., Stilgoe, J., Macnaghten, P., Gorman, M., Fisher, E. and Guston, D. (2013). A framework for responsible innovation, in *Responsible Innovation* (eds R. Owen, J. Bessant and M. Heintz), Wiley, Chichester, pp. 27–50.

Pesch, U. (2015). Engineers and active responsibility. *Science and Engineering Ethics*, 21, 925–938.

Peterson, M. (2017). *The Ethics of Technology: A Geometric Analysis of Five Moral Principles*, Oxford University Press, Oxford.

Petroski, H. (1982). *To Engineer Is Human. The Role of Failure in Successful Design*, St. Martin's Press, New York.

Piszkiewicz, D. (1995). *The Nazi Rocketeers. Dreams of Space and Crimes of War*, Praeger, Westport, CT.

Porter, M.E. and Kramer, M.R. (2006). Strategy and society: The link between competitive advantage and corporate social responsibility. *Harvard Business Review*, 84 (12), 78–92, 163.

Powley, C.R., Michalczyk, M.J., Kaiser, M.A. and Buxton, L.W. (2005). Determination of perfluorooctanoic acid (PFOA) extractable from the surface of commercial cookware under simulated cooking conditions by LC/MS/MS. *The Analyst*, 130 (9), 1299–1302.

Pritchard, M.S. (2001). Responsible engineering. The importance of character and imagination. *Science and Engineering Ethics*, 7 (3), 391–402.

Pritchard, M.S. (2009). Professional standards in engineering practice, in *Handbook of the Philosophy of Science. Vol. 9: Philosophy of Technology and Engineering Sciences* (ed A. Meijers), Elsevier, Amsterdam, pp. 953–971.

Radfar, A., Asgharzadeh, S.A.A., Quesada, F. and Filip, I. (2018). Challenges and perspectives of child labor. *Industrial Psychiatry Journal*, 27 (1), 17–20.

Raffensperger, C. and Tickner, J. (eds) (1999). *Protecting Public Health and the Environment: Implementing the Precautionary Principle*, Island Press, Washington, DC.

Rawls, J. (1971). *A Theory of Justice*, Harvard University Press, Cambridge, MA.

Rawls, J. (1993). *Political Liberalism*, Columbia University Press, New York.

Rawls, J. (2001). *Justice as fairness. A restatement*, The Belknap Press of Harvard University Press, Cambridge, MA.

Raz, J. (1986). *The Morality of Freedom*, Oxford University Press, Oxford.

Redman, B. and Caplan, A. (2015). No one likes a snitch. *Science and Engineering Ethics*, 21, 813–819.

Renn, O. (2005). *White Paper on Risk Governance – Towards an Integrative Approach*, International Risk Governance Council, Geneva.

Rip, A. (1987). Controversies as informal technology assessment. *Knowledge: Creation, Diffusion, Utilization*, 8 (2), 349–371.

Rip, A. and te Kulve, H. (2008). Constructive technology assessment and socio-technical scenarios, in *Yearbook of Nanotechnology in Society* (eds E. Fisher, C. Selin and J.M. Wetmore), Springer, Dordrecht, pp. 49–70.

Robeyns, I. (2020). The capability approach. Stanford Encyclopedia of Philosophy, https://plato.stanford.edu/entries/capability-approach.

Roeser, S. and Pesch, U. (2016). An emotional deliberation approach to risk. *Science, Technology, & Human Values*, 41 (2), 274–297.

Ross, W.D. (1930). *The Right and the Good*, Clarendon Press, Oxford.

Rowe, G. and Frewer, L.J. (2000). Public participation methods: A framework for evaluation. *Science, Technology & Human Values*, 25 (1), 3–29.

Royakkers, L.M.M., Timmer, J., Kool., L. and Van Est, O. (2018). Societal and ethical issues of digitization. *Ethics and Information Technology*, 20 (2), 127–142.

Royakkers, L.M.M. and Van Est, Q. (2016). *Just Ordinary Robots*, CRC Press, Boca Raton.

Royakkers, L.M.M. and van Est, R. (2010). The cubicle warrior: The marionette of digitalized warfare. *Ethics and Information Technology*, 12, 289–296.

Ryan, B.L.V. (1991). Conflicts inherent in corporate codes. *International Journal of Value-Based Management*, 4 (1), 119–136.

Sánchez, D., Proske, M. and Baur, S.J. (2022). *Life Cycle Assessment of the Fairphone 4*. Fraunhofer IZM, Berlin.

Sandin, P. (1999). Dimensions of the precautionary principle. *Human and Ecological Risk Assessment*, 5 (5), 889–907.

Santoni de Sio, F. and Van den Hoven, J. (2018). Meaningful human control over autonomous systems: A philosophical account. *Frontiers in Robotics and AI*, 5 (15), https://doi.org/10.3389/frobt.2018.00015.

Schivelbusch, W. (1986). *The Railway Journey, The Industrialization of Time and Space in the 19th Century*, University of California Press, Berkeley.

Scholten, V.E. and Van der Duin, P.A. (2015). Responsible innovation among academic spin-offs: How responsible practices help developing absorptive capacity. *Journal on Chain and Network Science*, 15 (2), 165–179.

Schönherr, N., Findler, F. and Martinuzzi, A. (2017). Exploring the Interface of CSR and the Sustainable Development Goals. *Transnational Corporations*, 24 (3), 33–47.

Schot, J.W. (1992). Constructive technology assessment and technology dynamics: The case of clean technologies. *Science, Technology & Human Values*, 17 (1), 36–57.

Schot, J.W. and Rip, A. (1997). The past and future of constructive technology assessment. *Technology Forecasting and Social Change*, 54 (2/3), 251–268.

Sedlmeir, J., Buhl, H.U., Fridgen, G. and Keller, R. (2020). The energy consumption of blockchain technology: Beyond myth. *Business & Information Systems Engineering*, 62 (6), 599–608.

Sen, A. (1992). *Inequality Re-examined*, The Clarendon Press, Oxford.

Sen, A. (1999). *Development as Freedom*, Knopf, New York.

Sen, A. (2009). *The Idea of Justice*, Harvard University Press, Cambridge.

Sharbatoghlie, A., Mosleh, M. and Shokatian, T. (2013). Exploring trends in the codes of ethics of the Fortune 100 and Global 100 corporations. *Journal of Management Development*, 32 (7), 675–689, https://doi.org/10.1108/JMD-04-2011-0044.

Sharkey, A. (2014). Robots and human dignity: A consideration of the effects of robot care on the dignity of older people. *Ethics and Information Technology*, 16 (1), 63–75.

Shrader-Frechette, K.S. (1985). *Risk Analysis and Scientific Method. Methodological and Ethical Problems with Evaluating Societal Hazards*, Reidel, Dordrecht.

Shrader-Frechette, K.S. (1991). *Risk and Rationality. Philosophical Foundations for Populist Reform*, University of California Press, Berkeley.

Shue, H. (1993). Subsistence emissions and luxury emissions. *Law & Policy*, 15 (1), 39–60.

Sidgwick, H. (1877). *Methods of Ethics*, Macmillan, London.

Simon, H.A. (1973). The structure of ill-structured problems. *Artificial Intelligence*, 4, 181–201.

Simon, H.A. (1974). Spurious correlation: A causal interpretation, in *Causal Models in the Social Sciences* (ed H.M. Blalock), Macmillan, London and Basingstoke, pp. 1–17.

Singer, P. (2002). *One World. The Ethics of Globalization*, Yale University Press, London.

Slovic, P., Fischhoff, B. and Lichtenstein, S. (1990). Rating the risks, in *Readings in Risk* (eds T.S. Glickman and M. Gough), Resources for the Future, Washington, pp. 61–74.

Smart, J.J.C. (1973). An outline of a system of utilitarian ethics, in *Utilitarianism for and Against* (eds J.J.C. Smart and B. Williams), Cambridge University Press, Cambridge, pp. 3–74.

Smith, A. (1776). *An Inquiry into the Nature and Causes of the Wealth of Nations*, Clarendon Press, Oxford.

Spapens, A. (2018). The 'Dieselgate' scandal: A criminological perspective, in *Green Crimes and Dirty Money* (eds A.C.M. Spapens, R.D. White, D.P. van Uhm and W. Huisman), Routledge, London and New York, pp. 91–112.

Sparrow, R. and Sparrow, L. (2006). In the hands of machines? The future of aged care. *Minds and Machines*, 16 (2), 141–161.

Spiekermann, S. (2015). *Ethical IT Innovation. A Value-Based System Design Approach*, Auerbach Publications, New York.

Starr, C. (1990). Social benefit versus technological risk, in *Readings in Risk* (eds T.S. Glickman and M. Gough), Resources for the Future, Washington, pp. 183–193.

Stern, P.C. and Feinberg, H.V. (1996). *Understanding Risk: Informing Decisions in a Democratic Society*, National Academy Press, Washington.

Stewart, F. and Deneulin, S. (2002). Amartya Sen's contribution to development thinking. *Studies in Comparative International Development*, 37 (2), 61–70.

Stilgoe, J., Owen, R. and Macnaghten, P. (2013). Developing a framework for responsible innovation. *Research Policy*, 42 (9), 1568–1580.

Stone, G.D. and Glover, D. (2017). Disembedding grain: Golden rice, the green revolution, and heirloom seeds in the Philippines. *Agriculture and Human Values*, 34 (1), 87–102.

Stuhlinger, E. and Ordway, F.I., III. (1994). *Wernher von Braun, Crusader for Space: A Biographical Memoir*, Krieger, Malabar.

Styron, W. (1979). *Sophie's Choice*, Random House, London.

Sullenberger, S. (2019). My letter to the editor of New York Times Magazine, https://www.sullysullenberger.com/my-letter-to-the-editor-of-new-york-times-magazine.

Sunstein, C.R. (2005). Cost-benefit analysis and the environment. *Ethics*, 115, 351–385.

Swierstra, T. (2000). *Kloneren in de polder. Het maatschappelijk debat over kloneren in Nederland Februari 1997-Oktober 1999* [Cloning in the polder. The societal debate on cloning in the Netherlands February 1997–October 1999], Rathenau Instituut (Studie no. 39), The Hague.

Swierstra, T. (2013). Nanotechnology and technomoral change. *Etica & Politica/Ethics & Politics*, XV (1), 200–219.

Swierstra, T. and Jelsma, J. (2006). Responsibility without moralism in techno-scienctific design practice. *Science, Technology & Human Values*, 31 (3), 309–332.

Szulecki, K. (2018). Conceptualizing energy democracy. *Environmental Politics*, 27 (1), 21–41.

Taebi, B. (2017). Bridging the gap between social acceptance and ethical acceptability. *Risk Analysis*, 37 (10), 1817–1827.

Tan, S.-H. (2022). Humanities education in the age of AI. Reflections from Deweyan and Confucian perspectives, in *John Dewey and Chinese Education* (eds H. Zhang and J. Garrison), Brill, Singapore, pp. 234–253.

Tavani, H.T. (2004). *Ethics & Technology. Ethical Issues in an Age of Information and Communication Technology*, John Wiley & Sons, Hoboken.

Taylor, F.W. (1911). *The Principles of Scientific Management*, Harper Bros, New York.

Ten Horn-van Nispen, M.-L. (2001). Johan van Veen. *Tijdschrift voor Waterstaatsgeschiedenis*, 10 (1), 16–20.

Tetlock, P.E. (2003). Thinking the unthinkable: Sacred values and taboo cognitions. *Trends in Cognitive Sciences*, 7 (7), 320–324.

Thaler, R.H. and Sunstein, C.R. (2009). *Nudge: Improving Decisions about Health, Wealth, and Happiness*, Penguin Books, New York.

The Royal Society & The Royal Academy of Engineering (2004). *Nanoscience and Nanotechnologies: Opportunities and Uncertainties*, The Royal Society & The Royal Academy of Engineering, London.

Thompson, D.F. (1980). Moral responsibility and public officials. *American Political Science Review*, 74, 905–916.

Thompson, D.F. (2014). Responsibility for failures of government: The PMH. *American Journal of Public Administration*, 44 (3), 259–273.

Tronto, J.C. (1993). *Moral Boundaries. A Political Argument for an Ethic of Care*, Routledge, New York.

Tundys, B. (2021). Corporate social responsibility and sustainable value creation, in *Sustainability in Bank and Corporate Business Models* (eds M. Ziolo, B. Zofia Filipiak and B. Tundys), Palgrave Macmillan, Cham, pp. 67–110, https://doi.org/10.1007/978-3-030-72098-8_4.

Tversky, A. and Kahneman, D. (1981). The framing of decisions and the psychology of choice. *Science*, 211, 453–458.

Ujomudike, P.O. (2016). Ubuntu ethics, in *Encyclopedia of Global Bioethics* (ed H. ten Have), Springer, Cham, pp. 2869–2881.

Unger, S.H. (1994). *Controlling Technology: Ethics and the Responsible Engineer*, John Wiley & Sons, New York.

Valenti, J. and Wilkins, L. (1995). An ethical risk communication protocol for science and mass communication. *Public Understanding of Science*, 4 (18), 177–194.

Van Beers, C., Knorringa, P. and Leliveld, A. (2020). Can frugal innovations be responsible innovations? in *Responsible Innovation in Large Technological Systems* (eds J.R. Ortt, D. van Putten, L.M. Kamp and I. van de Poel), Routledge – Taylor & Francis Group, London, https://www.taylorfrancis.com/chapters/edit/10.4324/9781003019930-6/frugal-innovations-responsible-innovations-cees-van-beers-peter-knorringa-andr%C3%A9-leliveld.

Van den Hoven, J. (2013). Value sensitive design and responsible innovation, in *Responsible Innovation* (eds R. Owen, J. Bessant and M. Heintz), Wiley, Chichester, pp. 75–84.

Van den Hoven, J., Lokhorst, G.-J. and Van de Poel, I. (2012). Engineering and the problem of moral overload. *Science and Engineering Ethics*, 18 (1), 143–155.

Van den Hoven, J. and Vermaas, P.E. (2007). Nano-technology and privacy: On continuous surveillance outside the panopticon. *Journal of Medicine and Philosophy*, 32 (3), 283–297.

Van den Hoven, J., Vermaas, P.E. and Van de Poel, I. (2015). *Handbook of Ethics, Values, and Technological Design*, Springer, Dordrecht.

Van de Poel, I. (1998). *Changing Technologies. A Comparative Study of Eight Processes of Transformation of Technological Regimes*, Universiteit Twente, Enschede.

Van de Poel, I. (2001). Investigating ethical issues in engineering design. *Science and Engineering Ethics*, 7 (3), 429–446.

Van de Poel, I. (2007). Ethics in engineering practice, in *Philosophy in Engineering* (eds S.H. Christensen, M. Meganck and B. Delahousse), Academica, Aarhus, Denmark, pp. 245–262.

Van de Poel, I. (2009a). Values in engineering design, in *Handbook of the Philosophy of Science. Vol. 9: Philosophy of Technology and Engineering Sciences* (ed A. Meijers), Elsevier, Amsterdam, pp. 973–1006.

Van de Poel, I. (2009b). The introduction of nanotechnology as a societal experiment, in *Technoscience in Progress: Managing the Uncertainty of Nanotechnology* (eds S. Arnaldi, A. Lorenzet and F. Russo), IOS Press, Amsterdam, pp. 129–142.

Van de Poel, I. (2013). Translating values into design requirements, in *Philosophy and Engineering: Reflections on Practice, Principles and Process* (eds D. Mitchfelder, N. McCarty and D.E. Goldberg), Springer, Dordrecht, pp. 253–266.

Van de Poel, I. (2015a). Design for values, in *Social Responsibility and Science in Innovation Economy* (eds P. Kawalec and R.P. Wierzchoslawski), KUL, Lublin, pp. 115–164.

Van de Poel, I. (2015b). Conflicting values in design for values, in *Handbook of Ethics, Values, and Technological Design* (eds J. van den Hoven, P.E. Vermaas and I. van de Poel), Springer, Dordrecht, pp. 89–116.

Van de Poel, I. (2016). A coherentist view on the relation between social acceptance and moral acceptability of technology, in *Philosophy of Technology after the Empirical Turn* (eds M. Franssen, P. Vermaas, P. Kroes and A.W.M. Meijers), Springer International Publishing, Cham, pp. 177–193.

Van de Poel, I. (2017a). Design for sustainability, in *Philosophy, Technology, and the Environment* (ed D.M. Kaplan), MIT Press, Cambridge, pp. 121–142.

Van de Poel, I. (2017b). Dealing with moral dilemmas through design, in *Designing in Ethics* (eds J. van den Hoven, S. Miller and T. Pogge), Cambridge University Press, Cambridge, pp. 57–77.

Van de Poel, I. (2020). Embedding values in Artificial Intelligence (AI) systems. *Minds and Machines*, 30 (3), 385–409, https://doi.org/10.1007/s11023-020-09537-4.

Van de Poel, I. (2021). Values and design, in *Routledge Handbook to Philosophy of Engineering* (eds D.P. Michelfelder and N. Doorn), Routledge, New York, pp. 300–314.

Van de Poel, I., Asveld, L., Flipse, S., Klaassen, P., Scholten, V. and Yaghmaei, E. (2017). Company strategies for responsible research and innovation (RRI): A conceptual model. *Sustainability*, 9 (11), 2045, https://doi.org/10.3390/su9112045.

Van de Poel, I. and Kudina, O. (2022). Understanding technology-induced value change: A pragmatist proposal. *Philosophy & Technology*, 35 (2), Article number 40.

Van de Poel, I. and Royakkers, L. (2007). The ethical cycle. *Journal of Business Ethics*, 71 (1), 1–13.

Van de Poel, I. and Royakkers, L. (2015). Introduction, in *Moral Responsibility and the Problem of Many Hands* (eds I. van de Poel, L. Royakkers and S.D. Swart), Routledge, New York, pp. 1–11.

Van de Poel, I. and Sand, M. (2021). Varieties of responsibility: Two problems of responsible innovation. *Synthese*, 198, 4769–4787.

Van de Poel, I. and Taebi, B. (2022). Value change in energy systems. *Science, Technology, & Human Values*, 47 (3), 371–379.

Van de Poel, I., Zandvoort, H. and Brumsen, M. (2001). Ethics and engineering courses at Delft University of Technology: Contents, educational setup and experiences. *Science and Engineering Ethics*, 7 (2), 267–282.

Van de Poel, I. and Zwart, S.D. (2010). Reflective equilibrium in R&D networks. *Science, Technology & Human Values*, 35, 174–199.

Van der Burg, S. and Van de Poel, I.R. (2005). Teaching ethics and technology with Agora, an electronic tool. *Science and Engineering Ethics*, 11 (2), 277–297.

Van der Ham, W. (2003). *Meester van de zee: Johan van Veen, waterstaatsingenieur 1893–1959* [Master of the sea: Johan van Veen, civil engineer], Balans, Amsterdam.

Van Gorp, A. (2005). *Ethical Issues in Engineering Design. Safety and Sustainability*, Simon Stevin Series in the Philosophy of Technology, Delft.

Van Gorp, A. and van de Poel, I.R. (2001). Ethical considerations in engineering design processes. *IEEE Technology and Society Magazine*, 20 (3), 15–22.

Van Norren, D. and Verbeek, P.P. (2021). The ethics of AI and Ubuntu. Paper presented at *Africa Knows! Conference*, Leiden, Netherlands, https://nomadit.co.uk/conference/africaknows/paper/57969.

Van Poortvliet, A. (1999). *Risks, Disasters and Management*, Eburon, Delft.

Van Veen, J. (1962). *Dredge, Drain, Reclaim. The Art of a Nation*, 5th edn, Martinus Nijhoff, The Hague.

Van Wynsberghe, A. (2021). Sustainable AI: AI for sustainability and the sustainability of AI. *AI and Ethics*, 1 (3), 213–218.

Vaughan, D. (1996). *The Challenger Launch Decision*, The University of Chicago Press, Chicago.

Verbeek, P.P. (2005). *What Things Do – Philosophical Reflections on Technology, Agency, and Design*, Pennsylvania State University Press, University Park.

Verbeek, P.P. (2008). Obstetric ultrasound and the technological mediation of morality: A postphenomenological analysis. *Human Studies*, 31, 11–26.

Vincenti, W. (1990). *What Engineers Know and How They Know It*, Johns Hopkins University Press, Baltimore.

Voegtlin, C., Scherer, A.G., Stahl, G.K. and Hawn, O. (2022). Grand societal challenges and responsible innovation. *Journal of Management Studies*, 59 (1), 1–28.

Von Schomberg, R. (2012). Prospects for Technology Assessment in a framework of responsible research and innovation, in *Technikfolgen abschätzen lehren: Bildungspotenziale transdisziplinärer Methoden* (eds M. Dusseldorp and R. Beecroft), Springer, Wiesbaden, pp. 39–61.

Von Schomberg, R. (2019). Why responsible innovation? in *International Handbook on Responsible Innovation; A Global Resource* (eds R. von Schomberg and J. Hankins), Edwar Elgar Publishing, Cheltenham, UK, pp. 12–33.

Von Schomberg, R. and Hankins, J. (2019). *International Handbook on Responsible Innovation: A Global Resource*, Edward Elgar Publishing, Cheltenham, UK.

Warren, S.D. and Brandeis, L.D. (1890). The right to privacy. *Harvard Law Review*, 4 (5), 193–220.

WCED (1987). *Our Common Future. Report of the World Commission on Environment and Development*, Oxford University Press, Oxford.

Weckert, J. and Moor, J. (2007). The precautionary principle in nanotechnology, in *Nanoethics: The Social and Ethical Implications of Nanotechnology* (eds F. Allhoff, P. Lin, J. Moor and J. Weckert), John Wiley & Sons, Hoboken, pp. 133–146.

Wei, P. (2022). Confucius and the whistleblower. Palladium Magazine, March 4, 2022, https://www.palladiummag.com//2022/03/04/confucius-and-the-whistleblower.

West, M., Kraut, R. and Chew, H.E. (2019). *I'd blush if I could: Closing gender divides in digital skills through education*. Unesco, https://unesdoc.unesco.org/ark:/48223/pf0000367416.

Westacott, E. (2012). Moral relativism. Internet Encyclopedia of Philosophy, https://iep.utm.edu/moral-re.

Whitbeck, C. (1998). *Ethics in Engineering Practice and Research*, Cambridge University Press, Cambridge.

Whitehead, T. (2008). 13,000 people wrongly branded criminals. Telegraph, November 12, https://www.telegraph.co.uk/news/uknews/law-and-order/3449207/13000-people-wrongly-branded-criminals.html.

Winner, L. (1980). Do artifacts have politics? *Daedalus*, 109, 121–136.

Wong, P.-H. (2012). Dao, harmony and personhood: Towards a Confucian ethics of technology. *Philosophy & Technology*, 25, 67–86.

Wong, P.-H. (2020). Why Confucianism matters in ethics of technology, in *Oxford Handbook of Philosophy of Technology* (ed S. Vallor), Oxford University Press, Oxford, pp. 609–628.

World Commission on Environment and Development (1987). *Our Common Future*, Oxford University Press, Oxford.

World Health Organization (2006). Constitution of the World Health Organization – Basic Documents. 45th edn. Supplement, https://www.afro.who.int/sites/default/files/pdf/generic/who_constitution_en.pdf.

Wynne, B. (1975). The rhetorics of consensus politics: A critical review of technology assessment. *Research Policy*, 4 (2), 108–158.

Wynne, B. (1989). Frameworks of rationality in risk management: Towards the testing of naive sociology, in *Environmental Threats: Perception Analysis, and Management* (ed J. Brown), Bellhaven, London, pp. 33–47.

Zah, R., Böni, H., Gauch, M., Hischier, R., Lehmann, M. and Wäger, P. (2007). *Life Cycle Assessment of Energy Products: Environmental Assessment of Biofuels* (Technical Report), Empa, St. Gallen (Switzerland).

Zandvoort, H. (2000). Codes of conduct, the law, and technological design and development, in *The Empirical Turn in the Philosophy of Technology. Vol. 20 Research in Philosophy and Technology* (eds P. Kroes and A. Meijers), Elsevier/JAI Press, London, pp. 193–205.

Zhang, M., Atwal, G. and Kaiser, M. (2021). Corporate social irresponsibility and stakeholder ecosystems: The case of Volkswagen Dieselgate scandal. *Strategic Change*, 30 (1), 79–85.

Zhu, Q. (2020). Ethics, society, and technology: A Confucian role ethics perspective. *Technology in Society*, 63, https://doi.org/10.1016/j.techsoc.2020.101424.

Zhu, Q., Williams, T. and Wen, R. (2019). Confucian robot ethics. *Computer Ethics – Philosophical Enquiry (CEPE) Proceedings*, article number 12, https://digitalcommons.odu.edu/cepe_proceedings/vol2019/iss1/12.

Zyglidopoulos, S.C. (2002). The social and environmental responsibilities of multinationals: Evidence from the Brent Spar case. *Journal of Business Ethics*, 36 (1/2), 141–151.

Index of Cases

2,4,5-T herbicide, 328–9

animal welfare, battery cage design, 157–8
atomic bomb, scientists' petition, 41–2
Austin, Inez, whistle-blowing, 26
automatic seatbelts, 168

Bay Area Rapid Transport Project, 38–9
biofuels, 242–3, 254–5
biometrics, Super Bowl XXXV, 78
Boeing 737 MAX crashes, 183–5
Brent Spar oil platform, 66

Challenger Space Shuttle, 22–3
child labor, 96
Cubicle Warrior, military robots, 175–6

Deepwater Horizon drilling rig, 214–16
Dieselgate scandal, 7–9
dioxin risks, 193–4

facial recognition systems, 78
Fairphone, 267–8
Ford Pinto, 72–4

Gillbane Gold, waste water processing, 323–5
Golden Gate Bridge, suicide barrier, 188–9
golden rice, abating vitamin A deficiency, 281–2
Google in China, 56–8

heat pump boiler design, 253
heavy metals in waste water, 324–5
Herald of Free Enterprise, sinking of, 321–3
Herbicide 2,4,5-T, 328–9
high speed train disaster, Germany, 325–7
Highway safety, 122–3

household refrigerators, CFC alternatives, 159–61
housing systems for laying hens, 157–8
Hyatt Regency Hotel walkway collapse, 329–30

intelligent speed adaption system, cars, 333–4

LeMessurier, William, Citicorp Center, 101–2

military robots, moralization of, 175–6
Moses, Robert, racist overpasses, 146

nanoparticles, 205–6

obstetric ultrasound, 332–3
Onshore Carbon Dioxide storage in Barendrecht, 334–5

refrigerants, alternatives to CFCs, 159–61
RFID (Radio Frequency Identity) chips, 330–1
rice, genetically modified, 281–2
ride-sharing platforms, 176–8

Shell in Nigeria, unsustainable development, 50–3
storm surge barrier, Eastern Scheldt, 170–1

Tacoma Narrows Bridge, collapse of, 185
Teflon invention, 31–2
Tozer, John, silencing of, 55–6
traffic accident reduction, 122–3
train wheel design, German railways, 326–7

Uber, bias issues, 176–8

V-chip, violence on television, 232–4
values hierarchy for biofuels, 254–5

Ethics, Technology, and Engineering: An Introduction, Second Edition. Ibo van de Poel and Lambèr Royakkers.
© 2023 John Wiley & Sons Ltd. Published 2023 by John Wiley & Sons Ltd.

Index

Note: Page numbers in *italics* refer to figures; page numbers in **bold** refer to key term definitions.

2,4,5-T herbicide, 328–9

"ability to pay" principle, 251–2
acceptability, moral, of innovation, **272**
acceptable risk, **187**, 194
  and availability of alternatives, 199–200
  and calculated risk size, 195–6
  and informed consent, 197–8
  and objections to cloning, 195
acceptance, social, of innovation, **272**
accountability, **11**
Accreditation Board for Engineering and
  Technology (ABET) Engineering Criteria
  2000, 1
act utilitarianism, **90**
  application of, 132
action(s) *see* ethical cycle; ethical theories;
  normative ethics
active responsibility, **14**
  and the effectiveness requirement, 222
  features of, 14–15
  and the moral fairness requirement, 222
  and responsible innovation, 289–90
  see also ideals of engineers
actors, **30**
  and the problem of many hands, 217–22
  and technological development, 30–1
  and virtue ethics, 81
  see also stakeholders
advisory code of conduct, **40**
Africa
  Niger Delta oil spills case, 50–3
  Ubuntu philosophy, 112–13

algorithms
  ethical issues in design of, 116–17
  face recognition, racial bias, 152
  for rating videos and shows, 235
  re-design for social media, 149
ambiguity, **187**
American Society of Civil Engineers (ASCE),
  codes of conduct, 19–20, 43
American Society of Mechanical Engineering
  (ASME), codes of conduct, 19–20, 43
amicus curiae letter, 38–9
animal tests, **192**, 194
animal welfare, battery cage design, 157–8
anthropocentrism, **245**
anticipation criterion for responsible innovation,
  274–5
applied ethics, 115
  design of algorithms, 116–17
Aquinas, Thomas (1225–1274), 100
Aristotle (384–322 BC), 98–100
artefacts, values embedded in, 147–9
artificial intelligence (AI)
  ethical code for, Google, 14
  guiding value of "explainability", 288
  protest against use of, Google employees, 13
  see also robots
asbestos, 185
aspirational code of conduct, **40**
atomic bomb case, 41–2
auditing, external, **65**
Austin, Inez, whistle-blowing, 26
Australia
  compulsory registration of engineers, 65

Ethics, Technology, and Engineering: An Introduction, Second Edition. Ibo van de Poel and Lambèr Royakkers.
© 2023 John Wiley & Sons Ltd. Published 2023 by John Wiley & Sons Ltd.

organizations, 298
titles and qualifications, 297–8
Tozer case study, 55–6
automatic correction system, Boeing 737 MAX, 105, 184, 191, 198
automatic pilots, design decisions, 231–2
automatic seatbelts, 168
autonomous cars, 110
autonomy
    and active responsibility, Bovens, 15
    in Kantian ethics, 91, 93
    privacy as essential aspect of, 78
    see also moral autonomy
Ayala, Dr Alberto, 25

back-up systems, redundant design, 190
Bay Area Rapid Transport Project (BART), 38–9
"beings and doings", capability approach, 107
Bentham, Jeremy (1748–1832), 82–5
best available technology, **199–200**
bias
    design, 151–2
    offender profiling, 116
    ride-sharing platforms, 176–8
biocentrism, **245**
biofuels, 242–3
    innovation and value conflict, 262
    and intergenerational justice, 250–1
    and respecification, 261
    value hierarchy for, 254–5
    vs fossil fuels, aggregated environmental impact, *257*
biometrics, 78, 152
black-and-white strategy, **130**, 131, 136
blame, apportioning, 216, 217
blameworthiness, **11**
    four conditions for, 11–13
Boeing 737 MAX crashes, 183–5
    and engineer's responsibility for safety, 187–8, 191
    engineers' moral obligation to inform pilots, 198
    precautionary principle violation, 203–4
Boeing Code of Conduct, 312–13
Boisjoly, Roger, 23
Bovens, Mark, active responsibility features, 14–15
Brambell committee, UK, 158
Brent Spar oil platform, 66
bridge design
    Golden Gate Bridge, 188–9
    Tacoma Narrows Bridge, 185
British Board of Film Classification (BBFC), 234–5
Brumsen, M., accident reduction, 327
Brundtland definition of sustainable development, 246–8, 253–4, 262–3
building design

CitiCorp Center, New York, 101–3, 218–21
    Hyatt Hotel walkway collapse, Kansas City, 329–30
building inspection, conflicting ethical frameworks, 133

calculated risk size and acceptability of, 195–6
Canada, dioxin concentrations, 194
capability approach, **106–8**
    applications of, 109–10
    criticism of, 110–11
carbon capture and storage (CCS), 334–5
carcinogens
    asbestos, 185
    dioxin, 193–4
    Teflon constituents, 32
care ethics, **103**
    criticism of, 104
    in engineering, 104–6
    relationships, importance of, 103–4
categorical imperative, 137–8
categorical imperative, Kant, **92–3**
    applied to Ford Pinto case, 97
    and equal treatment, 200
    and informed consent, 197
    and intergenerational justice, 250
causal contribution, 11–12
    CitiCorp building, 218–19, 220
censorship in China, Google's role in, 56–8
certification, **173**
Chadwick, James, discovery of neutron, 41
chain reaction, atomic bomb, 41–2
Challenger Space Shuttle, 22–3
cheating software *see* Dieselgate scandal
Chevron Corporation, Code of Conduct, 59, 310–11
child labor, 96–7, 106, 110
China
    censorship, Google's participation in, 56–8, 86
    Confucianism, 113–15
chlorofluorocarbons (CFCs), alternatives to, 159–61
CitiCorp Center, New York, 101–3
    problem of many hands, 218–21
Civic Integrity team, Facebook, 28
Clancy, R.F., cross-cultural values, 43
Clean Air Act, US, 12, 199
Clean Water Act, US, 199, 216
climate justice, 251–2
Clinton, Bill, *232*
cloning, ethical objections to, 195
cobalt mining, 267–8
codes of conduct, **39**
    arguments against, 60–2, 66
    corporate, 47–55
    difficult to live by, 62–4

enforcement of, 64–6
examples of, 308–13
and loyalty, 58–9
and obedience, 59–60
professional, 40–7
and self-interest, 55–8
website links, 295, 297, 298, 300
collective responsibility, **218**
collective responsibility model, **228–9**
collective risks, **201**
Collingridge dilemma, **32–3**
  overcoming, 275, 279
common sense method, ethical evaluation, **131**, 137
COMPAS algorithm, offender risk-profiling, 116–17
competitive advantage, **270–1**
competitive bidding, 305
computer ethics, threats to personal privacy, 78
conceptual disruption, 288
conceptualization of values, **154–6**
confidentiality duties, **62**
  limits to, 62–3
conflict minerals, Fairphone case, 267–8
conflict of values *see* value conflicts
conflicts of interest, **46**
Confucianism, 113–15
consensus, overlapping, **141**
consent, informed, 197–8, 208–9
consequence (risk) assessment, 192–3
consequentialism, 81, **82**, 118
  and risk communication, 202
  see also utilitarianism
Constructive Technology Assessment (CTA), **33**, 141–2
contingent validation, **162**
coolants in refrigerators, 159–61, 164–5, 168–9
cooperation strategy, **130**, 131, 136
Copernicus, Nicolaus (1473–1543), 95
core values, 53–4
  Chevron, 310–11
  Pfizer, 308–9
corporate codes, **39**, 47
  enforcement of, 65
  mission and vision statements, 53
  responsibilities to a stakeholders, 54
corporate liability, **227**
Corporate Social Responsibility (CSR), **48–9**
  and responsible innovation, 283, 284
cost-benefit analysis (CBA), **161–2**
  Ford Pinto case, 73, 74, 90–1
courage
  as a middle course, 99
  and Enlightenment, Kant, 81
  of whistleblowers, 114
critical loyalty, **58–9**

cross-cultural values, two moral principles related to, 43
Cubicle Warrior, military robots, 175–6
Czeskis, A., PhoneTracker value scenario, 276

Davis, Michael, 59, 60, 61
Deepwater Horizon drilling rig, 214–16
deliberation, moral, 140–1
deliberation with stakeholders, guidelines for, 278
deontological ethics *see* duty ethics
descriptive statements, 76
  vs normative judgments, 77
desiderata for allocation of responsibility for safety, 230–1
Design for Values, **33**, **147**
  case, Racist Overpasses, 146
  conceptualization and specification, 153–9
  design process, 149
  embedding values in artefact design, 147–9
  intended, embedded, and realized value, 148–9
  prototyping, testing and monitoring, 172–8
  stakeholder analysis, 150–2
  value conflicts, 159–72
  Value Sensitive Design (VSD) approach, 150
determinism, technological, 269, **270**
developments risks, **226**
Devon, Richard, values for group processes, 105–6
Didion, D.A., refrigerants, 160, 166, 167
Dieselgate scandal, 7–9
  ethical evaluation, 132–3
  moral problem formulation, 127–8
  options for actions, 130–1
  and passive responsibility, 11–13
  relevant values, 128–9
  some unknown/disputed facts, 130
  stakeholders, interests of, 129
  and window-dressing, 56
dioxin risks, 193–4
direct stakeholders, **150**
  values of, 153
disciplinary code of conduct, **40**
  enforcement of, 62, 64
discount rate, **161**, 250
disease detection algorithm, 117
disruptive innovation, 285
  conceptual and normative disruption, 288
  market disruption, 285–6
  regulatory disruption, 287–8
  social disruption, 286–7
distribution of responsibility, **221–2**
  Deepwater Horizon case, 214–17
  legal aspects, 222–7
  in organizations, 227–31
  problem of many hands, 217–22
  and technological design, 231–6

distributive justice, 89, 115
"doings and beings", capability approach, 107
dose-response relationship models, **193**
Dragonfly project, Google, 57–8
drugs, experimental phases for new, 279–80
Du Pont, Teflon, 31–2
duty of care, **225**
    negligence, 60
    violation of, Shell, 51–2
duty ethics, 81, **91–2**
    applied to Ford Pinto case, 97
    applied to Highway Safety case, 137–8
    and categorical imperative, 92–4
    criticism of, 95–7
    and sustainable development, 250–1

eco-indicators, 256, 258
effectiveness, **17**
    external goal, 19
effectiveness requirement, **222**
efficiency, **17**
    external goal, 19
efficiency ideal, 17–19
Einstein, Albert, 42
embedded values, **148**
emissions testing *see* Dieselgate scandal
energy transition towards low-carbon
    technologies, 260–1
Engineering Criteria (2000), ABET, 1–2
engineering design, **149**
    and distribution of responsibility, 231–6
    value conflicts in, 170–1
    Value Sensitive Design (VSD), 150
    see also Design for Values
Engineering Ethics: Concepts and Cases (Harris,
    Pritchard, and Rabins), 123
ENGINEERS EUROPE, 43–4, 101, 245,
    299–300, 306–7
engineers vs managers, 22–30
    Challenger space shuttle case, 22–3
    separatism, 24–5
    technocracy, 25
    whistle-blowing, 26–30
environmental damage, 204, 248–9
environmental degradation, 244
environmental problems, 243–4, 245
Environmental Protection Agency (EPA), US, 7,
    8, 32
environmental sustainability, 155
    alternative coolants case, 159–61, 167, 168–9
    see also sustainability
epidemiological research, **193**
equality postulate, **93**, 137–8
ethical cycle, 121–4, **125**, *126*, 127
    applied to Highway Safety case, 135–40
    ethical evaluation, 131–3
    fictional case: Highway Safety, 122–3

ill-structured problems, 124–5
    moral deliberation, 140–1
    options for actions, 130–1
    overlapping consensus, 141
    problem analysis, 128–30
    reflection, 134–5
ethical theories, 81
    applying, 115–17
    capability approach, 106–11
    care ethics, 103–6
    Kantian theory: duty ethics, 91–7
    non-Western, 111–15
    utilitarianism, 82–91
    virtue ethics, 97–103
ethics, **75–6**
    see also normative ethics
EU Council Directive 85/374/EEC for Product
    Liability, 225
European Expert Group on Science and
    Governance, 207
European Federation of National Engineering
    Associations (ENGINEERS EUROPE),
    Code of Conduct, 43–4, 306–7
European Union (EU)
    animal welfare directive, 158
    and best available technology, 200
    Expert Group on Science and Governance,
        207
    ENGINEERS EUROPE code of conduct,
        43–4, 306–7
    genetically modified organisms (GMOs) debate,
        204–5
    Horizon 2020 science funding program, 268
    product liability directive, 225, 226
    responsible innovation, 268
    titles and qualifications, 299
    Volkswagen diesel cars, 7–8
    Whistle-blower Directive (2019), 63–4
event tree, **192**
experimental subjects, principles for, 209
experiments, societal, 206–9
exposure (risk) assessment, 192
external auditing, **65**

face recognition technology, 78, 152
Facebook, 28–9
Fahlquist, J.N., heuristics when designing for
    responsibility, 235–6
failure modes, **192**
fairness, moral, 222, 235
Fairphone case, conflict minerals, 267–8
fake news, social media, 149, 287
fast fashion industry, child labor, 106
fault trees, **192**
Fermi, Enrico, atomic scientist, 41
filial piety, Confucianism, 113–14
first-order learning, **278**, 279

flooding, Zeeland, 20–1, 170–1
"flourishing", 107, 108
Ford Pinto case, 72–4
  applying Kant's theory to, 97
  applying utilitarianism to, 90–1
  and informed consent, 198
  and social ethics, 106
foreseeability, 12, 222, 225, 284, 289
fossil fuels, 244, 248, 255, *257*
Frankena, William, 100
free will, Kant, 93
freedom of action, 12–13
freedom principle, Mill, **87**
  and informed consent, 197
freedom of speech, 62–3
Friedman, Milton, 48
  objections against views of, 48–9
friendship, 163
  companion robots, 114–15
functionality in design process, 149
functionings, capability approach, 107
future generations, needs of, 247

genetically modified organisms (GMOs)
  golden rice, 281–2
  and the precautionary principle, 204–5
Germany
  and the atomic bomb, 41–2
  high speed train crash, 325–7
  professional code of engineers' association
    (VDI), 43
  refrigerator firms, 161, 169
Gillbane Gold, waste water processing, 323–5
global warming, 251–2
  contribution of biofuels to, 242–3
  and energy production technologies, 260
Global Warming Potential (GWP), refrigerator
  coolants, 160, 161, 168–9
Golden Gate Bridge, suicide barrier, 188–9
golden rice, and vitamin A deficiency, 281–2
"Golden Rule", 81, 113
"good life", Aristotle, **98**
good will, **92**
Google
  employees protest against AI project for the
    military, 13–14
  launch of censored version in China, 56–8
Google Earth, technological enthusiasm, 15
greenhouse effect, 186, 242, 245, 251
greenhouse gas emissions
  global emissions by sector, *260*
  mitigation/reduction of, 251
  necessary for subsistence, 251
Greenpeace, 66, 161, 169, 282
group processes, values for, 105–6

Habermas, Jürgen, moral judgments, 141
Hahn, Otto, nuclear fission, 41
happiness
  Bentham's utility principle, 84
  and cost-benefit analysis, 162
  and distributive justice, 89
  personal relationships, 89
  and reasoning, Aristotle's "good life", 98
  vs duty, Kant, 91
Hare, Richard, 89
hate speech, social media, 149
Haugen, Frances, Facebook whistleblower,
  28–9, 48
hazards, **186**
heat pump boiler, design for sustainability, 253
heavy metals in waste water, 324–5
hedonism, **82**, 154
Herald of Free Enterprise, capsizing of, 321–3
Herbicide 2,4,5-T, 328–9
HFC 134a vs isobutane, coolant in refrigerators,
  160–1, 164–5, 168–9
hierarchical responsibility model, **228**
hierarchical social structures, 232
hierarchy of values, **156–7**
  biofuels, 254–5
  housing systems for hens, 157–8
high speed train disaster, Germany, 325–7
highway safety case, 122–3
  ethical cycle applied to, 135–40
Hillerbrand, R., capability approach, 110
"hired guns", **24–5**
honesty, **46**
household refrigerators, CFC alternatives, 159–61
human capabilities, Nussbaum, 108–9, 111
human rights, 43, 54, 83, 90
  and free will, Kant, 93
  and Google in China, 57–8
human welfare ideal, 19–22
Hung, L., use of social robots, 112–13
Hyatt Regency Hotel walkway collapse, 329–30
hydrocarbons, 159–61, 214–15
hypothetical consent, **208**
hypothetical norm, **92**

ideals, **15**
ideals of engineers, 15–22
  effectiveness and efficiency, 17–19
  human welfare, 19–22
  technological enthusiasm, 15–17
ignorance, **187**, 210
  dealing with, 203–9
ill-structured problems, **124**
impact learning, 280
imperative, categorical, 92–4
inclusiveness, responsible innovation, **276–8**

inclusiveness, value for design, 155
income and marginal utility, 89
incommensurable values, **163**
indirect stakeholders, **150**
    source of value in design, 153
individual moral responsibility, conditions for, 219
individual responsibility model, **229**, 230–1
industry, responsible innovation in, 283–5
informed consent, **197–8**
    alternative set of principles, 209
    and communicating risk, 201–2
    leading principle for societal experiments, 208
inherently safe design, **190**
innovation
    application of strict liability to, 225, 226
    and regulation, 224
    strategic importance of, 270–1
    and value conflicts, 170, 262
    see also Responsible Innovation
Institute of Electrical and Electronic Engineers
    (IEEE), 38–9
institutional learning, 280
instrumental values, **77**
    anthropocentrism, 245
    and cost-benefit analysis, 162
    privacy as, 78
Integrated Pollution Prevention and Control
    Directive, EU, 199
integrity, **46**
intelligent speed adaption system, cars, 333–4
intended value, **148**
interests (of actors), **31**
intergenerational justice, **248**, 250–1, 255, 256,
    261, 263
    and values hierarchy for biofuels, *254*
International Council of Chemical associations
    (ICCA), 65
International Council on Clean Transportation
    (ICCT), 8
International Rice Research Institute (IRRI), 282
International Risk Governance Council (IRGC),
    ambiguity, 187
interval scale, *164*, **165**
intragenerational justice, **248**, 250–1, 255, 256,
    261, 263
    and values hierarchy for biofuels, *254*
intrinsic values, **77**
    biocentrism, 245
    and cost-benefit analysis, 162
    dominant values as, 131–2
    and privacy, 78
    and utilitarianism, 82
intuitivist framework, **131**
investigations, Value Sensitive Design (VSD), 150
isobutane coolant vs HFC 134a, 161, 164–5, 169
iterative processes
    ethical cycle as, *126*, 127, 135
    values hierarchy as, 156–7

Japan
    atomic bomb attacks on, 42
    professional engineering codes of conduct, 43
    Whistleblower Protection Act (WPA), 64
Johnson, Deborah, computer ethicist, 25, 78
Johnson, Stuart, Volkswagen, 8, 27
Joliot-Curie, Irene, 41–2
justice, 155
    climate, 251–2
    distributive, 89, 115, 155
    intergenerational, 248, 250, 255, 256, 261, 263

Kant, Immanuel (1724–1804), 94–5
    criticism of theory, 95–7
    on Enlightenment, 81
    theory of duty ethics, 91–4

labor
    child, 96–7, 110
    divisions of, 232, *233–4*
laws/legislation
    environmental, 199
    and innovation, 271–2
    new technologies disrupting existing, 287
    protecting whistle-blowers, 63–4
    responsibility and, 222–4
learning for responsible innovation
    experimental learning process, 279–80
    first-order vs second-order, 278–9
    three types of, 280
legal liability, 223
LeMessurier, William, Citicorp Center, 101–3
    problem of many hands, 218–20
liability, **222–3**, 236–7
    corporate, 227
    negligence versus strict liability, 224–6
    vs moral responsibility, 223–4
    vs regulation, 224
Liang, James Robert, Dieselgate scandal, 9, 10,
    24, 58
life cycle analysis (LCA), **256**
    cars, 257–8
    disadvantages of, 256–7
    mobile phones, 268
life phases of a product, **252–3**
    new product development, 283–4
limited liability, **227**
loyalty, 58–9, 113, 114
Luegenbiehl, H.C., cross-cultural values, 43

MacIntyre, Alasdair, on virtues, 79
malicious obedience, **59**
managers vs engineers, 22–30
Manhattan Project, scientists' petition to President
    Truman, 42
marginal utility, **89**
market disruption, 285–6
market pull, **274**

Martin, Mike, 202, 208
maxims, Kant's categorical imperative, 92, 97, 133
MCAS (Maneuvering Characteristics
    Augmentation System), 184–5, 191, 198, 204
McLinden, M.O., refrigerants, 160, 166, 167
measurement scales, choice of, 163–6
mediation, technological, **173–4**
medical drugs, experimental phases for, 279–80
methane, *257*
Microsoft, CSR initiatives, 49
middle course (between two extremes of evil),
    99–100
military robots, moralization of, 175–6
Mill, John Stuart (1806–1873), 87–8
Minerals Management Service (MMS), Deepwater
    Horizon case, 217
mining, Fairphone case, 267–8
misinformation, social media, 28
mission statements, 53
mistakes, 194
mobile phones
    and conflict materials, 267–8
    and market disruption, 285–6
    tracking scenarios, 276
monitoring effects of technology, 172–8
Moor, James, philosopher, 206
moral acceptability (of innovation), **272**
moral argumentation skills, 2, 3, *4*
moral autonomy, **97**
    and informed consent, 197, 209
    paternalism clashing with, 25, 174
moral balance sheet, **84–5**
    problems of, 86
moral competencies, 1–2
moral deliberation, **140–1**
moral dilemmas, **127**
moral fairness requirement, **222**
moral ideal of professionals, 61
moral judgments
    and categorical imperative, Kant, 92
    and codes of conduct, 60
    consequentialism/utilitarianism, 82, 139
    and the ethical cycle, 123–4, 125
    and ethical theories, 81, 86
    and moral deliberation, 141
    normative relativism, 80
    wide reflective equilibrium, 134, 139–40
moral norms, 79, 80, 91–2
moral problem, **127**
    statement of, 127–8, 135
moral responsibility, **10**
moral values, 77, 80
moral virtues, 79, 80, 99, 100
moralization of technology, **174**
Morton Thiokol Company, 22–3
Moses, Robert, racist overpasses, 146
Muilenberg, Denis, Boeing CEO, 203–4

multiple criteria analysis, **163**, 164–6, 172, 179
multiple independent safety barriers, **190**, 191

nanoparticles, 205–6
NASA, 22–3
National Safety Council (NSC), 123
National Society of Professional Engineers
    (NSPE), Code of Ethics, 43–4, 301–5
    Board of Ethical Review (BER) case, 44–5
    environmental responsibility, 245
    and integrity/honesty, 46
    and loyalty, 58, 59
    and obedience, 60
    and obligations toward clients and employers, 47
    professional ethics, 138
    and safety, responsibility for, 187–8
    and social responsibility, 47
natural resources
    exhaustion of, 244–5
    and polluter pays principle, 248–9
    and property rights, 248
needs, Brundtland definition, 247
negative feedback mechanism, **190**
negligence, **224**
    vs strict liability, 224–6
negligent obedience, **60**
negotiation strategy, **281**
Netflix, 234–5
Netherlands
    1953 flood disaster, 20–1
    accident prevention, 333
    code of conduct, 299
    Court of Appeal, The Hague, 51–2
    Eastern Scheldt storm surge barrier, 170–1
    Health Council, nanoparticles, 206
    onshore CO_2 storage, 334–5
    organizations, 299
    qualifications, 299
no harm principle, Mill, 87
Nokia, 54, 285–6
non-renewable resources, 244
non-Western ethical theories, 111–15
normative disruption, 288
normative ethics, 70–1, 76
    applied ethics, 115–17
    capability approach, 106–11
    care ethics, 103–6
    consequentialism: utilitarianism, 82–91
    deontology: duty ethics, 91–7
    descriptive statements, 76
    ethical theories, 81–116
    ethics and morality, 75–6
    Ford Pinto case study, 72–4
    and informed consent, 197
    non-Western ethical theories, 111–16
    normative judgments, 76–7
    norms, 78–9

relativism, 80–1
values, 77–8
virtue ethics, 97–103
virtues, 79–80
normative judgments, 76–7
normative learning, 280
normative relativism, **80**
norms, **78–9**
  in a values hierarchy, 156–8
  and corporate codes, 54–5
  hypothetical, 92
  legal, 79
  moral, 79, 80, 91–2
  prima facie, 95–6
  self-evident, 95, 96–7
  universal, 80, 118
  vs values, 79
  see also rules
NSPE *see* National Society of Professional
  Engineers
nuclear fission, atomic bomb, 41–2
nuclear power plants
  nuclear energy as societal experiment, 208
  radioactive waste vs lower CO_2 production, 248
  risk assessments, 195–6
nuclear reactors, redundant systems, 190
Nussbaum, Martha, capability theory, 106, 107,
  110–11
  ten central human capabilities, 108–9

O-rings, Challenger, redundant design, 190
obedience, forms of, 59–60
obligations, professional codes
  toward clients and employers, professional
    codes, 46–7
  toward the public, 47
obstetric ultrasound, 332–3
OECD, polluter pays principle, 249
Ogoni people, Nigeria, 50–2
oil, biofuels reducing reliance on for
    transportation, 242–3
oil spills
  BP Deepwater Horizon, 214–16
  Shell in Niger Delta, 50–3
Oosterlaken, E.T., "capability sensitive design", 109
ordinal scale, *164*, **165**
organizational responsibility, distribution of,
    227–8
  collective model, 228–9
  hierarchical model, 228
  individual model, 229
organizations, engineering, 295
overlapping consensus, **141**
ozone depletion, 159–60, *169*, *258*

pain, Bentham, 84–5
  measuring, 85–6
panopticon, *83*, 84

parental controls, V-chip, 232–5
participatory design, **152**
passive responsibility, **11**
paternalism, **25**
  standardization as, 200–1
"Peak Oil", 242
perception of risks, 202–3
perfluorooctanoicacid (PFOA) and Teflon, 32
persona, **151**
personal relationships
  and happiness, 89
  importance of, 103–4
personal risks, **201**
pesticides, 328–9
Pfizer code of conduct, 308–9
pleasure, 85
  Bentham's moral balance sheet, 84–5
  measuring, 85–6
  Mill's freedom principle, 87
  negative link to increased income, 89
Plunkett, Roy, Teflon invention, 31–2
politics of artefacts, Winner, 146–7, 232
polluter pays principle (PPP), **248–9**, 251, 252
pollution, 244
practical wisdom, **100**
precautionary principle, **203**
  applied to nanotechnology, 205–6
  Boeing's violation of, 203–4
  four dimensions of, 205
  and sustainable development, 204–5, 248–9
prediction
  and development risks, 226
  dose-response models, 193
  of recidivism, COMPAS algorithm, 116–17
  vs anticipation, 274–5
prima facie norms, **95**
  examples of, 96
  vs self-evident norms, 97
Pritchard, Michael, 61, 100–1, 102–3, 105, 135,
  221
privacy, 78, 155, 331
problem of many hands, **217–18**
  causes of, 221
  CitiCorp building, 218–21
  distribution of responsibility, 221–2
product development, life cycle phases in, 283–4
product dimension of responsible innovation, 273
product liability, **225**, 226
profession, defined, 61
professional codes, **39**, 40–5
  integrity, honesty and competence, 46
  obligations toward clients and employers, 46–7
  social responsibility and obligations toward the
    public, 47
professional ideals, **15**
  effectiveness and efficiency, 17–19
  human welfare, 19–22
  technological enthusiasm, 15–17

professional responsibility, **10–11**, 34
  and the BART train, 38–9
promises, breaking, 92–3
property rights, **248**
ProPublica, 116
prototype, **173**
public engagement methods, 277
Public Interest Disclosure Act (1998), UK, 63
"public utility", 136, 137, 139, 140

qualifications in engineering, 294–300

Radio Frequency Identity (RFID) chips, 330–1
radioactive waste, 208, 248
ratio scale, **165**, 166
rational foundationalist approach, Whitbeck, 124
rationality, 93
  of children, 96
  of consumers, 97
  of others, 138
Rawls, John, 3, 89, 141
realized values, **148**, 149
ReCiPe methodology for life cycle assessment, *258*
reciprocity principle, Kant, **93–4**, 118, 138
  applying to Ford Pinto case, 97
  and intergenerational justice, 250
recyclability, 253
redundant design, 190
reflection, ethical cycle, 134–5, 138–40
reflexivity (criterion for responsible innovation),
  278–9
refrigerants, alternatives to CFCs, 159–61
regulation, **224**
  new technologies disrupting, 287–8
  see also laws/legislation
regulators, **30**
relationships, importance of, 103–4
relativism, 80–1
release assessment, 191–2
renewable resources, 244, 260
respecification of values, **167–71**, 261–2
responsibilities of engineers, 6–35
  active responsibility, 14–15
  definitions, 9–11
  Dieselgate case, Volkswagen, 7–9
  effectiveness and efficiency, 17–19
  engineers vs managers, 22–30
  Google's AI project, employees' protest against,
    13–14
  ideals of engineers, 15–22
  passive responsibility, 11–13
  for responsible innovation, 289–90
  to solve environmental problems, 244–5
  technological development, social context of,
    30–3
  Teflon case, 31–2
responsibility distribution, 213–17
  and the law, 222–7

in organizations, 227–31
problem of many hands, 217–22
and technological designs, 231–6
Responsible Care, chemical industry, 285
Responsible Innovation, **33**, 265–7, **268**
  anticipation criterion, 274–6
  definitions of, 268–9, 272–3
  disruptive innovation, 285–8
  engineers' responsibility for, 289–90
  Fairphone case study, 267–8
  inclusiveness criterion, 276–8
  in industry, 283–5
  need for, 271–2
  as process, 273
  process criteria for, 274–80
  as product, 273
  reflexivity criterion, 278–9
  responsiveness criterion, 279–80
  science, technology, and society, 269–70
  and societal challenges, 274, 280–2
  strategic importance of, 270–1
responsiveness criterion for responsible innovation,
  279
RFID (Radio Frequency Identity) chips, 330–1
rice, genetically modified, 281–2
ride-sharing platforms, 176–8
Rio Declaration, precautionary principle, 204, 249
risk, **186**
risk assessments, **191**
  consequence assessment step, 192–3
  exposure assessment step, 192
  release assessment step, 191–2
  reliability of, 193–4
  risk estimation step, 193
risk-cost-benefit analysis, **199**
risk perception, 202–3
robots
  companions for the elderly, 110
  and conceptual disruption, 288
  Confucianism perspective, 114–15
  in the military, 175–6
  portrayal of, 152
  Ubuntu perspective, 112–13
rocket development, 16–17
role responsibilities, **10**
Ross, William David, criticism of Kantian theory,
  95–6
rule utilitarianism, **90**, 91
rules, 78–9
  social practices, 286
  see also codes of conduct; norms

safety, 155, **186–7**
  and automatic seatbelts, 168
  engineer's responsibility for, 187–91
  highway safety case, 122–3
  of products, strategies for ensuring, 190
  responsibility for, distribution, 230–1

safety factors, **190**
Sandin, Per, precautionary principle, 205
Sarbanes-Oxley Act (2002), US, 63
Saro-Wiwa, Ken, MOSOP leader, 52
scales of measurement, Multiple Criteria Analysis, 163–6
scenario techniques for responsible innovation, 275–6
Schinzinger, Roland, 202, 208
Schmidt, Oliver, 9, 25
Schönherr, N., benefits of SDGs to CSR, 49–50
science
    technology as applied, **269**
    technoscience, **270**
science, technology as applied, **269**
second-order learning, **278–9**
self-driving cars, 286–7
self-evident norms, 95, 96–7
self-interest, role in codes of conduct formulation, 55–8
Sen, Amartya, 106, 107
separatism, **24**
Sharkey, Amanda, use of robots in elderly care, 110
Shell
    Brent Spar oil platform, 66
    core values, 54
    oil spills in Nigeria, 50–3
    onshore CO_2 storage, Barendrecht, 334–5
Shell Petroleum Development Company (SPDC), 50–2
Sidgwick, Henry, 89
simulations, **173**
smartphones, introduction of, 285–6
social acceptance (of innovation), **272**
social context of technological development, 30–3
social disruption, 286–7
social ethics of engineering, **105**
social media
    creating new moral problems, 149, 287
    Facebook case, 28–9
    regulation difficulties, 287
social responsibility, 21, 22, 47
    Corporate Social Responsibility, 48, 283, 284
societal challenges, responsible innovation, 274, 280–2
societal experiment, **207**
    engineering as a, 206–9
society, relationship with science and technology, 269–70
solution strategy, **281**
Sophie's Choice (Styron), 127
specification of values, **154**
stakeholders, **31**, **129**
    analysis, 150–2
    conflicts, value dams and value flows, 262
    in Dieselgate case, 129

fruitful deliberation guidelines, 278
    responsibility to, 54
standardization, 153
    ethical objections against, 200–1
    setting thresholds, 166–7
standards, moral, 40, 272
standards, technical, **153**, 173
    and regulatory disruption, 287
standstill principle, **247**
storm surge barrier, Eastern Scheldt, 170–1
strict liability, **225**, **226**, 237
stupid obedience, **60**
Styron, William, *Sophie's Choice*, 127
suicide barrier for Golden Gate Bridge, 188–9
Supreme Court, US, 305
sustainability, 241–4
    design for, 252–3
    environmental ethics, 244–5
    life cycle analysis, 256–8
    specifying, 253–6
    and value conflicts in design for, 258–62
sustainable development, **246**
    Brundtland definition, 155, 246–8, 262–3
    moral justification, 248–52
Sustainable Development Goals (SDGs), 49–50, 54, 246
    societal challenges and, 280–1
Swierstra, Tsjalling, arguments against cloning, 195
"synthetic reasoning", Whitbeck, 125
Szilárd, Leó, atomic scientist, 41–2

Tacoma Narrows Bridge, *183*, 185
Taylor, Frederick W. (1856–1915), 17–19, 25
Taylor, Harriet, 88
technical codes, **153**
technical risks and ethics, 182–6
    acceptable risks, 194–201
    definitions, 186–7
    engineer's responsibility for safety, 187–91
    risk assessment, 191–4
    risk communication, 201–3
    uncertainty and ignorance, 203–9
technical standards, **153**
technocracy, **25**
technological determinism, 269, **270**
technological development, social context of, 30–3
technological enthusiasm, **15–16**
technological mediation, **173–4**
technology as applied science, **269**
Technology Assessment (TA), 32
technology push, **274**
technoscience, **270**
Teflon invention, case study, 31–2
testing, **173**
    new medical drugs, 280
thresholds for conflicting values, **166–7**, 172

total utility, 249–50
Tozer, John, silencing of, 55–6
trade-offs between values in design, 163, 172
traffic accidents, highway safety case, 122–3
    ethical cycle applied to, 135–40
traffic risks, calculating, 195–6
traffic safety, 78–9
    speed reduction system, 333–4
train derailment, Germany, 325–7
train wheel design, German railways, 326–7
transportation network companies (TNCs), 176–8
tripartite model, **24**
TV parental guidelines, 233–4
Type I errors, **194**
Type II errors, **194**

Uber, bias issues, 176–8
Ubuntu (humanness), 112–13
uncertainty, **187**
    precautionary principle, 203–6
    societal experiment, engineering as, 206–9
uncritical loyalty, **58**
United Kingdom (UK)
    Nuffield Council on Bioethics, 254, 255
    organizations, 296–7
    professional codes of conduct, 297
    titles and qualifications, 295–6
United States (US)
    compulsory registration of engineers, 65
    Ford Pinto case, 72–4
    organizations, 295
    professional codes of conduct, 43, 295
    titles and qualifications, 294–5
    Visa's corporate code of conduct, 309–10
universality principle, Kant, **92**, 118
    applying to Ford Pinto case, 97
use plan, **147**, 148
users, **30**
utilitarianism, **82**
    applying to the Ford Pinto case, 90–1
    criticism of, 89–90, 118
    ethical cycle example, 137, 138–9
    and intergenerational justice, 249–50
utility
    marginal, 89
    "public", 136, 137, 139, 140
    Sen's criticism of, 107
    total, 249–50
utility principle, Bentham, **84**, 118
    objection against, 134

V-chip, violence on television, 232–4
value conflicts, **159**
    comparison of methods for dealing with, 171–2
    cost-benefit analysis, 161–3
    in design for sustainability, 258–62

and innovation, 170–1, 262
    multiple criteria analysis, 163–6
    refrigerator coolants case, 159–61
    respecification, 167–70, 261–2
    thresholds, 166–7
value scenarios, 276
Value Sensitive Design (VSD), 147, 150
values, 77–8
    conceptualization, 154–6
    core values, 53–4
    designing for, 149–50
    specification, 154, 156–9
values hierarchy, **156**
    animal welfare, 157–8
    for biofuels, 254–5
van den Hoven, J., 273, 331
van Veen, Johan (1893–1959), 20–1, 26
Vermaas, P.E., RFID chips, 331
violence
    against women, ride-sharing platforms, 178
    on television, V-chip, *233–4*
virtue ethics, **97–8**
    Aristotle, 98–100
    criticism of, 100
virtues, **79–80**
    Aquinas' seven virtues, 100
    Aristotle's definition of, 98
    for morally responsible engineers, 100–3
Visa Code of Business Conduct and Ethics,
    309–10
vision statements, 53
    Volkswagen, 56
vitamin A deficiency and golden rice, 281–2
Volkswagen (VW) *see* Dieselgate scandal
von Braun, Wernher (1912–1977), 16–17, 24
Von Schomberg, R.,responsible innovation, 273

waste
    nuclear/radioactive, 26, 208, 248, 272
    water heavy metals in, 324
Weckert, John, precautionary principle, 206
welfare
    animal, 157–8
    human, 19–22
well-being, 107, 156
    three main theories of, 154
whistle-blowing, **26–7**
    Austin, Inez, nuclear waste, 26
    and confidentiality duties, 62
    disadvantages, 27, 30
    guidelines for, 27
    Haugen, Frances, Facebook, 28–9
    last resort strategy, 130, 131
    laws protecting, 63–4
    Royal DSM's Alert Policy, 314–20
    vs remonstration, Confucian ethics, 114

Whistleblower Protection Act (2006), Japan, 64
Whitbeck, Caroline, 124–5, 136
wide reflective equilibrium, **134**
window-dressing, **56**
Wingspread Statement, 204
Winner, Langdon, politics of artefacts, 146–7, 232
Winterkorn, Martin, VW CEO, 9, 12, 27, 56
wisdom, practical, **100**, 204

World Health Organization (WHO), 155, 281, 282
wrong-doing, 12, 26, 220–1
  Ford Pinto case as example of, 72–4

YouTube, 235

Z-Corp, 324